大数据技术与项目实战系列教材

Hadoop
核心技术与实战

◎ 潘正高 施霖 编著

清华大学出版社
北京

内 容 简 介

本书针对 Hadoop 体系的基本技术方法进行分析，并将理论与实际项目进行结合。本书层次清晰，结构合理，全书共分为 11 个章节，主要内容包括 Hadoop 简介、Hadoop 环境搭建、HDFS 文件系统操作、MapReduce 程序编写、Hive 安装与配置、Hive 操作、HBase 安装与配置、HBase 操作与客户端使用、Pig 安装与使用、Pig Latin 的使用、Sqoop 安装与数据操作、大数据项目案例分析等。

本书适合 Hadoop 的初学者进行阅读，也可作为高等院校相关课程的教学参考书。

本书封面贴有清华大学出版社防伪标签，无标签者不得销售。
版权所有，侵权必究。侵权举报电话：010-62782989　13701121933

图书在版编目(CIP)数据

Hadoop 核心技术与实战/潘正高，施霖编著. —北京：清华大学出版社，2022.3
大数据技术与项目实战系列教材
ISBN 978-7-302-52464-9

Ⅰ. ①H… Ⅱ. ①潘… ②施… Ⅲ. ①数据处理软件 Ⅳ. ①TP274

中国版本图书馆 CIP 数据核字(2019)第 043123 号

责任编辑：贾　斌　李　晔
封面设计：刘　键
责任校对：梁　毅
责任印制：朱雨萌

出版发行：清华大学出版社
　　　　　网　　址：http://www.tup.com.cn，http://www.wqbook.com
　　　　　地　　址：北京清华大学学研大厦 A 座　　　邮　编：100084
　　　　　社 总 机：010-83470000　　　　　　　　　邮　购：010-62786544
　　　　　投稿与读者服务：010-62776969，c-service@tup.tsinghua.edu.cn
　　　　　质量反馈：010-62772015，zhiliang@tup.tsinghua.edu.cn
　　　　　课件下载：http://www.tup.com.cn，010-83470236
印 装 者：三河市龙大印装有限公司
经　　销：全国新华书店
开　　本：185mm×260mm　　印　张：21　　　　　字　数：512 千字
版　　次：2022 年 4 月第 1 版　　　　　　　　　　印　次：2022 年 4 月第 1 次印刷
印　　数：1～1500
定　　价：59.80 元

产品编号：077978-01

FOREWORD 前言

2010年以后,信息产业进入了大数据时代,Hadoop成为大数据分析的首选平台和开发标准,数据分析软件纷纷向Hadoop靠拢。在Hadoop原有技术基础之上,涌现了Hadoop家族产品,它们正在配合"大数据"概念不断创新,推动科技进步。因此,新一代IT精英都必须顺应潮流,抓住机遇,随着Hadoop一起发展和成长!

Hadoop是一个能够对大量数据进行分布式处理的软件框架。但Hadoop是以一种可靠、高效、可伸缩的方式进行处理的。Hadoop是可靠的,因为它假设计算元素和存储会失败,因此它维护多个工作数据副本,确保能够针对失败的节点重新分布处理。Hadoop是高效的,因为它以并行的方式工作,通过并行处理加快处理速度。

Hadoop体系下包含了一系列进行数据处理、分析的组件,其中常见的组件包括Hive、HBase、Pig、Sqoop等。

Hive是一种底层封装了Hadoop的数据仓库处理工具,使用类SQL的HiveQL语言实现数据查询,所有Hive的数据都存储在Hadoop兼容的文件系统(例如,Amazon S3、HDFS)中。Hive在加载数据过程中不会对数据进行任何修改,只是将数据移动到HDFS中Hive设定的目录下。Hive构建在基于静态批处理的Hadoop之上,Hadoop通常都有较高的延迟并且在作业提交和调度的时候需要大量开销。因此,Hive并不能够在大规模数据集上实现低延迟快速的查询。

作为NoSQL家庭的一员,HBase的出现弥补了Hadoop只能离线批处理的不足,同时能够存储小文件,提供海量数据的随机检索,并保证一定的性能。而这些特性也完善了整个Hadoop生态系统,泛化了其大数据的处理能力,结合其高性能、稳定、扩展性好的特性,给使用大数据的企业带来了方便。

Pig是基于Hadoop的并行数据流处理开源引擎。通过Pig无须开发一个全功能的应用程序就可以在集群中进行数据批处理,这使得在新数据集上进行实验变得更加容易。

Sqoop是一个用来将Hadoop和关系型数据库中的数据进行相互转移的工具,可以将一个关系型数据库(例如,MySQL、Oracle、PostgreSQL等)中的数据导入Hadoop的HDFS中,也可以将HDFS的数据导入关系型数据库中。对于某些NoSQL数据库,它也提供了连接器。Sqoop使用元数据模型来判断数据类型并在数据从数据源转移到Hadoop时确保类型安全的数据处理。Sqoop专为大数据批量传输设计,能够分割数据集并创建Hadoop任务来处理每个区块。

本书针对Hadoop体系的基本技术方法进行了分析,并提供了相应的实例以帮助读者进一步加深了解。通过本书的学习,相信大家会在很短暂的时间内掌握Hadoop体系的相关技术，为以后的工作、学习提供指导与帮助。

本书特色

本书作者从实践出发，结合大量的教学经验以及工程案例，深入浅出地介绍大数据技术架构及相关组件。在章节的编排上，注重理论与实践相结合。首先提出相关的理论背景，并进行深入分析、讲解，然后着重介绍相关技术的环境搭建，最后通过实际操作，加深读者对技术的掌握及应用。通过项目实战案例介绍相关组件在实际大数据处理中的关键应用，本书介绍的 Hadoop 组件包括 HDFS、MapReduce、Hive、HBase、Pig、Sqoop。

为了方便读者对内容的理解以及满足相关教学、工作的需要，本书配套提供了真实的样本数据文件、PPT 课件以及实验视频，读者可以根据"勘误与支持"中提供的联系方式进行咨询或者获取。

本书适用对象

本书内容由浅入深，既适合初学者入门，也适合有一定基础的技术人员进一步提高技术水平。本书的读者对象包括：

- Hadoop 初学者
- Hadoop 开发人员
- Hadoop 管理人员
- 高等院校计算机相关专业的老师、学生
- 具有 Hadoop 相关经验，并希望进一步提高技术水平的读者

如何阅读本书

本书在章节的安排上，着眼于引导读者以最快的速度上手 Hadoop，本书一共包括 11 章，分为 3 个部分：基础篇、高级篇、实战篇。

基础篇（第 1～3 章）：第 1 章的主要内容包括 Hadoop 简介、Hadoop 项目及架构分析、Hadoop 计算模型、Hadoop 数据管理、Hadoop 环境搭建；第 2 章针对 Hadoop 分布式文件系统 HDFS 进行讲解，包括 HDFS 基本操作、WebHDFS 操作以及通过 Java API 进行 HDFS 文件操作；第 3 章针对 Hadoop 分布式计算 MapReduce 进行讲解，并通过实战案例帮助读者加深对相关知识的理解。

高级篇（第 4～10 章）：第 4 章针对 Hadoop 中的数据仓库 Hive 进行分析，包括 Hive 环境搭建、Hive 数据定义；第 5 章介绍了 Hive 的使用，包括数据操作、数据查询，并通过实战案例帮助读者进行深入的了解；第 6 章针对 HBase 进行详细的介绍，包括 HBase 的发展历史、HBase 的特性、HBase 与 Hadoop 的关系等，并讲解如何进行 HBase 环境的搭建；第 7 章针对 HBase 数据操作进行讲解，包括 Shell 工具的使用、基于 HBase 的 Java 客户端的使用；第 8 章介绍了 Pig 数据流引擎的使用，包括 Pig 概述、Pig 的安装与配置、Pig 命令行交互工具的使用；第 9 章针对 Pig Latin 进行讲解，主要内容包括 Pig Latin 介绍、关系操作和用户自定义函数的使用；第 10 章详细讲解 Sqoop 工具的使用，包括 Sqoop 概述、Sqoop 安装部署、Sqoop 常用命令介绍和 Sqoop 数据操作。

实战篇（第 11 章）：通过实际的大数据案例进行分析讲解，帮助读者进一步了解实际工程环境中 Hadoop 的应用，本章主要内容包括实战项目背景与数据源分析、环境搭建、数据清洗、数据统计分析以及定时任务处理。

致谢

在本书的编写过程中,得到了许多企事业单位人员的大力支持;在此谨向北京科技大学陈红松教授致以深深的谢意。在本书的编辑和出版过程中还得到了清华大学出版社相关人员的无私帮助与支持,在此一并表示感谢。

编　者

2021 年 12 月

目录 CONTENTS

基 础 篇

第 1 章 Hadoop 基础 …………………………………………………………………… 3

- 1.1 Hadoop 简介 …………………………………………………………………… 3
  - 1.1.1 什么是 Hadoop ……………………………………………………… 3
  - 1.1.2 Hadoop 项目及其结构 ……………………………………………… 5
  - 1.1.3 Hadoop 体系结构 …………………………………………………… 7
  - 1.1.4 Hadoop 与分布式开发 ……………………………………………… 8
  - 1.1.5 Hadoop 计算模型——MapReduce on Yarn ……………………… 10
  - 1.1.6 Hadoop 数据管理 …………………………………………………… 12
  - 1.1.7 Hadoop 集群安全策略 ……………………………………………… 14
- 1.2 Hadoop 的安装与配置 ………………………………………………………… 17
  - 1.2.1 安装 JDK 1.8 与配置 SSH 免密码登录 …………………………… 17
  - 1.2.2 安装并运行 Hadoop ………………………………………………… 20

第 2 章 Hadoop 存储：HDFS ………………………………………………………… 27

- 2.1 HDFS 的基本操作 ……………………………………………………………… 27
  - 2.1.1 HDFS 的命令行操作 ………………………………………………… 27
  - 2.1.2 HDFS 的 Web 界面 ………………………………………………… 28
  - 2.1.3 通过 distcp 进行并行复制 ………………………………………… 29
  - 2.1.4 使用 Hadoop 归档文件 ……………………………………………… 31
- 2.2 WebHDFS ……………………………………………………………………… 35
  - 2.2.1 WebHDFS 的配置 …………………………………………………… 35
  - 2.2.2 WebHDFS 命令 ……………………………………………………… 36
- 2.3 HDFS 常见的 Java API 介绍 ………………………………………………… 41
  - 2.3.1 使用 Hadoop URL 读取数据 ………………………………………… 41
  - 2.3.2 使用 FileSystem API 读取数据 …………………………………… 42
  - 2.3.3 创建目录 ……………………………………………………………… 44
  - 2.3.4 写数据 ………………………………………………………………… 45

2.3.5　删除数据 ·· 46
　　2.3.6　文件系统查询 ·· 46

第 3 章　Hadoop 计算：MapReduce ··· 51

3.1　MapReduce 应用程序编写 ·· 51
　　3.1.1　实例描述 ·· 51
　　3.1.2　设计思路 ·· 52
　　3.1.3　代码数据流 ·· 52
　　3.1.4　程序代码 ·· 53
　　3.1.5　代码解读 ·· 55
　　3.1.6　程序执行 ·· 56
　　3.1.7　代码结果 ·· 58

3.2　使用 MapReduce 求每年最低温度 ··· 58
　　3.2.1　作业描述 ·· 58
　　3.2.2　程序代码 ·· 58
　　3.2.3　准备输入数据 ·· 60
　　3.2.4　运行程序 ·· 60

高　级　篇

第 4 章　数据仓库：Hive ··· 63

4.1　Hive 的安装和配置 ·· 65
　　4.1.1　安装详细步骤 ·· 65
　　4.1.2　Hive 内部是什么 ··· 72

4.2　数据定义 ·· 73
　　4.2.1　Hive 中的数据库 ··· 73
　　4.2.2　修改数据库 ·· 77
　　4.2.3　创建表 ·· 78
　　4.2.4　分区表 ·· 85
　　4.2.5　删除表 ·· 94
　　4.2.6　修改表 ·· 95

第 5 章　Hive 数据操作与查询 ·· 101

5.1　数据操作 ·· 101
　　5.1.1　向管理表中装载数据 ·· 101
　　5.1.2　通过查询语句向表中插入数据 ··· 105
　　5.1.3　单个查询语句中创建表并加载数据 ··· 109
　　5.1.4　导出数据 ·· 110

5.2 数据查询 ·· 112
    5.2.1 SELECT⋯FROM 语句 ··· 112
    5.2.2 WHERE 语句 ·· 128
    5.2.3 GROUP BY 语句 ·· 132
    5.2.4 HAVING 语句 ·· 132
    5.2.5 JOIN 语句 ··· 133
    5.2.6 ORDER BY 和 SORT BY ··· 143
    5.2.7 含有 SORT BY 的 DISTRIBUTE BY ·· 145
    5.2.8 CLUSTER BY ··· 146
    5.2.9 类型转换 ··· 146
    5.2.10 抽样查询 ··· 147
    5.2.11 UNION ALL ·· 150
5.3 Hive 实战 ··· 150
    5.3.1 背景 ··· 150
    5.3.2 实战数据及要求 ·· 150
    5.3.3 实验步骤 ··· 152

第 6 章 Hadoop 数据库：HBase ··· 159

6.1 HBase 概述 ··· 159
    6.1.1 HBase 的发展历史 ··· 159
    6.1.2 HBase 的发行版本 ··· 160
    6.1.3 HBase 的特性 ··· 161
    6.1.4 HBase 与 Hadoop 的关系 ··· 162
    6.1.5 HBase 的核心功能模块 ·· 162
6.2 HBase 的安装和配置 ·· 166
    6.2.1 HBase 的运行模式 ··· 167
    6.2.2 HBase 的 Web UI ··· 173
    6.2.3 Hbase Shell 工具使用 ··· 174
    6.2.4 停止 HBase 集群 ·· 175

第 7 章 HBase 数据操作 ·· 176

7.1 Shell 工具的使用 ··· 176
    7.1.1 命令分类 ··· 176
    7.1.2 常规命令 ··· 178
    7.1.3 DDL 命令 ·· 178
    7.1.4 DML 命令 ·· 180
    7.1.5 工具命令 Tools ··· 183

7.1.6　复制命令 ··· 184
　　　7.1.7　安全命令 ··· 185
　7.2　Java 客户端的使用 ·· 185
　　　7.2.1　客户端配置 ··· 185
　　　7.2.2　创建表 ··· 194
　　　7.2.3　删除表 ··· 195
　　　7.2.4　插入数据 ··· 195
　　　7.2.5　查询数据 ··· 197
　　　7.2.6　删除数据 ··· 200

第 8 章　并行数据流处理引擎：Pig ·· 202

　8.1　Pig 概述 ··· 202
　　　8.1.1　Pig 是什么 ·· 202
　　　8.1.2　Pig 的发展简史 ·· 208
　8.2　Pig 的安装和使用 ·· 209
　　　8.2.1　下载和安装 Pig ··· 209
　　　8.2.2　命令行使用以及配置选项介绍 ··· 213
　　　8.2.3　返回码 ··· 215
　8.3　命令行交互工具 ··· 216
　　　8.3.1　Grunt 概述 ··· 216
　　　8.3.2　在 Grunt 中输入 Pig Latin 脚本 ·· 217
　　　8.3.3　在 Grunt 中使用 HDFS 命令 ··· 218
　　　8.3.4　在 Grunt 中控制 Pig ·· 220

第 9 章　Pig Latin 的使用 ·· 222

　9.1　Pig Latin 概述 ··· 222
　　　9.1.1　基础知识 ··· 222
　　　9.1.2　输入和输出 ··· 223
　9.2　关系操作 ··· 225
　　　9.2.1　foreach ··· 226
　　　9.2.2　Filter ··· 234
　　　9.2.3　Group ·· 236
　　　9.2.4　Order by ··· 239
　　　9.2.5　distinct ··· 242
　　　9.2.6　Join ·· 242
　　　9.2.7　Limit ··· 247
　　　9.2.8　Sample ··· 248

  9.2.9 Parallel ………………………………………………………………………… 249

 9.3 用户自定义函数 UDF ………………………………………………………………… 252

  9.3.1 注册 UDF ………………………………………………………………… 252

  9.3.2 define 命令和 UDF ……………………………………………………… 255

  9.3.3 调用静态 Java 函数 ……………………………………………………… 258

第 10 章　SQL to Hadoop：Sqoop …………………………………………………… 260

 10.1 Sqoop 概述 ………………………………………………………………………… 260

  10.1.1 Sqoop 的产生背景 ……………………………………………………… 260

  10.1.2 Sqoop 是什么 …………………………………………………………… 260

  10.1.3 为什么选择 Sqoop ……………………………………………………… 261

  10.1.4 Sqoop1 和 Sqoop2 的异同 ……………………………………………… 261

  10.1.5 Sqoop1 与 Sqoop2 的架构图 …………………………………………… 261

  10.1.6 Sqoop1 与 Sqoop2 的优缺点 …………………………………………… 262

 10.2 Sqoop 安装部署 …………………………………………………………………… 263

  10.2.1 下载 Sqoop ……………………………………………………………… 263

  10.2.2 设置 /etc/profile 参数 …………………………………………………… 265

  10.2.3 设置 bin/configure-sqoop 配置文件 …………………………………… 265

  10.2.4 设置 conf/sqoop-env.sh 配置文件 ……………………………………… 265

  10.2.5 验证安装完成 …………………………………………………………… 266

 10.3 Sqoop 常用命令介绍 ……………………………………………………………… 267

  10.3.1 如何列出帮助 …………………………………………………………… 267

  10.3.2 Export …………………………………………………………………… 268

  10.3.3 Import …………………………………………………………………… 268

  10.3.4 Job 作业 ………………………………………………………………… 270

 10.4 数据操作 …………………………………………………………………………… 273

  10.4.1 MySQL 数据导入到 HDFS 中 ………………………………………… 273

  10.4.2 HDFS 数据导入到 MySQL 中 ………………………………………… 281

实 战 篇

第 11 章　项目实战 …………………………………………………………………………… 285

 11.1 项目背景与数据情况 ……………………………………………………………… 285

  11.1.1 项目概述 ………………………………………………………………… 285

  11.1.2 项目分析指标 …………………………………………………………… 286

  11.1.3 项目开发步骤 …………………………………………………………… 287

11.1.4 表结构设计 ·· 287
11.2 环境搭建 ··· 287
　11.2.1 MySQL 的安装 ·· 287
　11.2.2 Eclipse 的安装 ··· 291
11.3 数据清洗 ··· 293
　11.3.1 数据分析 ··· 293
　11.3.2 数据清洗流程 ·· 294
11.4 数据统计分析 ·· 314
　11.4.1 建立分区表 ·· 314
　11.4.2 使用 HQL 统计关键指标 ····························· 315
　11.4.3 使用 Sqoop 将数据导入到 MySQL 数据表 ········ 318
11.5 定时任务处理 ·· 319
　11.5.1 日志数据定时上传 ····································· 320
　11.5.2 日志数据定期清理 ····································· 320
　11.5.3 数据定时统计分析 ····································· 321

参考文献 ··· 323

基础篇

第 1 章

Hadoop基础

Hadoop 是 Apache 软件基金会旗下的一个开源分布式计算平台。以 Hadoop 分布式文件系统(Hadoop Distributed File System,HDFS)和 MapReduce(Google MapReduce 的开源实现)为核心的 Hadoop 为用户提供了系统底层细节透明的分布式基础架构。HDFS 的高容错性、高伸缩性等优点允许用户将 Hadoop 部署在低廉的硬件上,形成分布式系统;MapReduce 分布式编程模型允许用户在不了解分布式系统底层细节的情况下开发并行应用程序。所以用户可以利用 Hadoop 轻松地组织计算机资源,从而搭建自己的分布式计算平台,并且可以充分利用集群的计算和存储能力,完成海量数据的处理。经过业界和学术界长达十几年的锤炼,目前 Hadoop 2.7.1 是最新的稳定版本,在实际的数据处理和分析任务中担当着不可替代的角色。

1.1 Hadoop 简介

1.1.1 什么是 Hadoop

1. Hadoop 的历史

Hadoop 的源头是 Apache Nutch,该项目始于 2002 年,是 Apache Lucene 的子项目之一。2004 年,Google 在"操作系统设计与实现"(Operating System Design and Implementation,OSDI)会议上公开发表了题为 *MapReduce：Simplified Data Processing on Large Clusters*(《MapReduce：简化大规模集群上的数据处理》)的论文之后,受到启发的 Doug Cutting 等人开始尝试实现 MapReduce 计算框架,并将它与 NDFS(Nutch Distributed File System)结合,用来支持 Nutch 引擎的主要算法。由于 NDFS 和 MapReduce 在 Nutch 引擎中有着良好的应用,所以它们于 2006 年 2 月被分离出来,成为一套完整而独立的软件,并命名为 Hadoop。到了 2008

年年初，Hadoop 已成为 Apache 的顶级项目，包含众多子项目。它被包括 Yahoo！在内的很多互联网公司应用。现在的 Hadoop 2.7.1 版本已经发展成为包含 HDFS、MapReduce 子项目，与 Pig、ZooKeeper、Hive、HBase 等项目相关的大型应用工程。

2. Hadoop 的功能与作用

我们为什么需要 Hadoop 呢？众所周知，现代社会信息增长速度很快，这些信息中又积累着大量数据，其中包括个人数据和工业数据。预计到 2023 年，每年产生的数字信息中将会有超过 1/3 的内容驻留在云平台或借助云平台处理。我们需要对这些数据进行分析处理，以获取更多有价值的信息。那么如何高效地存储管理这些数据、如何分析这些数据呢？这时可以选用 Hadoop 系统。在处理这类问题时，它采用分布式存储方式来提高读写速度和扩大存储容量；采用 MapReduce 整合分布式文件系统上的数据，保证高速分析处理数据；与此同时，还采用存储冗余数据来保证数据的安全性。

Hadoop 中的 HDFS 具有高容错性，并且是基于 Java 语言开发的，这使得 Hadoop 可以部署在成本低廉的计算机集群中，同时不限于某个操作系统。Hadoop 中 HDFS 的数据管理能力、MapReduce 处理任务时的高效率以及它的开源特性，使其在同类分布式系统中大放异彩，并在众多行业和科研领域中被广泛应用。

3. Hadoop 的优势

Hadoop 是一个能够让用户轻松架构和使用的分布式计算平台。用户可以轻松地在 Hadoop 上开发运行处理海量数据的应用程序。它主要有以下几个优点：

（1）高可靠性。Hadoop 按位存储和处理数据的能力值得人们信赖。

（2）高扩展性。Hadoop 是在可用的计算机集簇间分配数据完成计算任务的，这些集簇可以方便地扩展到数以千计的节点中。

（3）高效性。Hadoop 能够在节点之间动态地移动数据，以保证各个节点的动态平衡，因此其处理速度非常快。

（4）高容错性。Hadoop 能够自动保存数据的多份副本，并且能够自动将失败的任务重新分配。

4. Hadoop 应用现状和发展趋势

由于 Hadoop 优势突出，基于 Hadoop 的应用已经遍地开花，尤其是在互联网领域。Yahoo！通过集群运行 Hadoop，用于支持广告系统和 Web 搜索的研究；Facebook 借助集群运行 Hadoop 来支持其数据分析和机器学习；搜索引擎公司百度则使用 Hadoop 进行搜索日志分析和网页数据挖掘工作；淘宝的 Hadoop 系统用于存储并处理电子商务交易的相关数据；中国移动研究院基于 Hadoop 的"大云"（BigCloud）系统对数据进行分析并对外提供服务。

2008 年 2 月，作为 Hadoop 最大贡献者的 Yahoo！构建了当时最大规模的 Hadoop 应用。他们在 2000 个节点上面执行了超过 1 万个 Hadoop 虚拟机器来处理超过 5PB 的网页内容，分析大约 1 兆个网络连接之间的网页索引资料。这些网页索引资料压缩后超过 300TB。Yahoo！正是基于这些为用户提供了高质量的搜索服务。

Hadoop 目前已经取得了非常突出的成绩。随着互联网的发展，新的业务模式还将不断涌现，Hadoop 的应用也会从互联网领域向电信、电子商务、银行、生物制药等领域拓展。相信在未来，Hadoop 将会在更多的领域中扮演幕后英雄，为用户提供更加快捷优质的服务。

1.1.2 Hadoop 项目及其结构

现在 Hadoop 已经发展成为包含很多项目的集合。虽然其核心内容是 MapReduce 和 Hadoop 分布式文件系统，但与 Hadoop 相关的 Common、Avro、Chukwa、Hive、HBase 等项目也是不可或缺的。它们提供了互补性服务或在核心层上提供了更高层的服务。图 1.1 是 Hadoop 的项目结构图。

下面将对 Hadoop 的各个关联项目进行更详细的介绍。

Common：Common 是为 Hadoop 其他子项目提供支持的常用工具，它主要包括 FileSystem、RPC 和串行化库。它们为在廉价硬件上搭建云计算环境提供基本的服务，并且会为运行在该平台上的软件开发提供所需的 API。

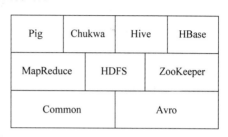

图 1.1　Hadoop 的项目结构图

Avro：Avro 是用于数据序列化的系统。它提供了丰富的数据结构类型、快速可压缩的二进制数据格式、存储持久性数据的文件集、远程调用 RPC 的功能和简单的动态语言集成功能。其中代码生成器既不需要读写文件数据，也不需要使用或实现 RPC 协议，它只是一个可选的对静态类型语言的实现。

Avro 系统依赖于模式 (Shcema)，数据的读和写是在模式之下完成的。这样可以减少写入数据的开销，提高序列化的速度并缩减其大小；同时，也可以方便动态脚本语言的使用，因为数据连同其模式都是自描述的。

在 RPC 中，Avro 系统的客户端和服务端通过握手协议进行模式的交换，因此当客户端和服务端拥有彼此全部的模式时，不同模式下相同命名字段、丢失字段和附加字段等信息的一致性问题就得到了很好的解决。

MapReduce：MapReduce 是一种编程模式，用于大规模数据集 (大于 1TB) 的并行运算。映射 (Map)、化简 (Reduce) 的概念及其主要思想都是从函数式编程语言中借鉴而来的。它极大地方便了编程人员——即使在不了解分布式并行编程的情况下，也可以将自己的程序运行在分布式系统上。MapReduce 在执行时先指定一个 Map (映射) 函数，把输入键值对映射成一组新的键值对，经过一定处理后交给 Reduce，Reduce 对相同 key 下的所有 value 进行处理后再输出键值对作为最终的结果。

图 1.2 是 MapReduce 的任务处理流程图，它展示了 MapReduce 程序将输入划分到不同的 Map 上，再将 Map 的结果合并到 Reduce，然后进行处理的输出过程。

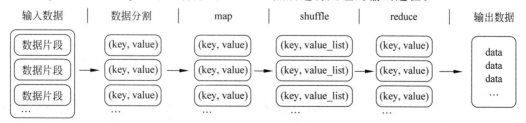

图 1.2　MapReduce 的任务处理流程图

HDFS：HDFS 是一个分布式文件系统。因为 HDFS 具有高容错性（fault-tolerent）的特点，所以它可以设计部署在成本低廉（low-cost）的硬件上。它可以通过提供高吞吐率（high throughput）来访问应用程序的数据，适合那些有着超大数据集的应用程序。HDFS 放宽了对可移植操作系统接口（Portable Operating System Interface，POSIX of UNIX）的要求，这样可以实现以流的形式访问文件系统中的数据。HDFS 原本是开源的 Apache 项目 Nutch 的基础结构，最后它却成为了 Hadoop 基础架构之一。

以下几个方面是 HDFS 的设计目标：

（1）检测和快速恢复硬件故障。硬件故障是计算机常见的问题。整个 HDFS 系统由数百甚至数千个存储着数据文件的服务器组成。如此多的服务器则意味着高故障率，因此，故障的检测和快速自动恢复是 HDFS 的一个核心目标。

（2）流式的数据访问。HDFS 使应用程序流式地访问它们的数据集，HDFS 被设计成适合进行批量处理，而不是用户交互式处理。所以它重视数据吞吐量，而不是数据访问的反应速度。

（3）简化一致性模型。大部分的 HDFS 程序对文件的操作需要一次写入，多次读取。一个文件一旦经过创建、写入、关闭就不需要修改了。这个假设简化了数据一致性问题和高吞吐量的数据访问问题。

（4）通信协议。所有的通信协议都是在 TCP/IP 协议之上的，一个客户端和明确配置了端口的名字节点（Name Node）建立连接之后，它和名字节点的协议便是客户端协议（Client Protocal）。数据节点（Data Node）和名字节点之间则用数据节点协议（Data Node Protocal）。

Chukwa：Chukwa 是开源的数据收集系统，用于监控和分析大型分布式系统的数据。Chukwa 是在 Hadoop 的 HDFS 和 MapReduce 框架之上搭建的，它继承了 Hadoop 的可扩展性和健壮性。Chukwa 通过 HDFS 来存储数据，并依赖 MapReduce 任务处理数据。Chukwa 中也附带了灵活且强大的工具，用于显示、监视和分析数据结果，以便更好地利用所收集的数据。

Hive：Hive 最早是由 Facebook 设计的，是一个建立在 Hadoop 基础之上的数据仓库，它提供了一些用于对 Hadoop 文件中数据集进行数据整理、特殊查询和分析存储的工具。Hive 提供的是一种结构化数据的机制，它支持类似于传统 RDBMS 中的 SQL 语言的查询语言，来帮助那些熟悉 SQL 的用户查询 Hadoop 中的数据，该查询语言称为 Hive QL。与此同时，传统的 MapReduce 编程人员也可以在 Mapper 或 Reducer 中通过 Hive QL 查询数据。Hive 编译器会把 Hive QL 编译成一组 MapReduce 任务，从而方便 MapReduce 编程人员进行 Hadoop 系统开发。

HBase：HBase 是一个分布式的、面向列的开源数据库，该技术来源于 Google 论文《Bigtable：一个结构化数据的分布式存储系统》。如同 Bigtable 利用了 Google 文件系统（Google File System）提供的分布式数据存储方式一样，HBase 在 Hadoop 之上提供了类似于 Bigtable 的能力。HBase 不同于一般的关系数据库，原因有两个：其一，HBase 是一个适合于非结构化数据存储的数据库；其二，HBase 是基于列而不是基于行的模式。HBase 和 Bigtable 使用相同的数据模型。用户将数据存储在一个表里，一个数据行拥有一个可选的键和任意数量的列。由于 HBase 表是疏松的，用户可以为行定义各种不同的列。HBase 主要用于需要随机访问、实时读写的大数据(Big Data)。

Pig：Pig 是一个对大型数据集进行分析、评估的平台。Pig 最突出的优势是它的结构能够经受住高度并行化的检验,这个特性使得它能够处理大型的数据集。目前,Pig 的底层由一个编译器组成,它在运行的时候会产生一些 MapReduce 程序序列,Pig 的语言层由一种叫做 Pig Latin 的正文型语言组成。

ZooKeeper：ZooKeeper 是一个为分布式应用所设计的开源协调服务。它主要为用户提供同步、配置管理、分组和命名等服务,减轻分布式应用程序所承担的协调任务。ZooKeeper 的文件系统使用了我们所熟悉的目录树结构。ZooKeeper 是使用 Java 编写的,但是它支持 Java 和 C 两种编程语言。

1.1.3　Hadoop 体系结构

如前所述,HDFS 和 MapReduce 是 Hadoop 的两大核心。而整个 Hadoop 的体系结构主要是通过 HDFS 来实现分布式存储的底层支持的,并且它会通过 MapReduce 来实现分布式并行任务处理的程序支持。

下面首先介绍 HDFS 的体系结构。HDFS 采用了主从(Master/Slave)结构模型,一个 HDFS 集群是由多个 NameNode 和若干个 DataNode 组成的,可以同时部署多个 NameNode,这些 NameNodes 之间是相互独立。其中 NameNode 作为主服务器,管理文件系统的命名空间和客户端对文件的访问操作。集群中的 DataNode 管理存储的数据,HDFS 允许用于以文件的形式存储数据。从内部来看,文件被分成若干个数据块,而且这若干个数据块存放在一组 DataNode 上。NameNode 执行文件系统的命名空间操作,比如打开、关闭、重命名文件或目录等,它也负责数据块到具体 DataNode 的映射。DataNode 负责处理文件系统客户端的文件读写请求,并在 NameNode 的统一调度下进行数据块的创建、删除和复制工作。图 1.3 介绍了 HDFS 单个 NameNode 的体系结构。

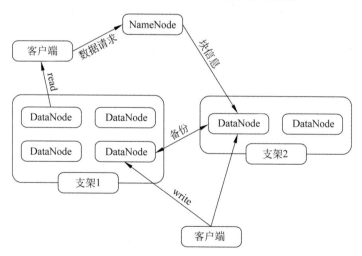

图 1.3　HDFS 单个 NameNode 的体系结构

NameNode 和 DataNode 都可以在普通商用计算机上运行。这些计算机通常运行的是 GNU/Linux 操作系统。HDFS 采用 Java 语言开发,因此任何支持 Java 的机器都可以部署

NameNode 和 DataNode。一个典型的部署场景是集群中的一台机器运行一个 NameNode 实例，其他机器分别运行一个 DataNode 实例。当然，并不排除一台机器运行多个 DataNode 实例的情况。NameNode 是所有 HDFS 元数据的管理者，用户需要保存的数据不会经过 NameNode，而是直接流向存储数据的 DataNode。

MapReduce 是一种并行编程模式，利用这种模式软件开发者可以轻松地编写出分布式并行程序。在 Hadoop 的体系结构中，MapReduce 是一个简单易用的软件框架，基于它可以将任务分发到由上千台商用计算机组成的集群上，并以一种可靠容错的方式并行处理大量的数据集，实现 Hadoop 的并行任务处理功能。MapReduce 框架由 ResourceManager、ApplicationMaster、NodeManager 三个组件组成，可实现资源管理和任务调度/监控的功能。

从上面的介绍可以看出，HDFS 和 MapReduce 共同组成了 Hadoop 分布式系统体系结构的核心。HDFS 在集群上实现了分布式文件系统，MapReduce 在集群上实现了分布式计算和任务处理。HDFS 在 MapReduce 任务处理过程中提供了对文件操作和存储等的支持，MapReduce 在 HDFS 的基础上实现了任务的分发、跟踪、执行等工作，并收集结果，二者相互作用，完成了 Hadoop 分布式集群的主要任务。

1.1.4 Hadoop 与分布式开发

我们通常所说的分布式系统其实是分布式软件系统，即支持分布式处理的软件系统。它是在通信网络互联的多处理机体系结构上执行任务的系统，包括分布式操作系统、分布式程序设计语言及其编译（解释）系统、分布式文件系统和分布式数据库系统等。Hadoop 是分布式软件系统中文件系统层的软件，它实现了分布式文件系统和部分分布式数据库系统的功能。Hadoop 中的分布式文件系统 HDFS 能够实现数据在计算机集群组成的云上高效地存储和管理，Hadoop 中的并行编程框架 MapReduce 能够让用户编写的 Hadoop 并行应用程序运行得以简化。下面简单介绍基于 Hadoop 进行分布式并发编程的相关知识，详细的介绍请参看 3.1 节。

Hadoop 上并行应用程序的开发是基于 MapReduce 编程模型的。MapReduce 编程模型的原理是：利用一个输入的键值对集合来产生一个输出的键值对集合。MapReduce 库的用户用两个函数来表达这个计算：Map 和 Reduce。

用户自定义的 Map 函数接收一个输入的键值对，然后产生一个中间键值对的集合。MapReduce 把所有具有相同 key 的 value 集合在一起，然后传递给 Reduce 函数。

用户自定义的 Reduce 函数接收 key 和相关 value 集合。Reduce 函数合并这些 value，形成一个较小的 value 集合。一般来说，每次调用 Reduce 函数只产生 0 或 1 个输出 value。通常通过一个迭代器把中间 value 提供给 Reduce 函数，这样就可以处理无法全部放入内存中的大量的 value 集合了。

图 1.4 是 MapReduce 的数据流图，体现了 MapReduce 处理大数据集的过程。简言之，这个过程就是将大数据集分解为成百上千个小数据集，每个（或若干个）数据集分别由集群中的一个节点（一般就是一台普通的计算机）进行处理并生成中间结果，然后这些中间结果又由大量节点合并，形成最终结果。图 1.4 也说明了 MapReduce 框架下并行程序中的两个主要函数：Map、Reduce。在这个结构中，用户需要完成的工作是根据任务编写 Map 和

Reduce 两个函数。

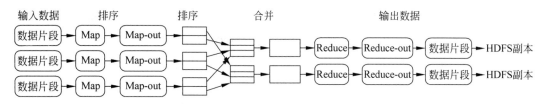

图 1.4　MapReduce 的数据流图

MapReduce 计算模型非常适合在大量计算机组成的大规模集群上并行运行。图 1.4 中的每一个 Map 任务和每一个 Reduce 任务均可以同时运行于一个单独的计算节点上，可想而知，其运算效率是很高的，那么这样的并行计算是如何做到的呢？下面将简单介绍其原理。

1．数据分布存储

Hadoop 分布式文件系统（HDFS）由一个名字节点（NameNode）和多个数据节点（DataNode）组成，每个节点都是一台普通的计算机。在使用方式上 HDFS 与我们熟悉的单机文件系统非常类似，利用它可以创建目录，创建、复制、删除文件，并且可以查看文件内容等。但文件在 HDFS 底层被切割成了 Block，这些 Block 分散地存储在不同的 DataNode 上，每个 Block 还可以复制数份数据存储在不同的 DataNode 上，达到容错容灾的目的。NameNode 则是整个 HDFS 的核心，它通过维护一些数据结构来记录每一个文件被切割成了多少个 Block，这些 Block 可以从哪些 DataNode 中获得，以及各个 DataNode 的状态等重要信息。

2．分布式并行计算

Hadoop 中有一个作为主控的 ResourceManger，用于调度和管理其他的 ApplicationMaster、NodeManger。类似于 Google 的 Map/Reduce 计算框架，它能够把应用程序分割成许多很小的工作单元，每个单元可以在任何集群节点上执行或重复执行。可见，Map/Reduce 是一种简化的分布式编程模式，可以让程序自动在由普通计算机组成的集群中以并行方式分布执行。

3．本地计算

数据存储在哪一台计算机上，就由哪台计算机进行这部分数据的计算，这样可以减少数据在网络上的传输，降低对网络带宽的需求。在 Hadoop 这类基于集群的分布式并行系统中，计算节点可以很方便地扩充，因此它所能够提供的计算能力近乎无限。但是数据需要在不同的计算机之间流动，故而网络带宽变成了瓶颈。"本地计算"是一种最有效的节约网络带宽的手段，业界将此形容为"移动计算比移动数据更经济"。

4．任务粒度

在把原始大数据集切割成小数据集时，通常让小数据集小于或等于 HDFS 中一个 Block 的大小（默认是 64MB），这样能够保证一个小数据集是位于一台计算机上的，便于本地计算。假设有 M 个小数据集待处理，就启动 M 个 Map 任务，注意这 M 个 Map 任务分布于 N 台计算机上，它们将并行运行，Reduce 任务的数量 R 则可由用户指定。

5. 数据分割（Partition）

把 Map 任务输出的中间结果按 key 的范围划分成 R 份（R 是预先定义的 Reduce 任务的个数），划分时通常使用 Hash 函数（如 hash(key) mod R），这样可以保证某一段范围内的 key 一定是由一个 Reduce 任务来处理的，可以简化 Reduce 的过程。

6. 数据合并（Combine）

在数据分割之前，还可以先对中间结果进行数据合并，即将中间结果中有相同 key 的键值对合并成一对。Combine 的过程与 Reduce 的过程类似，在很多情况下可以直接使用 Reduce 函数，但 Combine 是作为 Map 任务的一部分、在执行完 Map 函数后紧接着执行的。Combine 能够减少中间结果中键值对的数目，从而降低网络流量。

7. Reduce

Map 任务的中间结果在执行完 Combine 和 Partition 之后，以文件形式存储于本地磁盘上。当所有 Map 的任务结束后，ApplicationMaster 通过心跳机制（heartbeat mechanism），由它知道 mapping 的输出结果与机器 host，所以 reducer 会定时通过一个线程访问 Applicationmaster 请求 Map 的输出结果，将 Map 的输出结果作为 Reduce 的输入。注意，所有的 Map 任务产生的中间结果均按其 key 通过同一个 Hash 函数划分成了 R 份，R 个 Reduce 任务各自负责一段 key 区间。每个 Reduce 需要向许多个 Map 任务节点取得落在其负责的 key 区间内的中间结果，然后执行 Reduce 函数，形成一个最终的结果文件。

8. 任务管道

有 R 个 Reduce 任务，就会有 R 个最终结果。很多情况下这 R 个最终结果并不需要合并成一个最终结果，因为这 R 个最终结果又可以作为另一个计算任务的输入，开始另一个并行计算任务，这也就形成了任务管道。

以上简要介绍了在并行编程方面 Hadoop 中 MapReduce 编程模型的原理、流程、程序结构和并行计算的实现，MapReduce 程序的详细流程、编程接口、程序实例等请参见第 3 章。

1.1.5 Hadoop 计算模型——MapReduce on Yarn

MapReduce 是 Google 公司的核心计算模型，它将运行于大规模集群上的复杂的并行计算过程高度地抽象为两个函数：Map 和 Reduce。Hadoop 是 Doug Cutting 受到 Google 发表的关于 MapReduce 的论文启发而开发出来的。Hadoop 中 MapReduce 是一个使用简单的软件框架，基于它写出来的应用程序能够运行在由上千台商用计算机组成的大型集群上，并以一种可靠容错的方式并行处理上 T 级别的数据集，实现了 Hadoop 在集群上的数据和任务的并行计算与处理。

一个 MapReduce 作业（Job）通常会把输入的数据集切分为若干独立的数据块，由 Map 任务（Task）以完全并行的方式处理它们。框架会先对 Map 的输出进行排序，然后把结果输入给 Reduce 任务。通常作业的输入和输出都会被存储在文件系统中。整个框架负责任务的调度和监控，以及重新执行已经失败的任务。

通常，Map/Reduce 框架和分布式文件系统是运行在一组相同的节点上的，也就是说，计算节点和存储节点在一起。这种配置允许框架在那些已经存好数据的节点上高效地调度任务,这样可以使整个集群的网络带宽得到非常高效的利用。

基于 Yarn 的 MapReduce 框架由 ApplicationMaster、NodeManager、ResourceManager、Container 组成,可实现资源管理和任务调度/监控的功能。

1. 资源管理器(ResourceManager)

根据功能不同将资源管理器分成两个组件:调度器(Scheduler)和应用管理器。调度器根据集群中容量、队列和资源等限制,将资源分配给各个正在运行的应用。虽然被称为调度器,但是它仅负责资源的分配,而不负责监控各个应用的执行情况和任务失败、应用失败或硬件失败时的重启任务。调度器根据各个应用的资源需求和集群各个节点的资源容器(Resource Container,是集群节点将自身内存、CPU、磁盘等资源封装在一起的抽象概念)进行调度。应用管理器负责接收作业,协商获取第一个资源容器用于执行应用的任务主题并为重启失败的应用主题分配容器。

2. 节点管理器(NodeManager)

节点管理器是每个节点的框架代理。它负责启动应用的容器,监控容器的资源使用(包括 CPU、内存、硬盘和网络带宽等),并把这些有用信息汇报给调度器。应用对应的应用主体负责通过协商从调度器处获取资源容器,并跟踪这些容器的状态和应用执行的情况。

集群每个节点上都有一个节点管理器,它的主要责任是:

(1) 为应用启用调度器已分配给应用的容器。

(2) 保证已启动的容器不会使用超过分配的资源量。

(3) 为 task 构建容器环境,包括二进制可执行文件,如.jars 等。

(4) 为所在的节点提供一个管理本地存储资源的简单服务。

应用程序可以继续使用本地存储资源,即使它没有从资源管理器处申请。比如:MapReduce 可以利用这个服务存储 Map 任务的中间输出结果并将其混洗、排序给 Reduce 任务。

3. 应用主体(ApplicationMaster)

应用主体和应用是一一对应的,它主要有以下职责:

(1) 与调度器协商资源。

(2) 与节点管理器合作,在合适的容器中运行对应的组件任务,并监控这些任务执行。

(3) 如果资源容器出现故障,应用主体会重新向调度器申请其他资源。

(4) 计算应用程序所需的资源量,并转化成调度器可识别的协议信息包。

(5) 在应用主体出现故障后,应用管理器会负责重启它,但由应用主体自己从之前保存的应用程序执行状态中恢复应用程序。

4. 资源容器(Container)

系统资源的组织形式是分割节点上的可用资源,每一份通过封装组织成系统的一个资源单元,即资源容器(比如固定大小的内存分片、CPU 核心数、网络带宽量和硬盘空间块等。所谓资源,是指内存资源,每个节点由多个 512MB 或 1GB 大小的内存容器组成)。而不像 MapReduce V1 中那样,将资源组织成 Map 池或 Reduce 池。应用主体可以申请任意多个该内存整数倍大小的容器。由于将每个节点上的内存资源分割成了大小固定、地位相同的容器,这些内存容器就可以在任务执行中进行互换,从而提高利用率,避免了在 MapReduce V1 中作业在 Reduce 池上的瓶颈问题和缺乏资源互换的问题。资源容器的主要职责就是运行、保存或传输应用主体提交的作业或需要存储和传输的数据。

具体架构如图 1.5 所示。

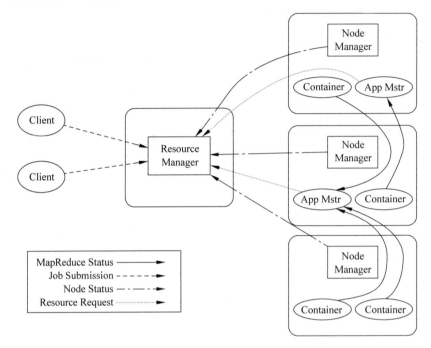

图 1.5　Hadoop MapReduce 框架（Yarn）架构

当 Client 提交一个任务后，首先 ResourceManger（RM）来调度出一个 Container，这个 Container 是在 NodeManger（NM）运作的，Client 直接和这个 Container 所在的 NM 进行通信，在这个 Container 中启动 Application Master（AM），启动成功之后，这个 AM 将全权负责此次任务的进度。AM 会计算此次任务所需的资源，然后向 RM 申请资源，得到一组供 MapReduce task 运行的 Container，然后协同 NM 一起对每个 Container 执行一些必要的任务，在任务执行过程中，AM 会一直监视着任务的运行进度，若中间某个 NM 上的 Container 中的任务失败，那么 AM 会重新找一个节点来运行此任务。

1.1.6　Hadoop 数据管理

前面重点介绍了 Hadoop 及其体系结构与计算模型 MapReduce，现在介绍 Hadoop 的数据管理。HDFS 是分布式计算的存储基石，Hadoop 分布式文件系统和其他分布式文件系统有很多类似的特性：

（1）对于整个集群有单一的命名空间。

（2）具有数据一致性，适合一次写入、多次读取的模型，客户端在文件没有被成功创建之前是无法看到文件存在的。

（3）文件会被分割成多个文件块，每个文件块被分配存储到数据节点上，而且会根据配置由复制文件块来保证数据的安全性。

通过前面的介绍可以看出，HDFS 通过三个重要的角色来进行文件系统的管理：NameNode、DataNode 和 Client。NameNode 可以看作分布式文件系统中的管理者，主要负责管理文件系统的命名空间、集群配置信息和存储块的复制等。NameNode 会将文件系统的 Metadata 存储在内存中,这些信息主要包括文件信息、每一个文件对应的文件块的信息和每一个文件块在 DataNode 中的信息等。DataNode 是文件存储的基本单元,它将文件块(Block)存储在本地文件系统中,保存了所有 Block 的 Metadata,同时周期性地将所有存在的 Block 信息发送给 NameNode。Cilent 就是需要获取分布式文件系统文件的应用程序。接下来通过三个具体的操作来说明 HDFS 对数据的管理。

1. 文件写入

(1) Client 向 NameNode 发起文件写入的请求。

(2) NameNode 根据文件大小和文件块配置情况,返回给 Client 所管理的 DataNode 的信息。

(3) Client 将文件划分为多个 Block,根据 DataNode 的地址信息,按顺序将其写入到每一个 DataNode 块中。

2. 文件读取

(1) Client 向 NameNode 发起文件读取的请求。

(2) NameNode 返回文件存储的 DataNode 信息。

(3) Client 读取文件信息。

3. 文件块(Block)复制

(1) NameNode 发现部分文件的 Block 不符合最小复制数这一要求或部分 DataNode 失效。

(2) 通知 DataNode 相互复制 Block。

(3) DataNode 开始直接相互复制。

作为分布式文件系统,HDFS 在数据管理方面还有值得借鉴的几个功能。

- **文件块(Block)的放置**。一个 Block 会有三份备份:一份放在 NameNode 指定的 DataNode 上,另一份放在与指定 DataNode 不在同一台机器上的 DataNode 上,最后一份放在与指定 DataNode 同一 Rack 的 DataNode 上。备份的目的是为了数据安全,采用这种配置方式主要是考虑同一 Rack 失败的情况,以及不同 Rack 之间进行数据复制会带来的性能问题。
- **心跳检测**。用心跳检测 DataNode 的健康状况,如果发现问题就采取数据备份的方式来保证数据的安全性。
- **数据复制**(场景为 DataNode 失败、需要平衡 DataNode 的存储利用率和平衡 DataNode 数据交互压力等情况):使用 Hadoop 时可以用 HDFS 的 balancer 命令配置界限来平衡每一个 DataNode 的磁盘利用率。假设设置了 Threshold 为 10%,那么执行 balancer 命令时,首先会统计所有 DataNode 的磁盘利用率的平均值,然后判断如果某一个 DataNode 的磁盘利用率超过这个平均值,那么将会把这个 DataNode 的 Block 转移到磁盘利用率低的 DataNode 上,这对于新节点的加入十分有用。
- **数据校验**:采用 CRC32 做数据校验。在写入文件块的时候,除了会写入数据外还会写入校验信息,在读取的时候则需要先校验后读入。
- **单个 NameNode**:如果单个 NameNode 失败,那么任务处理信息将会记录在本地文

- **数据管道性地写入**：当客户端要写入文件到 DataNode 上时，首先会读取一个 Block，然后将其写到第一个 DataNode 上，接着由第一个 DataNode 将其传递到备份的 DataNode 上，直到所有需要写入这个 Block 的 DataNode 都成功写入后，客户端才会开始写下一个 Block。
- **安全模式**：分布式文件系统启动时会进入安全模式（系统运行期间也可以通过命令进入安全模式），当分布式文件系统处于安全模式时，文件系统中的内容不允许修改也不允许删除，直到安全模式结束。安全模式主要是为了在系统启动的时候检查各个 DataNode 上数据块的有效性，同时根据策略进行必要的复制或删除部分数据块。在实际操作过程中，如果在系统启动时修改和删除文件会出现安全模式不允许修改的错误提示，此时只需要等待一会儿即可。

1.1.7　Hadoop 集群安全策略

众所周知，Hadoop 的优势在于其能够将廉价的普通 PC 组织成能够高效稳定处理事务的大型集群，企业正是利用这一特点来构架 Hadoop 集群，获取海量数据的高效处理能力的。但是，Hadoop 集群搭建起来后如何保证它安全稳定地运行呢？旧版本的 Hadoop 中没有完善的安全策略，导致 Hadoop 集群面临很多风险，例如，用户可以以任何身份访问 HDFS 或 MapReduce 集群，可以在 Hadoop 集群上运行自己的代码来冒充 Hadoop 集群的服务，任何未被授权的用户都可以访问 DataNode 的数据块等。经过 Hadoop 安全小组的努力，在 Hadoop 2.7.1 版本中已经加入最新的安全机制和授权机制（Simple 和 Kerberos），使 Hadoop 集群更加安全和稳定。下面从用户权限管理、HDFS 安全策略和 MapReduce 安全策略三个方面简要介绍 Hadoop 的集群安全策略。有关安全方面的基础知识（如 Kerberos 认证等）读者可自行查阅相关资料。

1. 用户权限管理

Hadoop 上的用户权限管理主要涉及用户分组管理，为更高层的 HDFS 访问、服务访问、Job 提交和配置 Job 等操作提供认证和控制基础。

Hadoop 上的用户和用户组名均由用户自己指定，如果用户没有指定，那么 Hadoop 会调用 Linux 的 whoami 命令获取当前 Linux 系统的用户名和用户组名作为当前用户的对应名，并将其保存在 Job 的 user.name 和 group.name 两个属性中。这样用户所提交 Job 的后续认证和授权以及集群服务的访问都将基于此用户和用户组的权限及认证信息进行。

Hadoop 集群的管理员是创建和配置 Hadoop 集群的用户，它可以配置集群，使用 Kerberos 机制进行认证和授权。同时管理员可以在集群的服务（集群的服务主要包括 NameNode、DataNode、ResourceManager、NodeManager、ApplicationMaster）授权列表中添加和更改某确定用户和用户组，系统管理员同时负责 Job 队列和队列的访问控制矩阵的创建。

2. HDFS 安全策略

用户和 HDFS 服务之间的交互主要有两种情况：用户机和 NameNode 之间的 RPC 交互获取待通信的 DataNode 位置；客户机和 DataNode 交互传输数据块。

RPC 交互可以通过 Kerberos 或授权令牌来认证。在认证与 NameNode 的连接时，用户需要使用 Kerberos 证书来通过初试认证，获取授权令牌。授权令牌可以在后续用户 Job 与 NameNode 连接的认证中使用，而不必再次访问 Kerberos Key Server。授权令牌实际上是用户机与 NameNode 之间共享的密钥。授权令牌在不安全的网络上传输时，应给予足够的保护，防止被其他用户恶意窃取，因为获取授权令牌的任何人都可以假扮成认证用户与 NameNode 进行不安全的交互。需要注意的是，每个用户只能通过 Kerberos 认证获取唯一一个新的授权令牌。用户从 NameNode 获取授权令牌之后，需要告诉 NameNode：谁是指定的令牌更新者。指定的更新者在为用户更新令牌时应通过认证确定自己就是 NameNode。更新令牌意味着延长令牌在 NameNode 上的有效期。为了使 MapReduce Job 使用一个授权令牌，ApplicationMaster 需要保证这一令牌在整个任务的执行过程中都是可用的，在任务结束之后，它可以选择取消令牌。

数据块的传输可以通过块访问令牌来认证，每一个块访问令牌都由 NameNode 生成，它们都是特定的。块访问令牌代表着数据访问容量，一个块访问令牌保证用户可以访问指定的数据块。块访问令牌由 NameNode 签发被用在 DataNode 上，其传输过程就是将 NameNode 上的认证信息传输到 DataNode 上。块访问令牌是基于对称加密模式生成的，NameNode 和 DataNode 共享了密钥。对于每个令牌，NameNode 基于共享密钥计算一个消息认证码（Message Authentication Code，MAC）。接下来，这个消息认证码就会作为令牌验证器成为令牌的主要组成部分。当一个 DataNode 接收到一个令牌时，它会使用自己的共享密钥重新计算一个消息认证码，如果这个认证码同令牌中的认证码匹配，那么认证成功。

3. MapReduce 安全策略

MapReduce 安全模式主要涉及 ResourceManager Delegation Token、YARN Application Token、YARN NodeManager Token、YARN Container Token、YARN Localizer Token、MapReduce Client Token、MapReduce Job Token、MapReduce Shuffle Secret 八个方面。

（1）ResourceManager Delegation Token。

ResourceManager Delegation Token 是 ResourceManager 授权令牌，持有该令牌的应用程序及其发起的任务可以安全地与 ResourceManager 交互，比如持有该令牌的 MapReduce 作业可以在任务中再次向 ResourceManager 提交一个或者多个作业，进而形成一个 MapReduce 工作流，Hadoop 生态系统中的工作流引擎 Oozie 正是采用了该策略。该令牌由 ResourceManager 中的组件 RMDelegationTokenSecretManager 管理和维护。

（2）YARN Application Token。

Application Token 用于保证 ApplicationMaster 与 ResourceManager 之间的通信安全。该 Token 的密钥（masterKey）由 ResourceManager 传递给 NodeManager，并保存到 ApplicationMaster Container 的私有目录下。当 NodeManager 启动 ApplicationMaster 时，所有的 Token 将被加载到 ApplicationMaster 的 UGI 中（NodeManager 通过环境变量 HADOOP_TOKEN_FILE_LOCATION 将 Token 所在目录传递给 UGI，这样 UGI 可以直接从文件中读取 Token 信息，所有其他 Token 的传递过程也是一样的），以在与 ResourceManager 通信时进行安全认证，需要注意的是，该 Token 的生命周期与 ApplicationMaster 实例一致。该 Token 由 ResourceManager 中的 AMRMTokenSecretManager 管理和维护。

(3) YARN NodeManager Token。

ApplicationMaster 与 NodeManager 通信时，需出示 NodeManager Token 以表明 ApplicationMaster 自身的合法性。该 Token 是由 ResourceManager 通过 RPC 函数 ApplcationMasterProtocol♯allocate 的应答中发送给 ApplicationMaster 的，它的密钥是各个 NodeManager 向 ResourceManager 注册（ResourceTracker♯registerNodeManager）和发送心跳信息（ResourceTracker♯nodeHeartbeat）时领取的。ApplicationMaster 通过 ContainerManagementProtocol 协议与 NodeManager 通信时，需要出示该 Token。该 Token 由 ResourceManager 中的 NMTokenSecretManagerInRM 管理和维护。

(4) YARN Container Token。

ApplicationMaster 与 NodeManager 通信启动 Container 时，需出示 Container Token 以表明 Container 的合法性。该 Token 是由 ResourceManager 通过 RPC 函数 ApplcationMasterProtocol♯allocate 的应答存放到 Container 中发送给 ApplicationMaster 的，它的密钥是各个 NodeManager 向 ResourceManager 注册和发送心跳信息时领取的。ApplicationMaster 通过 RPC 函数 ContainerManagementProtocol♯startContainer 与 NodeManager 通信启动 Container 时，需要出示相应的 ContainerToken。该 Token 由 ResourceManager 中的 RMContainerTokenSecretManager 管理和维护。

(5) YARN Localizer Token。

Localizer Token 用于保证 ContainerLocalizer 与 NodeManager 之间的通信安全。ContainerLocalizer 负责在任务运行之前从 HDFS 上下载各类所需的文件资源，以构建一个本地执行环境。在文件下载过程中，ContainerLocalizer 通过 RPC 协议 LocalizationProtocol 不断向 NodeManager 汇报状态信息。

(6) MapReduce Client Token。

MapReduce Client Token 用于保证 MapReduce JobClient 与 MapReduce Application Master 之间的通信安全。它由 ResourceManager 在作业提交时创建，并通过 RPC 函数 ApplicationClientProtocol♯getApplicationReport 发送给 JobClient。该 Token 由 ResourceManager 中的 ClientToAMTokenSecretManagerInRM 管理和维护。

(7) MapReduce Job Token。

MapReduce Job Token 用于保证 MapReduce 的各个 Task（包括 Map Task 和 Reduce Task）与 MapReduce Application Master 之间的通信安全。它由 ApplicationMaster 创建，通过 RPC 函数 ContainerManagementProtocol♯startContainer 传递给 NodeManager，并由 NodeManager 写入 Container 的私有目录中，以在任务启动时加载到 UGI 中，从而使得任务可以安全地通过 RPC 协议 TaskUmbilicalProtocol 与 ApplicationMaster 通信。

(8) MapReduce Shuffle Secret。

MapReduce Shuffle Secret 用于保证运行在各个 NodeManager 上的 ShuffleHandler（内部封装了一个 Netty Server）与 Reduce Task 之间的通信安全，即只有同一个作业的 Reduce Task 才允许读取该作业 Map Task 产生的中间结果，该安全机制是借助 Job Token 完成的。

1.2 Hadoop 的安装与配置

学习 Hadoop 之前首先要做的便是 Hadoop 环境的搭建,在一些主流的操作系统如 UNIX、Windows 上 Hadoop 都运行良好。在实际工程环境中,建议在 Linux 环境进行 Hadoop 的安装和使用。

本书中所有的操作都是基于 Linux 环境进行的,操作系统的版本为 CentOS 6.5。本书中的 Hadoop 使用的是伪分布式环境,伪分布式环境与集群环境的工作原理类似。

在 Linux 上安装 Hadoop 之前,需要先安装两个程序:

(1) JDK 1.7(或更高版本)。Hadoop 是用 Java 编写的程序,Hadoop 的编译及 MapReduce 的运行都需要使用 JDK。因此在安装 Hadoop 前,必须安装 JDK 1.7 或更高版本。

(2) SSH(安全外壳协议),推荐安装 OpenSSH。Hadoop 需要通过 SSH 来启动 Slave 列表中各台主机的守护进程,因此 SSH 也是必须安装的,即使是安装伪分布式版本(因为 Hadoop 并没有区分开集群式和伪分布式)。对于伪分布式,Hadoop 会采用与集群相同的处理方式,即按次序启动文件 conf/slaves 中记载的主机上的进程,只不过在伪分布式中 Slave 为 localhost(即为自身),所以对于伪分布式 Hadoop,SSH 一样是必需的。

1.2.1 安装 JDK 1.8 与配置 SSH 免密码登录

可以在 Oracle JDK 的官网下载相应版本的 JDK,本例以 JDK 1.8 为例,官网地址为 http://www.oracle.com/technetwork/java/javase/downloads/index.html。

1. 安装 JDK 1.8

1) 解压安装包

本书的 JDK 安装包 jdk-8u60-linux-x64.tar.gz 下载到了/usr/software/,首先需要创建工作目录,在/usr 目录下建立 cx 文件夹。

将安装包解压到/usr/cx/目录下,执行如下命令:

```
tar -zxvf /usr/software/jdk-8u60-linux-x64.tar.gz -C /usr/cx
```

2) 配置环境变量

打开文件~/.bashrc。

```
vi ~/.bashrc
```

在文件中增加下列内容:

```
# set java environment
export JAVA_HOME=/usr/cx/jdk1.8.0_60
export PATH=$JAVA_HOME/bin:$PATH
export CLASSPATH=.:$JAVA_HOME/jre/lib/rt.jar:$JAVA_HOME/jre/lib/tools.jar
```

保存退出,通过下面的命令更新环境变量:

```
source ~/.bashrc
```

3)验证 JDK 是否安装成

输入命令

```
java - version:
```

得到如图 1.6 所示的结果。

```
[root@master ~]# java -version
java version "1.8.0_60"
Java(TM) SE Runtime Environment (build 1.8.0_60-b27)
Java HotSpot(TM) 64-Bit Server VM (build 25.60-b23, mixed mode)
```

图 1.6 查看 Java 版本信息

如果出现上述 JDK 版本信息,则说明当前系统的 JDK 已设置成 CentOS 系统默认的 JDK。

2. 配置 SSH 免密码登录

(1)查看本机是否安装了 SSH,输入命令:ssh 127.0.0.1,如果可以登录本机说明已经安装了 SSH;否则输入 yes,下载并安装 SSH,输入命令:yum install ssh。

(2)配置为可以免密码登录本机。

CentOS 启用了 SElinux,它在网络服务方面权限要求比较严格,需要编辑配置文件关闭 SElinux:

```
vi /etc/selinux/config
```

把 SELINUX=enforcing 改成 SELINUX=disabled。配置后的文件如图 1.7 所示。

```
# This file controls the state of SELinux on the system.
# SELINUX= can take one of these three values:
#     enforcing - SELinux security policy is enforced.
#     permissive - SELinux prints warnings instead of enforcing.
#     disabled - No SELinux policy is loaded.
SELINUX=enforcing
# SELINUXTYPE= can take one of these two values:
#     targeted - Targeted processes are protected,
#     mls - Multi Level Security protection.
SELINUXTYPE=targeted
```

图 1.7 编辑配置文件

然后重启操作系统,输入命令:reboot。

(3)生成密钥文件,输入命令:

```
ssh-keygen -t dsa
```

通过一直按回车便可以成功生成密钥文件。

切换到～/.ssh 目录下，输入命令：

```
cd ~/.ssh
```

把 id_dsa.pub 的内容输出追加到 authorized_keys 的末尾，输入命令：

```
cat id_dsa.pub >> authorized_keys
```

配置过程如图 1.8 所示。

```
[root@CentOS1 桌面]# ssh-keygen -t dsa
Generating public/private dsa key pair.
Enter file in which to save the key (/root/.ssh/id_dsa):
Enter passphrase (empty for no passphrase):
Enter same passphrase again:
Your identification has been saved in /root/.ssh/id_dsa.
Your public key has been saved in /root/.ssh/id_dsa.pub.
The key fingerprint is:
33:76:34:8a:9f:41:6e:84:e3:ba:df:81:a4:f2:df:95 root@CentOS1
The key's randomart image is:
+--[ DSA 1024]----+
|                 |
|        .        |
|      o o o      |
|     . * o .     |
|      + S .      |
|     + = * .     |
|  . o . + E      |
|   o . o o       |
|    ooo o        |
+-----------------+
[root@CentOS1 桌面]# cd ~/.ssh
[root@CentOS1 .ssh]# cat id_dsa.pub>>authorized-keys
```

图 1.8 配置免密码登录

(4) 验证 SSH 是否已安装成功，以及是否可以免密码登录本机。

```
ssh -version
```

如图 1.9 所示，表示 SSH 已经安装成功了。

```
[root@CentOS1 ~]# ssh -version
OpenSSH_5.3p1, OpenSSL 1.0.1e-fips 11 Feb 2013
Bad escape character 'rsion'.
```

图 1.9 查看 SSH 信息

如果没有安装成功，可能是缺少部分必须要安装的软件，标准的需要安装的软件如图 1.10 所示，如果没有安装成功，可以对照着进行安装。

```
[root@VM-cc6f9475-47b6-4cc3-91f9-760ee60191ba .ssh]# rpm -qa| grep ssh
openssh-server-5.3p1-94.el6.x86_64
libssh2-1.4.2-1.el6.x86_64
openssh-clients-5.3p1-94.el6.x86_64
openssh-5.3p1-94.el6.x86_64
openssh-askpass-5.3p1-94.el6.x86_64
```

图 1.10 SSH 标准组件

输入如下命令,验证是否可以免密码登录本机,如图 1.11 所示。

```
ssh localhost
```

```
[root@VM-5aa7c8bb-d820-47f8-a79c-7695aa49c594 .ssh]# ssh localhost
Last login: Thu Aug 18 10:43:21 2016 from vm-7f135ec5-afc1-4ffb-a8
ef-db2d35c9b536.cs2cloud.internal
```

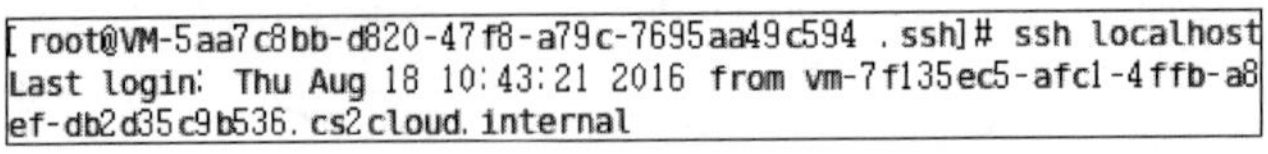

图 1.11　验证免密码登录

第一次登录时会询问是否继续链接,输入"yes"即可进入。

实际上,在 Hadoop 的安装过程中,是否免密码登录是无关紧要的,但是如果不配置免密码登录,每次启动 Hadoop 都需要输入密码以登录到每台计算机的 DataNode 上,考虑到一般的 Hadoop 集群动辄拥有数百或上千台计算机,因此一般来说都会配置 SSH 的免密码登录。

1.2.2　安装并运行 Hadoop

Hadoop 分别从三个角度将主机划分为两种角色。第一,最基本的划分为 Master 和 Slave,即主人与从属;第二,从 HDFS 的角度,将主机划分为 NameNode 和 DateNode(在分布式文件系统中,目录的管理很重要,管理目录相当于主人,而 NameNode 就是目录管理者);第三,从 MapReduce 的角度,将主机划分为 ResourceManager、ApplicationMaster 和 NodeManager。

Hadoop 有官方发行版与 Cloudera 版,其中 Cloudera 版是 Hadoop 的商用版本,这里介绍 Hadoop 官方发行版的安装方法。

Hadoop 有三种运行方式:单机模式、伪分布式与完全分布式。乍看之下,前两种方式并不能体现云计算的优势,但是它们便于程序的测试与调试,所以还是很有意义的。

通过以下地址可以获得 Hadoop 的官方发行版:

http://www.apache.org/dyn/closer.cgi/Hadoop/core/

下载 hadoop-2.7.1.tar.gz 并将其解压,本书中将安装文件下载到了/usr/software 目录下。

1. 解压该压缩文件到工作目录

输入命令:

```
tar - zxvf /usr/software/hadoop-2.7.1.tar.gz -C /usr/c
```

2. 配置 Hadoop 环境变量

(1) 编辑 ~/.bashrc 文件。

```
vim ~/.bashrc
```

(2) 在 ~/.bashrc 文件中增加以下内容:

```
export HADOOP_INSTALL = /usr/cx/hadoop-2.7.1
export PATH = $PATH:$HADOOP_INSTALL/bin:$PATH
export PATH = $PATH:$HADOOP_INSTALL/sbin:$PATH
```

```
export HADOOP_MAPRED_HOME = $ HADOOP_INSTALL
export HADOOP_COMMON_HOME = $ HADOOP_INSTALL
export HADOOP_HDFS_HOME = $ HADOOP_INSTALL
export YARN_HOME = $ HADOOP_INSTALL
```

结果显示如图 1.12 所示。

图 1.12　编辑配置文件

通过以上配置就可以让系统找到 Hadoop 的安装路径。

3. 单机模式配置方式

安装单机模式的 Hadoop 无须配置，在这种方式下，Hadoop 被认为是一个单独的 Java 进程，这种方式经常用来调试。

4. 伪分布式 Hadoop 配置

可以把伪分布式的 Hadoop 看作只有一个节点的集群，伪分布式的配置过程也很简单，只需要修改几个配置文件。

（1）进入 hadoop 所在目录。

```
cd /usr/cx/hadoop-2.7.1/etc/hadoop
```

（2）配置 hadoop-env.sh。

```
export JAVA_HOME = /usr/cx/jdk1.8.0_60
```

（3）配置 core-site.xml。

```
<configuration>
```

```
/*这里的值指的是默认的HDFS访问路径*/
<property>
<name>fs.defaultFS</name>
<value>hdfs://master:9000</value>
</property>
/*缓冲区大小:io.file.buffer.size 默认是4KB*/
<property>
<name>io.file.buffer.size</name>
<value>131072</value>
</property>
/*临时文件夹路径*/
<property>
<name>hadoop.tmp.dir</name>
<value>file:/usr/tmp</value>
<description>Abase for other temporary directories.</description>
</property>
<property>
<name>hadoop.proxyuser.hduser.hosts</name>
<value>*</value>
</property>
<property>
<name>hadoop.proxyuser.hduser.groups</name>
<value>*</value>
</property>
</configuration>
```

（4）配置 yarn-site.xml 文件。

```
<configuration>
<property>
<name>yarn.nodemanager.aux-services</name>
<value>mapreduce_shuffle</value>
</property>
<property>
<name>yarn.nodemanager.aux-services.mapreduce.shuffle.class</name>
<value>org.apache.hadoop.mapred.ShuffleHandler</value>
</property>
/*resourcemanager 的地址*/
<property>
<name>yarn.resourcemanager.address</name>
<value>master:8032</value>
</property>
/*调度器的端口*/
<property>
<name>yarn.resourcemanager.scheduler.address</name>
<value>master:8030</value>
</property>
/*resource-tracker 端口*/
<property>
<name>yarn.resourcemanager.resource-tracker.address</name>
<value>master:8031</value>
```

```
</property>
/* resourcemanager 管理器端口 */
<property>
<name>yarn.resourcemanager.admin.address</name>
<value>master:8033</value>
</property>
/* ResourceManager 的 Web 端口,监控 job 的资源调度 */
<property>
<name>yarn.resourcemanager.webapp.address</name>
<value>master:8088</value>
</property>
</configuration>
```

(5) 配置 mapred-site.xml 文件。

首先将 mapred-site.xml.template 复制为 mapred-site.xml 文件。执行命令如下:

```
cp mapred-site.xml.template mapred-site.xml
```

再对 mapred-site.xml 文件进行配置:

```
<configuration>
/* hadoop 对 map-reduce 运行框架一共提供了 3 种实现,在 mapred-site.xml 中通过"mapreduce.
framework.name"这个属性来设置为"classic""yarn"或者"local" */
<property>
<name>mapreduce.framework.name</name>
<value>yarn</value>
</property>
/* MapReduce JobHistory Server 地址 */
<property>
<name>mapreduce.jobhistory.address</name>
<value>master:10020</value>
</property>
/* MapReduce JobHistory Server Web UI 地址 */
<property>
<name>mapreduce.jobhistory.webapp.address</name>
<value>master:19888</value>
</property>
</configuration>
```

5. 创建 namenode 和 datanode 目录,并配置其相应路径

(1) 在根目录下创建 hdfs 文件夹,创建 namenode 和 datanode 目录(如果存在,则需要先删除原来的目录然后再创建),执行以下命令:

```
mkdir /hdfs/namenode
mkdir /hdfs/datanode
```

(2) 执行命令后,再次回到目录/usr/cx/hadoop-2.7.1/etc/hadoop,配置 hdfs-site.xml 文件,在文件中添加如下内容:

```
<configuration>
<property>
<name>dfs.namenode.secondary.http-address</name>
<value>master:9001</value>
</property>
<property>
<name>dfs.namenode.name.dir</name>
<value>file:/hdfs/namenode</value>
</property>
/*配置datanode的数据存储目录*/
<property>
<name>dfs.datanode.data.dir</name>
<value>file:/hdfs/datanode</value>
</property>
/*配置副本数*/
<property>
<name>dfs.replication</name>
<value>1</value>
</property>
/*将 dfs.webhdfs.enabled 属性设置为 true,否则就不能使用 webhdfs 的 LISTSTATUS、LIST
FILESTATUS 等需要列出文件、文件夹状态的命令,因为这些信息都是由 namenode 保存的*/
<property>
<name>dfs.webhdfs.enabled</name>
<value>true</value>
</property>
</configuration>
```

6. 配置/etc/sysconfig/network 和 slaves 文件

(1) 编辑 slaves 文件,在文件中写入 slave 主机名,在此填写为本机 localhost。

(2) /etc/sysconfig/network 文件负责配置主节点的主机名。例如,主节点名为 master,则需要在/etc/sysconfig/network 中修改内容。打开文件:

```
vim /etc/sysconfig/network
```

修改内容,结果显示如图 1.13 所示。

```
NETWORKING=yes
HOSTNAME=master
```

图 1.13 编辑配置文件

保存退出,重启计算机后查看结果,执行命令如下:

```
hostname
```

结果显示如图 1.14 所示。

```
[root@master hadoop]# hostname
master
```

图 1.14 查看主机名

(3) 修改 hosts 中的主机名,执行命令如下:

```
vim /etc/hosts
```

结果显示如图 1.15 所示。

```
127.0.0.1    master localhost localhost.localdomain localhost4 localhost4.localdomain4
::1          localhost localhost.localdomain localhost6 localhost6.localdomain6
```

图 1.15　编辑配置文件

保存退出,检测结果,执行命令如下:

```
ping master
```

若修改成功,将会显示如图 1.16 所示的结果。

```
[root@master hadoop]# ping master
PING master (127.0.0.1) 56(84) bytes of data.
64 bytes from master (127.0.0.1): icmp_seq=1 ttl=64 time=0.032 ms
64 bytes from master (127.0.0.1): icmp_seq=2 ttl=64 time=0.038 ms
64 bytes from master (127.0.0.1): icmp_seq=3 ttl=64 time=0.044 ms
64 bytes from master (127.0.0.1): icmp_seq=4 ttl=64 time=0.052 ms
64 bytes from master (127.0.0.1): icmp_seq=5 ttl=64 time=0.046 ms
64 bytes from master (127.0.0.1): icmp_seq=6 ttl=64 time=0.044 ms
64 bytes from master (127.0.0.1): icmp_seq=7 ttl=64 time=0.046 ms
64 bytes from master (127.0.0.1): icmp_seq=8 ttl=64 time=0.048 ms
64 bytes from master (127.0.0.1): icmp_seq=9 ttl=64 time=0.042 ms
64 bytes from master (127.0.0.1): icmp_seq=10 ttl=64 time=0.021 ms
```

图 1.16　测试网络

按 Ctrl+Z 键停止 ping。

7. 格式化 HDFS

启动 Hadoop 前,需要格式化 Hadoop 的文件系统 HDFS。

首先,通过 cd /usr/cx/hadoop-2.7.1 命令进入 Hadoop 的根目录,然后输入命令:

```
hadoop namenode -format
```

部分结果显示如图 1.17 所示。

8. 启动 Hadoop

输入命令:

```
start-all.sh
```

9. 查看是否配置和启动成功

通过 jps 命令,查看相应的 JVM 进程,执行如下命令:

```
jps
```

```
STARTUP_MSG:    java = 1.8.0_60
************************************************************/
16/08/12 17:32:17 INFO namenode.NameNode: registered UNIX signal handlers f
or [TERM, HUP, INT]
16/08/12 17:32:17 INFO namenode.NameNode: createNameNode [-format]
16/08/12 17:32:18 WARN util.NativeCodeLoader: Unable to load native-hadoop
library for your platform... using builtin-java classes where applicable
Formatting using clusterid: CID-72ede602-42e4-48a9-a273-3881b871d5bb
16/08/12 17:32:19 INFO namenode.FSNamesystem: No KeyProvider found.
16/08/12 17:32:19 INFO namenode.FSNamesystem: fsLock is fair:true
16/08/12 17:32:19 INFO blockmanagement.DatanodeManager: dfs.block.invalidat
e.limit=1000
16/08/12 17:32:19 INFO blockmanagement.DatanodeManager: dfs.namenode.datano
de.registration.ip-hostname-check=true
16/08/12 17:32:19 INFO blockmanagement.BlockManager: dfs.namenode.startup.d
elay.block.deletion.sec is set to 000:00:00:00.000
16/08/12 17:32:19 INFO blockmanagement.BlockManager: The block deletion wil
l start around 2016 八月 12 17:32:19
16/08/12 17:32:19 INFO util.GSet: Computing capacity for map BlocksMap
16/08/12 17:32:19 INFO util.GSet: VM type       = 64-bit
16/08/12 17:32:19 INFO util.GSet: 2.0% max memory 966.7 MB = 19.3 MB
16/08/12 17:32:19 INFO util.GSet: capacity      = 2^21 = 2097152 entries
16/08/12 17:32:19 INFO blockmanagement.BlockManager: dfs.block.access.token
.enable=false
16/08/12 17:32:19 INFO blockmanagement.BlockManager: defaultReplication
       = 1
16/08/12 17:32:24 INFO common.Storage: Storage directory /hdfs/namenode has
 been successfully formatted.
16/08/12 17:32:24 INFO namenode.NNStorageRetentionManager: Going to retain
1 images with txid >= 0
16/08/12 17:32:24 INFO util.ExitUtil: Exiting with status 0
16/08/12 17:32:24 INFO namenode.NameNode: SHUTDOWN_MSG:
/************************************************************
SHUTDOWN_MSG: Shutting down NameNode at master/127.0.0.1
************************************************************/
```

图 1.17　格式化文件系统

结果显示如图 1.18 所示。

```
[root@master hadoop-2.7.1]# jps
18048 Jps
1856 VmServer.jar
2834 DataNode
2997 SecondaryNameNode
17911 ResourceManager
2714 NameNode
18012 NodeManager
```

图 1.18　查看进程信息

10. 验证 Hadoop 是否安装成功

打开浏览器，分别输入网址：

- http://localhost:8088（MapReduce 的 Web 页面）
- http://localhost:50070（HDFS 的 Web 页面）

如果都能进行查看，则说明 Hadoop 已经配置成功。

第 2 章

Hadoop存储：HDFS

第 1 章介绍了如何搭建 Hadoop 安装环境，本章将对 Hadoop 的主要组件 HDFS 进行深入介绍，帮助大家提升对于 Hadoop 的认识。

2.1 HDFS 的基本操作

2.1.1 HDFS 的命令行操作

可以通过命令行接口来与 HDFS 进行交互。当然，命令行接口只是 HDFS 的访问接口之一，它的特点是更加简单直观，便于使用，可以进行一些基本操作。

下面具体介绍如何通过命令行访问 HDFS 文件系统。本节主要讨论一些基本的文件操作，比如读文件、创建文件存储路径、转移文件、删除文件、列出文件列表等。在终端中可以通过输入 hadoop fs-help 获得 HDFS 操作的详细帮助信息。

首先介绍 hadoop fs 和 hadoop dfs 的区别。fs 是一个比较抽象的层面，在分布环境中，fs 就是 dfs；但在本地环境中，fs 就是 local file system，这个时候 dfs 就不能使用了。

1．列出 HDFS 文件

通过-ls 命令列出 HDFS 下的文件：

```
# hadoop fs - ls
```

这里需要注意，在 HDFS 中，没有当前目录这样一个概念，也没有 cd 这个命令。

2．列出 HDFS 目录下某个文档中的文件

通过"-ls 文件名"命令浏览 HDFS 下名为 master1-file 的文档中文件：

```
# hadoop fs -ls /mater1-file
```

3. 上传文件到 HDFS

通过"-put 文件1 文件2"命令将 /usr/hadoop/hadoop 下的 data 文件上传到 HDFS 上并重命名为 master1-file：

```
# hadoop fs -put /usr/hadoop/hadoop/data /master1-file
```

在执行 -put 时只有两种可能，即执行成功和执行失败。在上传文件时，文件首先复制到 DataNode 上。只有所有的 DataNode 都成功接收完数据，文件上传才是成功的。其他情况（如文件上传终端等）对 HDFS 来说都是做了无用功。

4. 将 HDFS 中文件复制到本地系统中

通过"-get 文件1 文件2"命令将 HDFS 中 master1-file 文件复制到本地系统并命名为 getout，与 -put 命令一样，-get 操作既可以操作文件，也可以操作目录。

```
# hadoop fs -get /master1-file getout
```

5. 查看 HDFS 下的某个文件

通过"-cat 文件"命令查看 master1-file 文件中的内容：

```
# hadoop fs -cat /master1-file/data/current/VERSION*
```

6. 删除 HDFS 下的文档

通过"-rmr 文件"命令删除 HDFS 下的 master1-file 文档。

```
# hadoop fs -rmr /master1-file
```

hadoop fs 的命令远不止这些，本节介绍的命令已可以在 HDFS 上完成大多数常规操作。

2.1.2　HDFS 的 Web 界面

在部署好 Hadoop 集群之后，便可以直接通过 http://localhost：50070 访问 HDFS 的 Web 界面。HDFS 的 Web 界面提供了基本的文件系统信息，其中包括集群启动时间、版本号、编译时间及是否有升级。

HDFS 的 Web 界面还提供了文件系统的基本功能：Browse the filesystem（浏览文件系统），单击链接即可看到，它将 HDFS 的文件结构通过目录的形式展现出来，增加了对文件系统的可读性。此外，可以直接通过 Web 界面访问文件内容。同时，HDFS 的 Web 界面还将该文件所在的节点位置展现出来。可以通过设置 Chunk size to view 来设置一次读取并展开文件块的大小。

除了在本节中展示的信息之外，HDFS 的 Web 界面还提供了 NameNode 的日志列表、运行中的节点列表以及宕机的节点列表等信息。

2.1.3 通过 distcp 进行并行复制

Java API 等多种接口对 HDFS 访问模型都集中于单线程的存取,如果要对一个文件集进行操作,就需要编写一个程序进行并行操作。HDFS 提供了一个非常实用的程序——distcp,用来在 Hadoop 文件系统中并行地复制大量文件。distcp 一般适用于在两个 HDFS 集群间传送数据的情况。如果两个集群都运行在同一个 Hadoop 版本上,那么可以使用 HDFS 模式。

环境描述(需要两台主机)。
node1(节点)IP:192.168.30.44 主机名:master1
node2(节点)IP:192.168.30.144 主机名:master2
需要在/etc/hosts 中添加下列内容:

```
192.168.30.44 master1
192.168.30.144 master2
```

首先在两个集群中分别创建文件夹:master1 创建/master1-file,master2 创建/master2-file,如图 2.1 和图 2.2 所示。

```
[root@master1 ~]# hadoop fs -mkdir /master1-file
```

图 2.1 创建文件夹(一)

```
[root@master2 ~]# hadoop fs -mkdir /master2-file
```

图 2.2 创建文件夹(二)

将 master1 下的 master1-file 文件复制到 master2-file 文件下(端口默认为 8020)(注:检查两台主机是否可以 ping 通,且 hadoop 命令是否可以正常使用)

```
hadoop distcp hdfs://master1/master1-file hdfs://master2/master2-file
```

会出现很多如图 2.3 所示的信息。

打开 master2 的浏览器,输入 localhost:50070,可以查看到 master2 下增加了一个文件 master1-file,如图 2.4 所示。

这条命令会将第一个集群/maste1-file 文件夹以及文件夹下的文件复制到第二个集群/master2-file 目录下,即在第二个集群中会以/master2-file/master1-file 的目录结构出现。如果/master2-file 目录不存在,则系统会新建一个。也可以指定多个数据源,并且所有的内容都会被复制到目标路径。需要注意的是,源路径必须是绝对路径。

默认情况下,虽然 distcp 会跳过在目标路径上已经存在的文件,但是通过-overwrite 选项可以选择对这些文件进行覆盖重写,也可以使用-updata 选项仅对更新过的文件进行重写。

```
15/10/27 09:12:03 INFO mapreduce.Job: Counters: 26
        File System Counters
                FILE: Number of bytes read=104786
                FILE: Number of bytes written=392253
                FILE: Number of read operations=0
                FILE: Number of large read operations=0
                FILE: Number of write operations=0
                HDFS: Number of bytes read=0
                HDFS: Number of bytes written=0
                HDFS: Number of read operations=6
                HDFS: Number of large read operations=0
                HDFS: Number of write operations=2
                HFTP: Number of bytes read=0
                HFTP: Number of bytes written=0
                HFTP: Number of read operations=0
                HFTP: Number of large read operations=0
                HFTP: Number of write operations=0
        Map-Reduce Framework
                Map input records=2
                Map output records=0
                Input split bytes=168
                Spilled Records=0
                Failed Shuffles=0
                Merged Map outputs=0
                GC time elapsed (ms)=5
                Total committed heap usage (bytes)=16834560
        File Input Format Counters
                Bytes Read=352
        File Output Format Counters
                Bytes Written=8
        org.apache.hadoop.tools.mapred.CopyMapper$Counter
                COPY=2
```

图 2.3　运行命令

Browse Directory

/master2-file　　　　　　　　　　　　　　　　　　　　　　　　　　　Go!

Permission	Owner	Group	Size	Last Modified	Replication	Block Size	Name
drwxr-xr-x	root	supergroup	0 B	2015年10月27日 星期二 10时33分40秒	0	0 B	master1-file

图 2.4　查看运行结果

　　distcp 操作有很多选项可以设置，比如忽略失败、限制文件或者复制的数据量等。直接输入指令或者不附加选项可以查看此操作的使用说明。具体实现时，distcp 操作会被解析为一个 MapReduce 操作来执行，当没有 Reduce 操作时，复制被作为 Map 操作并行地在集群节点中运行。因此，每个文件都可以被当成一个 Map 操作来执行复制。而 distcp 会通过执行多个文件聚集捆绑操作，尽可能地保证每个 Map 操作执行相同数量的数据。那么执行 distcp 时，Map 操作如何确定呢？由于系统需要保证每个 Map 操作执行的数据量是合理的，来最大化地减少 Map 执行的开销，而按规定，每个 Map 最少要执行 256MB 的数据量（除非复制的全部数据量小于 256MB）。比如要复制 1GB 的数据，那么系统会分配 4 个 Map 任务，当数据量非常大时，就需要限制执行的 Map 任务数，以限制网络带宽和集群的使用率。默认情况下，每个集群的一个节点最多执行 20 个 Map 任务。比如，要复制 1000GB 数据到 100 节点的集群中，那么系统就会分配 2000 个 Map 任务（每个节点 20 个）。也就是说，每个节点会平均复制 512MB。还可以通过调整 distcp 的-m 参数减少 Map 任务量，比如-m 1000 就意味着分配 1000 个 Map，每个节点分配 1GB 数据量。

如果尝试使用 distcp 进行 HDFS 集群间的复制，使用 HDFS 模式之后，HDFS 运行在不同的 Hadoop 版本之上，复制将会因为 RPC 系统的不匹配而失败。为了纠正这个错误，可以使用基于 HTTP 的 HFTP 进行访问。因为任务要在目标集群中执行，所以 HDFS 的 RPC 版需要匹配，在 HFTP 模式下运行的代码如下：

```
hadoop distcp hftp://master1:50070/master1-file hdfs://master2/master2-file
```

需要注意的是，要定义访问源的 URI 中 NameNode 的网络接口，这个接口会通过 dfs.hftp.address 的属性值设定，默认值为 50070。

2.1.4 使用 Hadoop 归档文件

每个文件 HDFS 采用块方式进行存储，在系统运行时，文件块的元数据信息会被存在 NameNode 的内存中，因此，对 HDFS 来说，大规模存储小文件显然是低效的，很多小文件会耗尽 NameNode 的大部分内存。

Hadoop 归档文件和 HAR 文件可以将文件高效地放入 HDFS 块中的文件存档设备，在减少 NameNode 内存使用的同时，仍然允许对文件进行透明访问。具体来说，Hadoop 归档文件可以作为 MapReduce 的输入。这里需要注意的是，小文件并不会占用太多的磁盘空间，比如设定一个 128MB 的文件块来存储 1MB 的文件，实际上存储这个文件只需要 1MB 磁盘空间，而不是 128MB。

Hadoop 归档文件是通过 archive 命令工具根据文件集合创建的。因为这个工具需要运行一个 MapReduced 来并行处理输入文件，所以需要一个运行 MapReduce 的集群。而 HDFS 中的有些文件是需要进行归档的。

```
hadoop fs -lsr /master1-file/
-rw-r--r--   1 root supergroup 13 2015-11-26 14:09 /master1-file/1.txt
```

运行下面的命令进行文件归档：

```
hadoop archive -archiveName files.har -p /master1-file /
```

会出现如下信息：

```
15/11/26 16:13:09 WARN util.NativeCodeLoader: Unable to load native-hadoop library for your platform... using builtin-java classes where applicable
15/11/26 16:13:11 INFO Configuration.deprecation: session.id is deprecated. Instead, use dfs.metrics.session-id
15/11/26 16:13:11 INFO jvm.JvmMetrics: Initializing JVM Metrics with processName=JobTracker, sessionId=
```

```
15/11/26 16:13:11 INFO jvm.JvmMetrics: Cannot initialize JVM Metrics with processName=
JobTracker, sessionId= - already initialized
15/11/26 16:13:11 INFO jvm.JvmMetrics: Cannot initialize JVM Metrics with processName=
JobTracker, sessionId= - already initialized
15/11/26 16:13:12 INFO mapreduce.JobSubmitter: number of splits:1
15/11/26 16:13:12 INFO mapreduce.JobSubmitter: Submitting tokens for job: job_
local1968546553_0001
15/11/26 16:13:13 INFO mapreduce.Job: The url to track the job: http://localhost:8080/
15/11/26 16:13:13 INFO mapreduce.Job: Running job: job_local1968546553_0001
15/11/26 16:13:13 INFO mapred.LocalJobRunner: OutputCommitter set in config null
15/11/26 16:13:13 INFO mapred.LocalJobRunner: OutputCommitter is org.apache.hadoop.
mapred.FileOutputCommitter
15/11/26 16:13:13 INFO output.FileOutputCommitter: File Output Committer Algorithm version is 1
15/11/26 16:13:13 INFO mapred.LocalJobRunner: Waiting for map tasks
15/11/26 16:13:13 INFO mapred.LocalJobRunner: Starting task: attempt_local1968546553_0001_m
_000000_0
15/11/26 16:13:13 INFO output.FileOutputCommitter: File Output Committer Algorithm version is 1
15/11/26 16:13:13 INFO mapred.Task: Using ResourceCalculatorProcessTree : [ ]
15/11/26 16:13:13 INFO mapred.MapTask: Processing split: file:/usr/hadoop/hadoop/hadooptmp/
mapred/staging/root435803013/.staging/har_6j4tlb/_har_src_files:0+221
15/11/26 16:13:13 INFO mapred.MapTask: numReduceTasks: 1
15/11/26 16:13:13 INFO mapred.MapTask: (EQUATOR) 0 kvi 26214396(104857584)
15/11/26 16:13:13 INFO mapred.MapTask: mapreduce.task.io.sort.mb: 100
15/11/26 16:13:13 INFO mapred.MapTask: soft limit at 83886080
15/11/26 16:13:13 INFO mapred.MapTask: bufstart = 0; bufvoid = 104857600
15/11/26 16:13:13 INFO mapred.MapTask: kvstart = 26214396; length = 6553600
15/11/26 16:13:13 INFO mapred.MapTask: Map output collector class = org.apache.hadoop.
mapred.MapTask$MapOutputBuffer
15/11/26 16:13:13 INFO mapred.LocalJobRunner:
15/11/26 16:13:13 INFO mapred.MapTask: Starting flush of map output
15/11/26 16:13:13 INFO mapred.MapTask: Spilling map output
15/11/26 16:13:13 INFO mapred.MapTask: bufstart = 0; bufend = 122; bufvoid = 104857600
15/11/26 16:13:13 INFO mapred.MapTask: kvstart = 26214396(104857584); kvend = 26214392
(104857568); length = 5/6553600
15/11/26 16:13:13 INFO mapred.MapTask: Finished spill 0
15/11/26 16:13:13 INFO mapred.Task: Task:attempt_local1968546553_0001_m_000000_0 is done.
And is in the process of committing
15/11/26 16:13:13 INFO mapred.LocalJobRunner:
15/11/26 16:13:13 INFO mapred.Task: Task attempt_local1968546553_0001_m_000000_0 is allowed
to commit now
15/11/26 16:13:13 INFO output.FileOutputCommitter: Saved output of task 'attempt_
local1968546553_0001_m_000000_0' to hdfs://localhost:8020/files.har/_temporary/0/task_
local1968546553_0001_m_000000
15/11/26 16:13:13 INFO mapred.LocalJobRunner: Copying file hdfs://localhost:8020/master1-
file/1.txt to archive.
15/11/26 16:13:13 INFO mapred.Task: Task 'attempt_local1968546553_0001_m_000000_0' done.
15/11/26 16:13:13 INFO mapred.LocalJobRunner: Finishing task: attempt_local1968546553_0001_
m_000000_0
15/11/26 16:13:13 INFO mapred.LocalJobRunner: map task executor complete.
15/11/26 16:13:13 INFO mapred.LocalJobRunner: Waiting for reduce tasks
15/11/26 16:13:13 INFO mapred.LocalJobRunner: Starting task: attempt_local1968546553_0001_r
_000000_0
15/11/26 16:13:13 INFO output.FileOutputCommitter: File Output Committer Algorithm version is 1
```

```
15/11/26 16:13:13 INFO mapred.Task: Using ResourceCalculatorProcessTree : [ ]
15/11/26 16:13:13 INFO mapred.ReduceTask: Using ShuffleConsumerPlugin: org.apache.hadoop.
mapreduce.task.reduce.Shuffle@ff19d0b
15/11/26 16:13:13 INFO reduce.MergeManagerImpl: MergerManager: memoryLimit = 363285696,
maxSingleShuffleLimit = 90821424, mergeThreshold = 239768576, ioSortFactor = 10,
memToMemMergeOutputsThreshold = 10
15/11/26 16:13:13 INFO reduce.EventFetcher: attempt_local1968546553_0001_r_000000_0 Thread
started: EventFetcher for fetching Map Completion Events
15/11/26 16:13:14 INFO mapreduce.Job: Job job_local1968546553_0001 running in uber mode : false
15/11/26 16:13:14 INFO mapreduce.Job: map 100% reduce 0%
15/11/26 16:13:14 INFO reduce.LocalFetcher: localfetcher#1 about to shuffle output of map
attempt_local1968546553_0001_m_000000_0 decomp: 128 len: 132 to MEMORY
15/11/26 16:13:14 INFO reduce.InMemoryMapOutput: Read 128 bytes from map-output for attempt_
local1968546553_0001_m_000000_0
15/11/26 16:13:14 INFO reduce.MergeManagerImpl: closeInMemoryFile -> map-output of size:
128, inMemoryMapOutputs.size() -> 1, commitMemory -> 0, usedMemory -> 128
15/11/26 16:13:14 INFO reduce.EventFetcher: EventFetcher is interrupted.. Returning
15/11/26 16:13:14 INFO mapred.LocalJobRunner: 1 / 1 copied.
15/11/26 16:13:14 INFO reduce.MergeManagerImpl: finalMerge called with 1 in-memory map-
outputs and 0 on-disk map-outputs
15/11/26 16:13:14 INFO mapred.Merger: Merging 1 sorted segments
15/11/26 16:13:14 INFO mapred.Merger: Down to the last merge-pass, with 1 segments left of
total size: 122 bytes
15/11/26 16:13:14 INFO reduce.MergeManagerImpl: Merged 1 segments, 128 bytes to disk to
satisfy reduce memory limit
15/11/26 16:13:14 INFO reduce.MergeManagerImpl: Merging 1 files, 132 bytes from disk
15/11/26 16:13:14 INFO reduce.MergeManagerImpl: Merging 0 segments, 0 bytes from memory
into reduce
15/11/26 16:13:14 INFO mapred.Merger: Merging 1 sorted segments
15/11/26 16:13:14 INFO mapred.Merger: Down to the last merge-pass, with 1 segments left of
total size: 122 bytes
15/11/26 16:13:14 INFO mapred.LocalJobRunner: 1 / 1 copied.
15/11/26 16:13:14 INFO mapred.Task: Task:attempt_local1968546553_0001_r_000000_0 is done.
And is in the process of committing
15/11/26 16:13:14 INFO mapred.LocalJobRunner: 1 / 1 copied.
15/11/26 16:13:14 INFO mapred.Task: Task attempt_local1968546553_0001_r_000000_0 is allowed
to commit now
15/11/26 16:13:14 INFO output.FileOutputCommitter: Saved output of task 'attempt_
local1968546553_0001_r_000000_0' to hdfs://localhost:8020/files.har/_temporary/0/task_
local1968546553_0001_r_000000
15/11/26 16:13:14 INFO mapred.LocalJobRunner: reduce > reduce
15/11/26 16:13:14 INFO mapred.Task: Task 'attempt_local1968546553_0001_r_000000_0' done.
15/11/26 16:13:14 INFO mapred.LocalJobRunner: Finishing task: attempt_local1968546553_0001_
r_000000_0
15/11/26 16:13:14 INFO mapred.LocalJobRunner: reduce task executor complete.
15/11/26 16:13:15 INFO mapreduce.Job: map 100% reduce 100%
15/11/26 16:13:15 INFO mapreduce.Job: Job job_local1968546553_0001 completed successfully
15/11/26 16:13:15 INFO mapreduce.Job: Counters: 35
    File System Counters
    FILE: Number of bytes read = 45496
    FILE: Number of bytes written = 617838
    FILE: Number of read operations = 0
    FILE: Number of large read operations = 0
```

```
        FILE: Number of write operations = 0
        HDFS: Number of bytes read = 26
        HDFS: Number of bytes written = 163
        HDFS: Number of read operations = 31
        HDFS: Number of large read operations = 0
        HDFS: Number of write operations = 11
    Map - Reduce Framework
        Map input records = 2
        Map output records = 2
        Map output bytes = 122
        Map output materialized bytes = 132
        Input split bytes = 149
        Combine input records = 0
        Combine output records = 0
        Reduce input groups = 2
        Reduce shuffle bytes = 132
        Reduce input records = 2
        Reduce output records = 0
        Spilled Records = 4
        Shuffled Maps = 1
        Failed Shuffles = 0
        Merged Map outputs = 1
        GC time elapsed (ms) = 47
        Total committed heap usage (bytes) = 331489280
    Shuffle Errors
        BAD_ID = 0
        CONNECTION = 0
        IO_ERROR = 0
        WRONG_LENGTH = 0
        WRONG_MAP = 0
        WRONG_REDUCE = 0
    File Input Format Counters
        Bytes Read = 233
    File Output Format Counters
        Bytes Written = 0
```

在命令行中,第一个参数是归档文件的名称,这里是 file.har 文件;第二个参数是要归档的文件源,这里只归档一个源文件夹;最后一个参数即本例中的/是 HAR 文件的输出目录。

执行下面的命令:

```
hadoop fs - ls /
```

可以看到,已经创建了归档文件:

```
Found 3 items
drwxr-xr-x - root supergroup 0 2015-11-26 16:13 /files.har
drwxr-xr-x - root supergroup 0 2015-11-26 14:09 /master1-file
drwxr-xr-x - root supergroup 0 2015-11-26 11:08 /master1-file2
```

对于 HAR 文件需要了解一些它的不足。当创建一个归档文件时,还会创建原始文件的一个副本,这样就需要额外的磁盘空间(尽管归档完成后会删除原始文件)。而且当前还没有针对归档文件的压缩方案,只能对写入归档文件的原始文件进行压缩。归档文件一旦创建就不能改变,要增加或删除文件,就要重新创建。事实上,这对于那些写后不能更改的文件不构成问题,因为可以按日或者按周进行定期成批归档。

HAR 文件可以作为 MapReduce 的一个输入文件,但没有一个基于归档的 InputFormat 可以将多个文件打包到一个单一的 MapReduce 中去。所以,即使是 HAR 文件,处理小的文件时效率仍然不高。

2.2 WebHDFS

前面讲解了 HDFS 相关的内容,重点集中在如何使用 shell 下 Hadoop 的命令来管理 HDFS。本节将讲解 WebHDFS,即通过 Web 命令来管理 HDFS。

2.2.1 WebHDFS 的配置

WebHDFS 的原理是使用 curl 命令向指定的 Hadoop 集群对外接口发送页面请求,Hadoop 集群的网络接口接收到请求之后,会将命令中的 URL 解析成 HDFS 上对应的文件或者文件夹,URL 后面的参数解析成命令、用户、权限、缓存大小等参数。待完成相应的操作之后,将结果发还给执行 curl 命令的客户端,并显示执行信息或者错误信息。那么要使用 WebHDFS,首先就必须在期望使用 WebHDFS 的客户端安装 curl 软件包。待安装结束之后,在终端输入 curl-V 可以查看是否安装成功。如图 2.5 所示。

```
[root@master1 ~]# curl -V
curl 7.19.7 (x86_64-redhat-linux-gnu) libcurl/7.19.7 NSS/3.19.1 Basic ECC zlib/1
.2.3 libidn/1.18 libssh2/1.4.2
Protocols: tftp ftp telnet dict ldap ldaps http file https ftps scp sftp
Features: GSS-Negotiate IDN IPv6 Largefile NTLM SSL libz
```

图 2.5　查看版本

在客户端安装好 curl 软件包之后,还需要修改 Hadoop 集群的配置,使其开放 WebHDFS 服务,具体操作是:停止 Hadoop 所有服务之后,配置 hdfs-site.xml 中的 dfs.webhdfs.enabled、dfs.web.authentication.kerberos.principal、dfs.web.authentication.kerberos.keytab 这三个属性为适当的值,其中第一个属性值应配置为 true,代表启动 webHDFS 服务,后面两个代表使用 webHDFS 时采用的用户认证方法。配置结束之后重新启动 Hadoop 所有的服务,这样就可以使用 WebHDFS 来管理 Hadoop 的集群了。

在 hdfs-site.xml 中编辑配置文件,如图 2.6 所示。

```
<property>
  <name>dfs.webhdfs.enabled</name>
  <value>true</value>
</property>
<property>
  <name>dfs.permissions</name>
  <value>false</value>
</property>
</configuration>
```

图 2.6　编辑配置文件

启动之后,可以使用如下命令检测,结果如图 2.7 所示。

```
curl -i "http://master1:50070/webhdfs/v1/?user.name=hadoop&op=LISTSTATUS"
```

图 2.7　测试命令运行

2.2.2　WebHDFS 命令

2.2.1 节讲了如何配置 WebHDFS,本节将详细介绍 WebHDFS 命令的组织方式和具体的命令。

注意:在下面的实验中,主机名需要和配置 Hadoop 时的主机名一致,否则会出现错误。

1. WebHDFS 命令的一般形式

前面介绍 WebHDFS 实际上是用 curl 命令开发管理的命令,所以 WebHDFS 的命令组织类似。一般为下面的格式:

```
curl [-i/-X/-u/-T] [PUT] "HTTP://<HOST>:<PORT>/webhdfs/v1/<PATH>?[user.name
    =<user>&]&op=<operation>&[doas=<user>] …"
```

在这个命令中,引号前面的部分是 curl 自己的参数,需要大家自行了解;后面网页形式的内容代表着操作的命令、参数和路径。其中 http://<HOST>:<PORT>代表需要将命令发送的地址和端口,也就是 Hadoop 集群服务器的 IP 地址和 HDFS 端口(默认是 50070)。在这个地址之后的部分/webhdfs/v1/<PATH>代表需要操作的远程 HDFS 集群上的路径,比如/webhdfs/v1/master1-file,就代表着 HDFS 上/master1-file 这个目录。引号中再往后的内容就是操作的指令和参数了,其中最重要的是 op 参数,代表着具体的操作指令,下面会详细讲解。

2. 文件和路径操作

(1) 创建文件并写入内容：

```
curl -i -X PUT "http://<HOST>:<PORT>/webhdfs/v1/<PATH>?op=CREATE [overwrite=<true |
false>] [&blocksize=<LONG>][&replication=<SHORT>][&permission=<OCTAL>] [&buffersi
ze=<INT>]"
```

使用上述命令之后，会返回一个 location，如图 2.8 所示，它包括了已创建文件所在的 DataNode 地址及创建路径。下面就可以将文件内容发送到所显示 DataNode 对应路径下的文件内，命令如下，命令运行结果如图 2.9 所示。

```
[root@master1 hadoop]# curl -i -X PUT "http://master1:50070/webhdfs/v1/webhdfsFi
le?op=CREATE"
HTTP/1.1 307 TEMPORARY_REDIRECT
Cache-Control: no-cache
Expires: Thu, 05 Nov 2015 08:13:35 GMT
Date: Thu, 05 Nov 2015 08:13:35 GMT
Pragma: no-cache
Expires: Thu, 05 Nov 2015 08:13:35 GMT
Date: Thu, 05 Nov 2015 08:13:35 GMT
Pragma: no-cache
Content-Type: application/octet-stream
Location: http://master1:50075/webhdfs/v1/webhdfsFile?op=CREATE&namenoderpcaddre
ss=master1:8020&overwrite=false
Content-Length: 0
Server: Jetty(6.1.26)
```

图 2.8 命令运行结果（一）

```
curl -i -X PUT -T <LOCAL_FILE> "http://<DATANODE>:<PORT>/webhdfs/v1/<PATH>?op=
CREATE…"
```

```
[root@master1 ~]# curl -i -X PUT -T /usr/mytext/1.txt "http://master1:50075/web
hdfs/v1/webhdfsFile?op=CREATE&namenoderpcaddress=master1:8020&overwrite=false"
HTTP/1.1 100 Continue

HTTP/1.1 201 Created
Location: hdfs://master1:8020/webhdfsFile
Content-Length: 0
Connection: close
```

图 2.9 命令运行结果（二）

使用 cat 查看 WebHDFS 上传文件中的内容，如图 2.10 所示。

```
[root@master1 ~]# hadoop fs -cat /webhdfsFile
15/11/05 16:22:03 WARN util.NativeCodeLoader: Unable to load native-hadoop libra
ry for your platform... using builtin-java classes where applicable
hello world!
```

图 2.10 命令运行结果（三）

(2) 文件追加内容，首先使用下面的命令获取待追加内容文件所在地址，如图 2.11 所示。

```
curl -i -X POST "http://<HOST>:<PORT>/webhdfs/v1/<PATH>?op=APPEND[&buffersize=
<INT>]"
```

```
[root@master1 ~]# curl -i -X POST "http://master1:50070/webhdfs/v1/webhdfsFile?o
p=APPEND"
HTTP/1.1 307 TEMPORARY_REDIRECT
Cache-Control: no-cache
Expires: Thu, 05 Nov 2015 08:28:24 GMT
Date: Thu, 05 Nov 2015 08:28:24 GMT
Pragma: no-cache
Expires: Thu, 05 Nov 2015 08:28:24 GMT
Date: Thu, 05 Nov 2015 08:28:24 GMT
Pragma: no-cache
Content-Type: application/octet-stream
Location: http://master1:50075/webhdfs/v1/webhdfsFile?op=APPEND&namenoderpcaddre
ss=master1:8020
Content-Length: 0
Server: Jetty(6.1.26)
```

图 2.11 命令运行结果（四）

再结合返回内容的 location 信息，追加内容，命令如下，结果如图 2.12 所示。

`curl -i -X POST -T <LOCAL_FILE> "http://<DATANODE>:<PORT>/webhdfs/v1/<PATH>?op=APPEND"`

```
[root@master1 ~]# curl -i -X POST -T /usr/mytext/2.txt "http://master1:50075/we
bhdfs/v1/webhdfsFile?op=APPEND&namenoderpcaddress=master1:8020"
HTTP/1.1 100 Continue

HTTP/1.1 200 OK
Content-Length: 0
Connection: close
```

图 2.12 命令运行结果（五）

（3）打开并读取文件内容，使用下面的命令打开远程 HDFS 上的文件并读取内容：

`curl -i -L "http://<HOST>:<PORT>/webhdfs/v1/<PATH>?op=OPEN[&offset=<LONG>][&buffersize=<INT>]"`

需要注意的是，这个命令首先会返回文件所在的 location 信息，然后打印文件的具体内容。

（4）创建文件夹结果如图 2.13 所示。

`curl -i -X PUT "http://<HOST>:<PORT>/webhdfs/v1/<PATH>?op=MKDIRS[&permission=<OCTAL>]"`

（5）重命名文件夹或文件，结果如图 2.14 所示。

`curl -i -X PUT "http://<HOST>:<PORT>/webhdfs/v1/<PATH>?op=RENAME&destination=<PATH>"`

（6）删除文件夹或者文件，结果如图 2.15 所示。

`curl -i -X DELETE "http://<host>:<port>/webhdfs/v1/<PATH>?op=DELETE[&recursive=<true|false>]"`

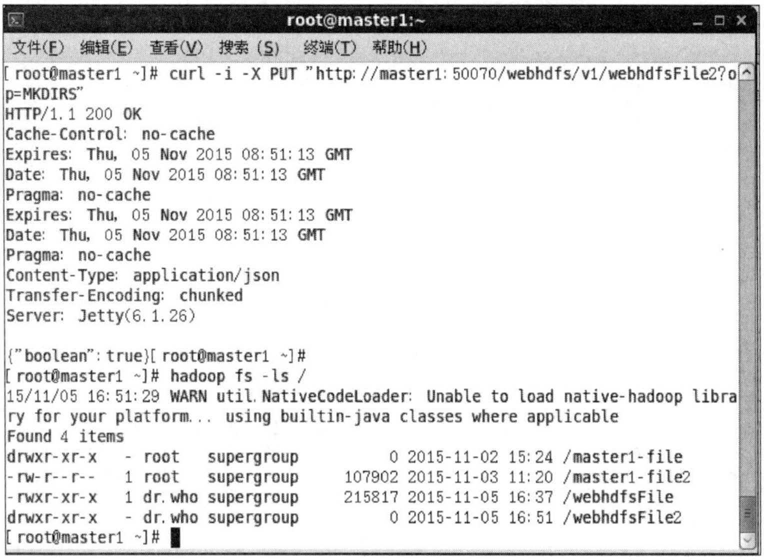

图 2.13　创建文件夹

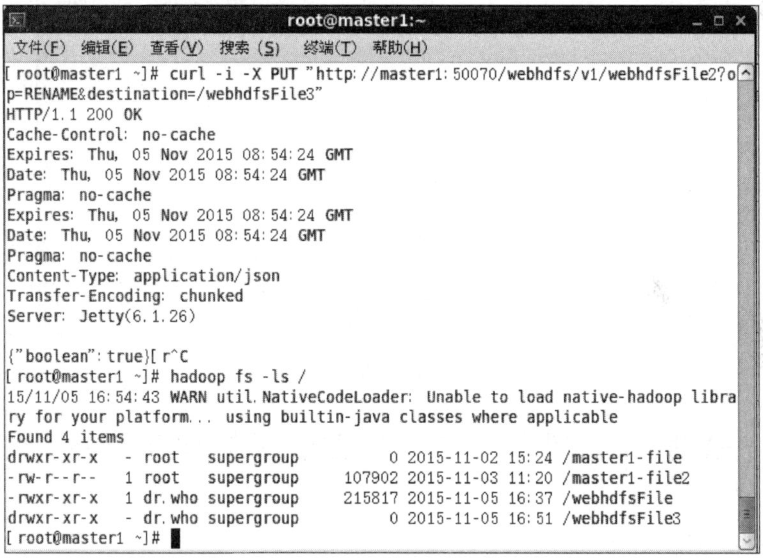

图 2.14　重命名文件

（7）查看文件夹或文件信息，结果如图 2.16 所示。

```
curl - i "http://< HOST >:< PORT >/webhdfs/v1/< PATH >?op = GETFILESTATUS"
```

（8）列举文件夹内容，结果如图 2.17 所示。

```
curl - i "http://< HOST >:< PORT >/webhdfs/v1/< PATH >?op = LISTSTATUS"
```

```
[root@master1 ~]# curl -i -X DELETE "http://master1:50070/webhdfs/v1/webhdfsFile3?op=DELETE"
HTTP/1.1 200 OK
Cache-Control: no-cache
Expires: Thu, 05 Nov 2015 08:58:07 GMT
Date: Thu, 05 Nov 2015 08:58:07 GMT
Pragma: no-cache
Expires: Thu, 05 Nov 2015 08:58:07 GMT
Date: Thu, 05 Nov 2015 08:58:07 GMT
Pragma: no-cache
Content-Type: application/json
Transfer-Encoding: chunked
Server: Jetty(6.1.26)

[root@master1 ~]# hadoop fs -ls /
15/11/05 16:59:59 WARN util.NativeCodeLoader: Unable to load native-hadoop library for your platform... using builtin-java classes where applicable
Found 3 items
drwxr-xr-x   - root   supergroup          0 2015-11-02 15:24 /master1-file
-rw-r--r--   1 root   supergroup     107902 2015-11-03 11:20 /master1-file2
-rwxr-xr-x   1 dr.who supergroup     215817 2015-11-05 16:37 /webhdfsFile
```

图 2.15　删除文件

```
[root@master1 ~]# curl -i "http://master1:50070/webhdfs/v1/webhdfsFile?op=GETFILESTATUS"
HTTP/1.1 200 OK
Cache-Control: no-cache
Expires: Thu, 05 Nov 2015 09:03:00 GMT
Date: Thu, 05 Nov 2015 09:03:00 GMT
Pragma: no-cache
Expires: Thu, 05 Nov 2015 09:03:00 GMT
Date: Thu, 05 Nov 2015 09:03:00 GMT
Pragma: no-cache
Content-Type: application/json
Transfer-Encoding: chunked
Server: Jetty(6.1.26)

{"FileStatus":{"accessTime":1446711540079,"blockSize":134217728,"childrenNum":0,"fileId":16409,"group":"supergroup","length":215817,"modificationTime":1446712678171,"owner":"dr.who","pathSuffix":"","permission":"755","replication":1,"storagePolicy":0,"type":"FILE"}}[root@master1 ~]#
```

图 2.16　查看文件

```
[root@master1 ~]# curl -i "http://master1:50070/webhdfs/v1/webhdfsFile?op=LISTSTATUS"
HTTP/1.1 200 OK
Cache-Control: no-cache
Expires: Thu, 05 Nov 2015 09:07:34 GMT
Date: Thu, 05 Nov 2015 09:07:34 GMT
Pragma: no-cache
Expires: Thu, 05 Nov 2015 09:07:34 GMT
Date: Thu, 05 Nov 2015 09:07:34 GMT
Pragma: no-cache
Content-Type: application/json
Transfer-Encoding: chunked
Server: Jetty(6.1.26)

{"FileStatuses":{"FileStatus":[
{"accessTime":1446711540079,"blockSize":134217728,"childrenNum":0,"fileId":16409,"group":"supergroup","length":215817,"modificationTime":1446712678171,"owner":"dr.who","pathSuffix":"","permission":"755","replication":1,"storagePolicy":0,"type":"FILE"}
]}}
```

图 2.17　列举文件夹内容

2.3 HDFS 常见的 Java API 介绍

上面介绍了 HDFS 的基本命令行操作。在实际应用时，也可以通过 HDFS 的 Java API 接口对文件进行操作。

2.3.1 使用 Hadoop URL 读取数据

如果想从 Hadoop 中读取数据，最简单的办法就是使用 java.net.URL 对象打开一个数据流，并从中读取数据，一般的调用格式如下：

```
InputStream in = null;
try {
    in = new URL("hdfs://host/path").openStream();
} finally {
    IOUtils.closeStream(in);
}
```

这里进行的处理是：通过 FsUrlStreamHandlerFactory 实例来调用在 URL 中的 setURLStreamHandlerFactory 方法。这种方法在一个 Java 虚拟机中只能调用一次，因此放在一个静态方法中执行。这意味着如果程序的其他部分也设置了一个 URLStreamHandlerFactory，那么会导致无法再从 Hadoop 中读取数据。

读取文件系统中的路径为 hdfs://master1/master1-file/1.txt 的文件 1.txt，如例 2-1 所示。这里 1.txt 的文件内容为"hello world!"。

例 2-1 使用 URLStreamHandler 以标准输出显示 Hadoop 文件系统文件。

```
package demoinput;
import org.apache.hadoop.io.*;
import org.apache.hadoop.fs.FsUrlStreamHandlerFactory;
import java.io.*;
import java.net.URL;
public class demoinputfile
{
    static{
        URL.setURLStreamHandlerFactory(new FsUrlStreamHandlerFactory());}
    public static void main(String[] args) throws Exception
    {
        InputStream in = null;
        try{
            in = new URL(args[0]).openStream();
            IOUtils.copyBytes(in,System.out, 4096,false);
        }finally{
            IOUtils.closeStream(in);
        }
    }
}
```

将 Java 文件导出成可执行 jar 包——Demo1.jar，运行结果如图 2.18 所示（注：需要将 Hadoop 的 jar 包导入工程中）。

```
[root@master1 ~]# hadoop jar usr/Demo1.jar hdfs://master1/master1-file/1.txt
15/11/02 17:18:10 WARN util.NativeCodeLoader: Unable to load native-hadoop library for your platform... using builtin-java classes where applicable
hello world!
```

图 2.18 运行结果（一）

运行程序即可看见 1.txt 中的文本内容。

需要说明的是，这里使用了 Hadoop 中简洁的 IOUtils 类来关闭 finally 子句中的数据流，同时复制输出流之间的字节（System.out）。例 2-1 中用到的 IOUtils.copyBytes() 方法，其中的两个参数，前者表示复制缓冲区的大小，后者表示复制后关闭数据流。

2.3.2 使用 FileSystem API 读取数据

文件在 Hadoop 文件系统中被视为一个 Hadoop Path 对象。可以把一个路径视为 Hadoop 的文件系统 URI，比如上面的 hdfs://master1/master1-file/1.txt。

FileSystemAPI 是一个高层抽象的文件系统 API，所以，首先要找到这里的文件系统实例 HDFS。取得 FileSystem 实例有两种静态工厂方法：

```
public static FileSystem get(Configuration conf) throws IOException
public static FileSystem get(URI uri, Configuration conf) throws IOException
```

Configuration 对象封装了一个客户端或服务器的配置，这是用路径读取的配置文件设置的，一般为 core-site.xml。第一个方法返回的是默认文件系统，如果没有设置，则为默认的本地文件系统。第二个方法使用指定的 URI 方案决定文件系统的权限，如果指定的 URI 中没有指定方案，则退回默认的文件系统。

有了 FileSystem 实例后，可通过 open() 方法得到一个文件的输入流：

```
public FSDataInputStream open(Path f) throws IOException
public abstract FSDataInputStream open(Path f, int bufferSize) throws IOException
```

第一个方法直接使用默认的 4KB 的缓冲区，如例 2-2 所示。

例 2-2 使用 FileSystem API 显示 Hadoop 文件系统中的文件。

```
package com.zhixueyun.edu.cn;
import org.apache.hadoop.io.*;
import org.apache.hadoop.fs.FsUrlStreamHandlerFactory;
import org.apache.hadoop.fs.Hdfs;
import org.apache.hadoop.fs.Path;
import org.apache.hadoop.filecache.DistributedCache;
import org.apache.hadoop.conf.*;
import org.apache.hadoop.fs.FileSystem;

import java.io.*;
import java.net.URI;
```

```
import java.net.URL;

public class FileSystemCat {

    public static void main(String[] args) throws Exception{
        String uri = args[0];
        Configuration conf = new Configuration();
        FileSystem fs = FileSystem.get(URI.create(uri),conf);
        InputStream in = null;
        try{
            in = fs.open(new Path(uri));
            IOUtils.copyBytes(in, System.out,4096,false);
        }finally{
            IOUtils.closeStream(in);
        }
    }
}
```

然后设置程序运行参数为 hdfs://master1/master1-file/1.txt，运行程序即可以看到 1.txt 中的文本内容"hello world!"，如图 2.19 所示。

```
[root@master1 ~]# hadoop jar usr/Demo2.jar hdfs://master1/master1-file/1.txt
15/11/02 17:16:49 WARN util.NativeCodeLoader: Unable to load native-hadoop libra
ry for your platform... using builtin-java classes where applicable
hello world!
```

图 2.19　运行结果（二）

FileSystem 中的 open 方法实际上返回的是一个 FSDataInputStream，而不是标准的 java.io 类。这个类是 java.io.DataInputStream 的一个子类，支持随机访问，并可以从流的任意位置读取，代码如下：

```
public class FSDataInputStream extends DataInputStream
implements Seekable,PositionedReadable{
          //implementation elided
}
```

Seekable 接口允许在文件定位并提供一个查询方法用于查询当前位置相对于文件开始的偏移量（getPos()），代码如下：

```
public interface Seekable{
    void seek(long pos) throws IOException;
      long getPos() throws IOException;
      Boolean seekToNewSource(long targetPos) throws IOException;
}
```

其中，调用 seek() 来定位大于文件长度的位置会导致 IOException 异常。开发人员并不常用 seekToNewSource() 方法，此方法倾向于切换到数据的另一个副本，并在新的副本中找寻 targetPos 定制的位置。HDFS 就采用这样的方法在数据节点出现故障时为客户端提供可靠的数据流访问。如例 2-3 所示。

例 2-3 扩展例 2-2,通过使用 seek()读取一次后,重新定位到文件头第三位,再次显示 Hadoop 文件系统的文件内容。

```java
package com.zhixueyun.edu.cn;
import org.apache.hadoop.io.*;
import org.apache.hadoop.fs.Path;
import org.apache.hadoop.conf.*;
import org.apache.hadoop.fs.FSDataInputStream;
import org.apache.hadoop.fs.FileSystem;

import java.net.URI;
public class DoubleCat {
    public static void main(String[] args) throws Exception{
        String uri = args[0];
        Configuration conf = new Configuration();
        FileSystem fs = FileSystem.get(URI.create(uri),conf);
        FSDataInputStream in = null;
        try {
            in = fs.open(new Path(uri));
            IOUtils.copyBytes(in, System.out,4096,false);
            in.seek(3);
            IOUtils.copyBytes(in, System.out,4096,false);
        } finally{
            IOUtils.closeStream(in);
        }
    }
}
```

然后设置程序运行参数为 hdfs://master1/master1-file/1.txt,运行程序即可看到 1.txt 中的文本内容"hello world! lo world!",如图 2.20 所示。

```
[root@master1 ~]# hadoop jar usr/Demo3.jar hdfs://master1/master1-file/1.txt
15/11/02 17:55:08 WARN util.NativeCodeLoader: Unable to load native-hadoop libra
ry for your platform... using builtin-java classes where applicable
hello world!
lo world!
```

图 2.20 运行结果(三)

需要注意的是,seek()是一个高开销的操作,需要慎重使用。通常我们是依靠流数据 MapReduce 构建应用访问模式,而不是大量地执行 seek()操作。

2.3.3 创建目录

FileSystem 显然也提供了创建目录的方法,代码如下:

```
public Boolean mkdir(Path f) throws IOException
```

这个方法会按照客户端请求创建未存在的父目录,就像 java.io.File 的 mkdirs()一样。 如果目录包括所有父目录且创建成功,那么它会返回 true。事实上,一般不需要特别地创建 一个目录,因为调用 creat()时写入文件会自动生成所有的父目录。

2.3.4 写数据

FileSystem 还有一系列创建文件的方法，最简单的就是给拟创建的文件指定一个路径对象，然后返回一个写输入流，代码如下：

```
public FSDataOutputStream creat(Path f) throws IOException
```

这个方法有很多重载方法，例如，可以设定是否强制覆盖源文件，设定文件副本数量，设置写入文件缓冲区大小、文件块大小及设置文件许可等。

还有一个用于传递回调接口的重载方法 Progressable，通过这个方法就可以获得数据节点写入进度，代码如下：

```
package org.apache.hadoop.util;
public interface Progressable{
    public void progress();
}
```

新建文件也可以使用 append() 在一个已有文件中追加内容，这个方法也有重载，代码如下：

```
public FSDataOutputStream append(Path f) throws IOException
```

这个方法对于写入日志文件很有用，比如在重启后可以在之前的日志中继续添加内容，但不是所有的 Hadoop 文件系统都支持此方法，比如 HDFS 支持，但 S3 不支持。

例 2-4 展示了如何将本地文件复制到 Hadoop 的文件系统，当 Hadoop 调用 progress() 方法时，打印星号来展示整个过程。

例 2-4　将本地文件复制到 Hadoop 文件系统。

```
package com.zhixueyun.edu.cn;
import org.apache.hadoop.io.*;
import org.apache.hadoop.util.Progressable;
import org.apache.hadoop.fs.FsUrlStreamHandlerFactory;
import org.apache.hadoop.fs.Hdfs;
import org.apache.hadoop.fs.Path;
import org.apache.hadoop.filecache.DistributedCache;
import org.apache.hadoop.conf.*;
import org.apache.hadoop.io.*;
import org.apache.hadoop.fs.FileSystem;

import java.io.*;
import java.net.URI;
import java.net.URL;
public class FileCopyWithProgress {
    public static void main(String[] args) throws Exception{
        String localSrc = args[0];
        String dst = args[1];
        InputStream in = new BufferedInputStream(new FileInputStream(localSrc));
        Configuration conf = new Configuration();
```

```
            FileSystem fs = FileSystem.get(URI.create(dst),conf);
            OutputStream out = fs.create(new Path(dst),new Progressable(){
                public void progress(){
                    System.out.print("*");
                }
            });
            IOUtils.copyBytes(in,out,4096,true);
        }
    }
```

然后配置应用参数,可以看到控制台输出"***",即上传显示进度*。目前其他文件系统写入时都不会调用 progress(),如图 2.21 所示。

```
[root@master1 ~]# hadoop jar usr/Demo3-4.jar /usr/mytext/2.txt  hdfs://master1/m
aster1-file2
15/11/03 11:19:52 WARN util.NativeCodeLoader: Unable to load native-hadoop libra
ry for your platform... using builtin-java classes where applicable
**[root@master1 ~]#
```

图 2.21 运行结果(四)

上面在介绍读取数据时提到 FSDataInputStream,这里 FileSystem 中的 creat()方法也返回一个 FSDataOutStream,它也有一个查询文件当前位置的方法,代码如下:

```
package com.zhixueyun.edu.cn
public class FSDataOutputStream extends DataOutputStream implements Syncable{
    public long getPos() throws IOException{
    //implementation elided
}
//implementation elided
}
```

但是它与 FSDataInputStream 不同,FSDataOutputStream 不允许定位。这是因为 HDFS 只对一个打开的文件顺序写入,或者向一个已有的文件添加。换句话说,它不支持对除文件尾部以外的其他位置进行写入,这样,写入时的定位就没有意义了。

2.3.5 删除数据

使用 FileSystem 的 delete()可以永久删除 Hadoop 中的文件或目录。

```
public boolean delete(Path f , Boolean recursive) throws IOException
```

如果传入的 f 为空文件或空目录,那么 recursive 的值会被忽略。只有当 recursive 的值为 true 时,非空的文件或目录才会被删除,否则抛出异常。

2.3.6 文件系统查询

同样,Java API 提供了文件系统的基本查询接口。通过这个接口,可以查询系统的元数据信息和文件目录结构,并可以进行更复杂的目录匹配等操作。下面将对其进行介绍。

1. 文件元数据 Filestatus

任何文件系统要具备的重要功能就是定位其目录结构以检索器存储的文件和目录信息。FileStatus 类封装了文件系统中文件和目录的元数据，其中包括文件长度、块大小、副本、修改时间、所有者和许可信息等。

FileSystem 的 getFileStatus() 方法提供了获取一个文件或目录的状态对象的方法，如例 2-5 所示。

例 2-5 获取文件状态信息。

```
package com.zhixueyun.edu.cn;

import org.apache.hadoop.conf.Configuration;
import org.apache.hadoop.fs.FileStatus;
import org.apache.hadoop.fs.FileSystem;
import org.apache.hadoop.fs.Path;
public class ShowFileStatusTest{
    public static void fileInfo(String path) throws IOException{
        Configuration conf = new Configuration();
        FileSystem fs = FileSystem.get(conf);
        Path p = new Path(path);
        //FileStatus 对象封装了文件和目录的元数据,包括文件长度、块大小、权限等信息
        FileStatus fileStatus = fs.getFileStatus(p);
        System.out.println("文件路径:" + fileStatus.getPath());
        System.out.println("块的大小:" + fileStatus.getBlockSize());
         System.out.println("文件所有者:" + fileStatus.getOwner() + ":" + fileStatus.getGroup());
        System.out.println("文件权限:" + fileStatus.getPermission());
        System.out.println("文件长度:" + fileStatus.getLen());
        System.out.println("备份数:" + fileStatus.getReplication());
        System.out.println("修改时间:" + fileStatus.getModificationTime());
    }
    public static void main(String[] args) throws IOException {
        String localSrc = args[0];
        fileInfo(localSrc);
    }
}
```

具体运行结果如图 2.22 所示。

```
[root@master1 ~]# hadoop jar usr/Demo3-5.jar  hdfs://master1/master1-file2
15/11/03 17:11:08 WARN util.NativeCodeLoader: Unable to load native-hadoop library for your platform... using builtin-java classes where applicable
文件路径: hdfs://master1/master1-file2
块的大小: 134217728
文件所有者: root: supergroup
文件权限: rw-r--r--
文件长度: 107902
备份数: 1
修改时间: 1446520800367
[root@master1 ~]#
```

图 2.22 运行结果（五）

2. 列出目录文件信息

列出目录的内容,这里使用 listStatus() 方法,代码如下:

```
public FileStatus[] listStatus(Path f) throws IOException
public FileStatus[] listStatus(Path f , PathFilter fliter) throws IOException
public FileStatus[] listStatus(Path[] files) throws IOException
public FileStatus[] listStatus(Path[] files,PathFilter filter) throws IOException
```

传入参数为一个目录时,它会返回 0 个或多个 FileStatus 对象,代表该目录所包含的文件和子目录。

我们看到 listStatus() 有很多重载方法,可以使用 PathFilter 来限制匹配的文件和目录。如果把路径数组作为参数来调用 listStatus() 方法,其结果与一次对多个目录进行查询、再将 FileStatus 对象数组收集到一个单一的数组的结果是相同的。当然,前者更为方便。例 2-6 是一个简单的示范。

例 2-6 显示 Hadoop 文件系统中的一个目录的文件信息。

```java
package com.zhixueyun;

import org.apache.hadoop.fs.Path;
import org.apache.hadoop.conf.*;
import org.apache.hadoop.fs.FileStatus;
import org.apache.hadoop.fs.FileSystem;
import org.apache.hadoop.fs.FileUtil;
import java.net.URI;

public class ListStatus {
    public static void main(String[] args) throws Exception{
        String uri = args[0];
        Configuration conf = new Configuration();
        FileSystem fs = FileSystem.get(URI.create(uri),conf);
        Path[] paths = new Path[args.length];
        for(int i = 0;i < paths.length;i++){
            paths[i] = new Path(args[i]);
        }
        FileStatus[] status = fs.listStatus(paths);
        Path[] listedPaths = FileUtil.stat2Paths(status);
        for(Path p:listedPaths){
            System.out.println(p);
        }
    }
}
```

具体运行结果如图 2.23 所示。

```
[root@master1 ~]# hadoop jar usr/Demo3-6.jar    hdfs://master1/master1-file
15/11/04 09:41:11 WARN util.NativeCodeLoader: Unable to load native-hadoop libra
ry for your platform... using builtin-java classes where applicable
hdfs://master1/master1-file/1.txt
```

图 2.23 运行结果(六)

配置应用参数可以查看文件系统的目录,可以查看 HDFS 中对应文件目录下的文件信息。

3. 通过通配符实现目录筛选

有时候我们需要批量处理文件,比如处理日志文件,这时可能要求 MapReduce 任务分析一个月的文件。这些文件包含在大量目录中,这就要求我们进行一个通配符操作,并使用通配符核对多个文件。Hadoop 为通配符提供了两个方法,可以在 FileSystem 中找到:

```
public FileStatus[] globStatus(Path pathPattern) throws IOException
public FileStatus[] globStatus(Path pathPattern,PathFilter filter) throws IOException
```

globStatus()返回了其路径匹配所提供的 FileStatus 对象数组,再按路径进行排序,其中可选的 PathFilter 命令可以进一步限定匹配。

表 2.1 是 Hadoop 支持的一系列通配符。

表 2.1　Hadoop 支持的通配符及其作用

通配符	名称	匹配功能
*	星号	匹配 0 个或多个字符
?	问号	匹配一个字符
[ab]	字符类别	匹配{a,b}中的一个字符
[^ab]	非此字符类别	匹配不属于{a,b}中的一个字符
[a—b]	字符范围	匹配在{a,b}范围内的字符(包括 a,b),a 在字典顺序上要小于或等于 b
[^a—b]	非此字符范围	匹配不在{a,b}范围内的字符(包括 a,b),a 在字典顺序上要小于或等于 b
{a,b}	或选择	匹配包含 a 或 b 中的语句
\c	转义字符	匹配元字符 c

下面通过例子进行详细说明。假设一个日志文件的存储目录是分层组织的,其中目录格式为年/月/日:/2014/12/30、/2014/12/31、/2010/01/01、/2010/01/02。表 2.2 是通配符的部分样例。

表 2.2　通配符使用样例

通配符	匹配结果
/*	/2014 /2015
/*/*	/2014/12 /2015/01
/*/12/*	/2014/12/30 2014/12/31
/201*	/2014 /2015
/201[4—5]	/2014 /2015
/201[^012356789]	/2014
/*/*/{31,01}	/2014/12/31 /2015/01/01
/*/{12/31,01/01}	/2014/12/31 /2015/01/01

4. PathFilter 对象

使用通配符有时也不一定能够精确地定位到要访问的文件集合,比如排除一个特定的文件，这时可以使用 FileSystem 中的 listStatus（）和 globStatus（）方法提供可选的 PathFilter 对象来通过编程的方法控制匹配结果，如下面的代码所示。

```
package com.zhixueyun;
public interface PathFilter{
Boolean accept (Path path);
}
```

下面看一个 PathFilter 的应用，如例 2-7 所示。

例 2-7　使用 PathFilter 排除匹配正则表达式的目录。

```
package cn.zhixueyun;
import org.apache.hadoop.fs.Path;

public class RegexExcludePathFilter implements PathFilter{
private final String regex;
public RegexExcludePathFilter (String regex) {
    this.regex = regex;
}
public boolean accept(Path path){
    return !path.toString().matches(regex);
}
public static void main(String args[]) throws Exception{
 String localSrc = args[0];
    RegexExcludePathFilter q = new RegexExcludePathFilter(localSrc);
    Path path = new Path(args[1]);
    System.out.println("" + q.accept(path));
 }
}
```

第 3 章

Hadoop 计算：MapReduce

3.1 MapReduce 应用程序编写

介绍了 HDFS 的基本操作之后，本章将会对 MapReduce 进行深入的讲解。如同 Java 中的"Hello World"经典程序一样，WordCount 是 MapReduce 的入门程序。虽然此例在本书中的其他章节也有涉及，但是本节主要从如何挖掘此问题中的并行处理可能性角度出发，让读者了解设计 MapReduce 程序的过程。

3.1.1 实例描述

计算出文件中每个单词的频数。要求输出结果按照单词的字母顺序进行排序。每个单词及其频数占一行，单词和频数之间有间隔。

例如，输入一个文件，其内容如下：

```
hello world
hello hadoop
hello mapreduce
```

对应上面给出的输入样例，其输出样例为：

```
hadoop 1
hello 3
mapreduce 1
world 1
```

3.1.2 设计思路

这个应用实例的解决方案很直接，就是将文件内容切分为单词，然后将所有相同的单词聚集在一起，最后计算单词出现的次数并输出。根据 MapReduce 并行程序设计原则可知，解决方案中的内容切分步骤和数据不相关，可以并行化处理，每个获得原始数据的机器只要将输入数据切分单词就可以了。所以可以在 Map 阶段完成单词切分任务。另外，相同单词的频数计算也可以并行化处理。从实例要求来看，不同单词之间的频数不相关，所以可以将相同的单词交给一台机器来计算频数，然后输出最终结果。这个过程可以在 Reduce 阶段完成。至于将中间结果根据不同单词分组再分发给 Reduce 机器，这正好是 MapReduce 过程中的 shuffle 能够完成的。至此，这个实例的 MapReduce 程序就设计出来了。Map 阶段完成由输入数据到单词切分的工作，shuffle 阶段完成相同单词的聚集和分发工作（这个过程是 MapReduce 的默认过程，不用具体配置），Reduce 阶段负责接收所有单词并计算其频数。MapReduce 中传递的数据都是< key，value >形式的，并且 shuffle 排序聚集分发都是按照 key 进行的，因此将 Map 的输出设计成由 word 作为 key、1 作为 value 的形式，这表示单词 word 出现了一次（Map 的输入采用 Hadoop 默认的输入方式：文件的一行作为 value，行号作为 key）。Reduce 的输出会设计成与 Map 输出相同的形式，只是后面的数字不再固定是 1，而是具体算出的 word 所对应的频数。

3.1.3 代码数据流

首先在 MapReduce 程序启动阶段，ApplicationMaster 先将 Job 的输入文件分割到每个 Map Task 上。假设现在有两个 Map Task，一个 Map Task 作为一个文件。

接下来 MapReduce 启动 Job，每个 Map Task 在启动之后会接收到自己所分配的输入数据，针对此例（采用默认的输入方式，每次读入一行，key 为行首在文件中的偏移量，value 为行字符串内容），两个 Map Task 的输入数据如下：

```
<0, "hello world">

<0, "hello hadoop">
<14, "hello mapreduce">
```

Map 函数会对输入内容进行切分，然后输出每个单词及其频次。第一个 Map Task 的 Map 输出如下：

```
<"hello", 1>
<"world", 1>
```

第二个 Map Task 的 Map 输出如下：

```
<"hello", 1>
<"hadoop", 1>
<"hello", 1>
<"mapreduce", 1>
```

由于在本例中设置了 Combiner 的类为 Reduce 的 class，所以每个 Map Task 将输出发送到 Reduce 时，会先执行一次 Combiner。这里的 Combiner 相当于将结果先局部进行合并，这样能够降低网络压力，提高效率。执行 Combiner 之后两个 Map Task 的输出如下：

```
Map Task1
<"hello", 1 >
<"world", 1 >
Map Task2
<"hello", 2 >
<"hadoop", 1 >
<"mapreduce", 1 >
```

接下来是 MapReduce 的 shuffle 过程，对 Map 的输出进行排序合并，并根据 Reduce 数量对 Map 的输出进行切分，将结果交给对应的 Reduce。经过 shuffle 过程的输出也就是 Reduce 的输入如下：

```
<"hadoop", 1 >
<"hello", <1, 2 >>
<"mapreduce", 1 >
<"world", 1 >
```

Reduce 接收到如上的输入之后，对每个< key，value >进行处理，计算每个单词也就是 key 的出现总数。最后输出单词和对应的频数，形成整个 MapReduce 的输出，内容如下：

```
<"hadoop", 1 >
<"hello", 3 >
<"mapreduce", 1 >
<"world", 1 >
```

3.1.4 程序代码

```java
package org.apache.hadoop.examples;
import java.io.IOException;
import java.util.StringTokenizer;

import org.apache.hadoop.conf.Configuration;
import org.apache.hadoop.fs.Path;
import org.apache.hadoop.io.IntWritable;
import org.apache.hadoop.io.Text;
import org.apache.hadoop.mapreduce.Job;
import org.apache.hadoop.mapreduce.Mapper;
import org.apache.hadoop.mapreduce.Reducer;
import org.apache.hadoop.mapreduce.lib.input.FileInputFormat;
import org.apache.hadoop.mapreduce.lib.output.FileOutputFormat;
import org.apache.hadoop.util.GenericOptionsParser;

public class WordCount {
```

```java
//继承 Mapper 接口,设置 map 的输入类型为<Object, Text>;输出类型为<Text, IntWritable>
public static class TokenizerMapper extends Mapper<Object, Text, Text, IntWritable>{
    //one 表示单词出现一次
    private final static IntWritable one = new IntWritable(1);
    //word 用于存储切下的单词
    private Text word = new Text();

    public void map(Object key, Text value, Context context) throws IOException, InterruptedException{
        StringTokenizer itr = new StringTokenizer(value.toString()); //对输入的行切词
        while(itr.hasMoreTokens()){
            //切下的单词存入 word
            word.set(itr.nextToken());
            context.write(word, one);
        }
    }
}

//继承 Reducer 接口,设置 Reduce 的输入类型为<Text, IntWritable>,输出类型为<Text, IntWritable>
public static class IntSumReducer extends Reducer<Text, IntWritable, Text, IntWritable>{
    //result 记录单词的频数
    private IntWritable result = new IntWritable();

    public void reduce(Text key, Iterable<IntWritable> values, Context context) throws IOException, InterruptedException{
        int sum = 0;
        //对获取的<key, value-list>计算 value 的和
        for(IntWritable val : values){
            sum += val.get();
        }
        //将频数设置到 result 中
        result.set(sum);
        //收集结果
        context.write(key, result);
    }
}

public static void main(String[] args) {
    // TODO Auto-generated method stub
    Configuration conf = new Configuration();
    conf.set("fs.defaultFS","hdfs://master:9000");
    //检查运行命令
    String[] otherArgs = null;
    try {
        otherArgs = new GenericOptionsParser(conf, args).getRemainingArgs();
    } catch (IOException e) {
        // TODO Auto-generated catch block
        e.printStackTrace();
    }
    //配置作业名
    Job job = null;
    try {
```

```
        job = new Job(conf, "word count");
    } catch (IOException e) {
        // TODO Auto-generated catch block
        e.printStackTrace();
    }
    //配置作业的各个类
    job.setJarByClass(WordCount.class);
    job.setMapperClass(TokenizerMapper.class);
    job.setCombinerClass(IntSumReducer.class);
    job.setReducerClass(IntSumReducer.class);
    job.setOutputKeyClass(Text.class);
    job.setOutputValueClass(IntWritable.class);
    try {
        FileInputFormat.addInputPath(job, new Path(args[0]));
    } catch (IllegalArgumentException e) {
        // TODO Auto-generated catch block
        e.printStackTrace();
    } catch (IOException e) {
        // TODO Auto-generated catch block
        e.printStackTrace();
    }
    FileOutputFormat.setOutputPath(job, new Path(args[1]));
    try {
        job.waitForCompletion(true);
    } catch (ClassNotFoundException e) {
        // TODO Auto-generated catch block
        e.printStackTrace();
    } catch (IOException e) {
        // TODO Auto-generated catch block
        e.printStackTrace();
    } catch (InterruptedException e) {
        // TODO Auto-generated catch block
        e.printStackTrace();
    }
    finally
    {
        System.out.println("mapreduce success!");
    }
    }
}
```

3.1.5 代码解读

WordCount 程序在 Map 阶段接收输入的<key,value>(key 是当前输入的行号,value 是对应行的内容),然后对此行内容进行切词,每切下一个词就将其组织成<word,1>的形式输出,表示 word 出现了一次。

在 Reduce 阶段,会接收到<word,{1,1,1,1…}>形式的数据,也就是特定单词及其出现次数的情况,其中 1 表示 word 的频数。所以 Reduce 每接收一个<word,{1,1,1,1…}>,就会在 word 的频数上加 1,最后组织成<word,sum>的形式直接输出。

3.1.6 程序执行

这里将介绍两种运行程序的方式(程序默认使用第一种方式运行,但是第二种方式也要掌握)。

1. 在 Eclipse 中运行

将两个输入文件上传到 DFS 文件系统中。

file1 的内容是:

```
hello world
```

file2 的内容是:

```
hello hadoop
hello mapreduce
```

在 Eclipse 中运行该程序,在运行之前,先来配置一下输入、输出文件,选择要运行的文件,然后单击 Run As→Run Configurations,选择要配置的文件,选择 Arguments 选项卡,然后在 "Program arguments:" 中输入 "/user/root/input/file ∗ /user/root/output/WordCount",单击右下角的 Apply 按钮,最后关闭该窗口。然后运行该文件。

2. 在命令行模式下运行

运行条件:将 WordCount.java 文件放在 Hadoop 安装目录下,并在目录下创建输入目录 input 和 WordCount,目录下有输入文件 file1 和 file2。

file1 的内容是:

```
hello world
```

file2 的内容是:

```
hello hadoop
hello mapreduce
```

准备好之后在命令行输入命令运行。下面对执行的命令进行介绍。

(1)进入 Hadoop 根目录,在集群上创建输入文件夹:

```
bin/hadoop fs -mkdir input
```

执行命令:

```
bin/hadoop fs -ls
```

运行结果如图 3.1 所示。

(2)将本地目录 input 下前四个字符为 file 的文件上传到集群上的 input 目录,如图 3.2 和图 3.3 所示。

第3章 Hadoop计算：MapReduce

```
[root@master hadoop-2.7.1]# bin/hadoop fs -ls
16/08/30 15:23:13 WARN util.NativeCodeLoader: Unable to load native-hadoop libra
ry for your platform... using builtin-java classes where applicable
Found 2 items
drwxr-xr-x   - root supergroup          0 2016-08-30 15:20 input
drwxr-xr-x   - root supergroup          0 2016-08-30 15:21 output
[root@master hadoop-2.7.1]# bin/hadoop fs -ls
16/08/30 15:24:05 WARN util.NativeCodeLoader: Unable to load native-hadoop libra
ry for your platform... using builtin-java classes where applicable
Found 1 items
drwxr-xr-x   - root supergroup          0 2016-08-30 15:20 input
```

图 3.1 查看文件

```
bin/hadoop fs - put input/file* input
```

```
[root@master hadoop-2.7.1]# bin/hadoop fs -put input/file* input
16/08/30 15:28:01 WARN util.NativeCodeLoader: Unable to load native-hadoop libra
ry for your platform... using builtin-java classes where applicable
[root@master hadoop-2.7.1]# bin/hadoop fs -ls
16/08/30 15:28:19 WARN util.NativeCodeLoader: Unable to load native-hadoop libra
ry for your platform... using builtin-java classes where applicable
Found 1 items
drwxr-xr-x   - root supergroup          0 2016-08-30 15:28 input
```

图 3.2 上传文件

```
[root@master hadoop-2.7.1]# bin/hadoop fs -ls input
16/08/30 15:52:21 WARN util.NativeCodeLoader: Unable to load native-hadoop libra
ry for your platform... using builtin-java classes where applicable
Found 2 items
-rw-r--r--   1 root supergroup         12 2016-08-30 15:50 input/file1.txt
-rw-r--r--   1 root supergroup         30 2016-08-30 15:50 input/file2.txt
```

图 3.3 查看文件

（3）将写好的 WordCount.java 放入 hadoop 的根目录下（即：/usr/cx/hadoop-2.7.1）。将 hadoop 的 classhpath 信息添加到 CLASSPATH 变量中，在 ~/.bashrc 中增加如图 3.4 所示的内容：

```
export HADOOP_HOME=/usr/cx/hadoop-2.7.1
export CLASSPATH=$($HADOOP_HOME/bin/hadoop classpath):$CLASSPATH
```

图 3.4 编辑配置文件

（4）执行 source ~/.bashrc 使变量生效。

（5）通过 javac 命令编译 WordCount.java，执行命令：javac WordCount.java。

编译时会有警告，可以忽略。编译后可以看到生成了几个 .class 文件，接着把 .class 文件打包成 jar，才能在 Hadoop 中运行。执行命令：

```
jar - cvf WordCount.jar ./WordCount * .class
```

（6）开始运行。执行命令

```
hadoop jar WordCount.jar org.apache.hadoop.examples.WordCount input output //hdfs 上的 input
文件夹,命令执行所在位置为 WordCount.jar 同一目录
```

(7) 把在 WordCount 目录中生成的 jar 包放到 WordCount 上一级目录中，在集群上运行 WordCount 程序，以 input 目录作为输入目录，output 目录作为输出目录：

```
bin/hadoop jar WordCount.jar org.apache.hadoop.five WordCount input output
```

(8) 查看输出结果：

```
bin/hadoop fs -cat output/part-00000
```

3.1.7 代码结果

运行结果如下：

```
hadoop 1
hello 3
mapreduce 1
word 1
```

3.2 使用 MapReduce 求每年最低温度

在了解了 MapReduce 的运行过程及原理之后，下面通过一个实际的应用案例分析来对 MapReduce 进行深入的理解。

3.2.1 作业描述

本例中的气象数据集包括十年的气象数据，这里将数据集放到了 /usr/software/ 下，编写一个 MapReduce 作业，求每年的最低温度，数据集的信息如图 3.5 所示。

图 3.5 数据信息

3.2.2 程序代码

编写 MinTemperature.java：

```java
package com.uicc.shizhan;

import org.apache.hadoop.fs.Path;
import org.apache.hadoop.io.IntWritable;
import org.apache.hadoop.io.Text;
import org.apache.hadoop.mapreduce.Job;
```

```java
import org.apache.hadoop.mapreduce.lib.input.FileInputFormat;
import org.apache.hadoop.mapreduce.lib.output.FileOutputFormat;

public class MinTemperature {
    public static void main(String[] args) throws Exception{
        if(args.length != 2){
            System.err.println("USage: MinTemperature<input path><output path>");
            System.exit(-1);
        }

        Job job = new Job();
        job.setJarByClass(MinTemperature.class);
        job.setJobName("Min temperature");
        FileInputFormat.addInputPath(job, new Path(args[0]));
        FileOutputFormat.setOutputPath(job, new Path(args[1]));
        job.setMapperClass(MinTemperatureMapper.class);
        job.setReducerClass(MinTemperatureReducer.class);
        job.setOutputKeyClass(Text.class);
        job.setOutputValueClass(IntWritable.class);
        System.exit(job.waitForCompletion(true)?0:1);
    }
}
```

编写 MinTemperatureMapper.java：

```java
package com.uicc.shizhan;

import java.io.IOException;

import org.apache.hadoop.io.IntWritable;
import org.apache.hadoop.io.LongWritable;
import org.apache.hadoop.io.Text;
import org.apache.hadoop.mapreduce.Mapper;
public class MinTemperatureMapper extends Mapper<LongWritable, Text, Text, IntWritable>{
    private static final int MISSING = 9999;

    public void map(LongWritable key, Text value, Context context)
        throws IOException, InterruptedException{
        String line = value.toString();           //按行读取数据
        String year = line.substring(15,19);      //获取年份

        int airTemperature;                        //温度
        if(line.charAt(87) == '+'){
            airTemperature = Integer.parseInt(line.substring(88,92));
        }else{
            airTemperature = Integer.parseInt(line.substring(87,92));
        }

        String quality = line.substring(92,93);
        if(airTemperature != MISSING && quality.matches("[01459]")){
            context.write(new Text(year), new IntWritable(airTemperature));
        }
    }
}
```

编写 MinTemperatureReducer.java：

```java
package com.uicc.shizhan;

import java.io.IOException;

import org.apache.hadoop.io.IntWritable;
import org.apache.hadoop.io.Text;
import org.apache.hadoop.mapreduce.Reducer;

public class MinTemperatureReducer extends Reducer<Text, IntWritable, Text, IntWritable>{
    public void reduce(Text key, Iterable<IntWritable> values ,Context context)
        throws IOException, InterruptedException{
        int minValue = Integer.MAX_VALUE;
        for(IntWritable value : values){
            minValue = Math.min(minValue, value.get());
        }
        context.write(key, new IntWritable(minValue));
    }
}
```

3.2.3　准备输入数据

解压气象数据集，命令如下：

```
cd /usr/software
tar - zxvf temperature.tar.gz
```

使用 zcat 命令把这些数据文件解压并合并到一个 sample.txt，命令如下：

```
cd temperature
zcat *.gz > /software/sample.txt
```

将 /software/sample.txt 上传到 HDFS 中的 /user/root/input 目录下作为程序的输入文件。

3.2.4　运行程序

在 Eclipse 中运行程序。

需要在 Arguments 选项卡填写 MinTemperature 运行的输入文件路径和作业输出路径两个参数，需要注意的是，输入、输出路径参数需要全路径，否则运行会报错。

- 输入日志文件路径：

```
/user/root/input/sample.txt
```

- 作业输出路径：

```
/user/root/out
```

高级篇

第 4 章

数据仓库：Hive

从早期的互联网主流大爆发开始，主要的搜索引擎公司和电子商务公司就一直在与不断增长的数据进行较量。最近，社交网站也遇到了同样的问题。如今，许多组织已经意识到他们所收集的数据是让他们了解用户、提升业务在市场上的表现以及提高基础架构效率的一个宝贵的资源。

Hadoop 生态系统就是为处理如此大的数据集而产生的一个合乎成本效益的解决方案。Hadoop 实现了一个特别的计算模型，也就是 MapReduce，其可以将计算任务分割成多个处理单元然后分散到一群家用的或服务器级别的硬件机器上，从而降低成本并提供水平可伸缩性。这个计算模型的下面是一个被称为 Hadoop 分布式文件系统（HDFS）的分布式文件系统。这个文件系统是"可插拔的"。不过，仍然存在一个挑战，那就是用户如何从一个现有的数据基础架构转移到 Hadoop 上，而这个基础架构是基于传统关系型数据库和结构化查询语句（SQL）的。对于大量的 SQL 用户（包括专业数据库设计师和管理员，也包括那些使用 SQL 从数据仓库中抽取信息的临时用户）来说，这个问题又将如何解决呢？

这就是 Hive 出现的原因。Hive 提供了一个被称为 Hive 查询语言（简称 HiveQL 或 HQL）的 SQL 方言，来查询存储在 Hadoop 集群中的数据。SQL 知识分布广泛的一个原因是：它是一个可以有效地、合理地且直观地组织和使用数据的模型。即使对于经验丰富的 Java 开发工程师来说，将这些常见的数据运算对应到底层的 MapReduce Java API 也是令人畏缩的。Hive 可以帮助用户来完成这些艰苦的任务，这样用户就可以集中精力关注查询本身了。Hive 可以将大多数的查询转换为 MapReduce 任务，进而在介绍一个令人熟悉的 SQL 抽象的同时，拓宽 Hadoop 的可扩展性。

Word Count 算法与基于 Hadoop 实现的大多数算法一样，过程是比较复杂的，尤其是真正使用 Hadoop 的 API 来实现这种算法时，甚至有更多的底层细节需要用户来控制。所以，这是一个只适用于有经验的 Java 开发人员的工作，这也就将 Hadoop 潜在地放在了一个非程序员用户无法触及的位置，即使这些用户了解他们想使用的算法。

事实上，许多这些底层细节实际上进行的是从一个任务到下一个任务的重复性工作，例如，将 Mapper 和 Reducer 一同写入某些数据操作构造这样的底层的繁重工作，通过过滤得到所需数据的操作，以及执行类似 SQL 中数据集键的连接（JOIN）操作等。不过幸运的是，存在一种方式，可以通过使用"高级"工具自动处理这些情况来重用这些通用的处理过程。

这就是引入 Hive 的原因。Hive 不仅提供了一个熟悉 SQL 的用户所熟悉的编程模型，还消除了大量的通用代码，甚至是那些必须使用 Java 编写的代码。这就是为什么 Hive 对于 Hadoop 如此重要的原因，无论用户是 DBA 还是 Java 开发工程师。Hive 可以让你花费相当少的精力就可以完成大量的工作。

图 4.1 显示了 Hive 的主要"模块"以及 Hive 是如何与 Hadoop 交互工作的。本书将主要关注利用 CLI（命令行界面）来与 Hive 进行交互。

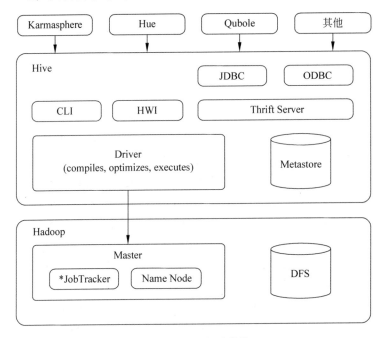

图 4.1　Hive 组成模块

Hive 发行版中附带的模块有 CLI，一个称为 Hive 网页界面（HWI）的简单网页界面，以及可通过 JDBC、ODBC 和一个 Thrift 服务器进行编程访问的几个模块。

所有的命令和查询都会进入到 Driver（驱动模块），通过该模块对输入进行解析编译，对需求的计算进行优化，然后按照指定的步骤执行（通常来说，都是启动多个 MapReduce 任务来执行）。当需要启动 MapReduce 任务时，Hive 本身是不会生成 Java MapReduce 算法程序的。相反，Hive 通过一个表示"任务执行计划"的 XML 文件驱动执行内置的、原生的 Mapper 和 Reducer 模块。换句话说，这些通用的模块函数类似于微型的语言翻译程序，而这个驱动计算的"语言"是以 XML 形式编码的。

Hive 通过和 JobTracker 通信来初始化 MapReduce 任务，而不必部署在 JobTracker 所在的管理节点上执行。在大型集群中，通常会有网关机专门用于部署像 Hive 这样的工具。在这些网关机上可远程和管理节点上的 JobTracker 通信来执行任务。通常，要处理的数据文件是存储在 HDFS 中的，而 HDFS 是由 NameNode 进行管理的。Metastore（元数据存储)是一个独立的关系型数据库(通常是 MySQL 实例),Hive 会在其中保存表模式和其他系统元数据。

Hive 最适合于数据仓库程序,对于数据仓库程序不需要实时响应查询,不需要记录级别的插入、更新和删除。当然,Hive 也适合于有一定 SQL 知识的用户,而且用户的某些工作可能采用其他的工具会更容易进行处理。

4.1 Hive 的安装和配置

接下来开始学习 Hive 的安装,之后将讨论如何配置 Hive 以了解如何在 Hadoop 集群上使用 Hive。

4.1.1 安装详细步骤

Hive 依赖于 Hadoop,而 Hadoop 依赖于 Java。所以在安装 Hive 之前,需要先安装好 Java 和 Hadoop 环境(**Hadoop** 和 **JDK** 在前面已经安装过了,但是为了更加直观地了解 **Hive** 的安装配置,在此重新进行 **Hadoop** 和 **JDK** 安装配置,具体的配置思路与上面的介绍相同,在配置细节上可能会有所简略)。

1. 安装 Java

(1) 在官网上下载 Java 的 JDK 安装包 jdk-8u60-linux-x64.tar.gz。

(2) 开始安装 JDK。

在 shell 命令行输入下面的命令,在/usr 目录下新建一个 java 文件夹,将 jdk-8u60-linux-x64.tar.gz 复制到新建目录下,完成解压后对其重命名。

```
# mkdir /usr/java
# cp /root/下载/jdk-8u60-linux-x64.tar.gz /usr/java
# cd /usr/java
# tar -zxvf jdk-8u60-linux-x64.tar.gz        --解压操作
# mv jdk1.8.0_60 jdk                         --重命名仅仅是为了后面配置环境变量的
                                               时候方便操作
```

(3) 开始配置环境变量。

在 shell 命令行中输入"vi ~/.bashrc"命令打开文件。这时,我们并不能对文件进行编辑,按下键盘上的 Esc 键后按 I 键进入插入模式,然后需要在该文件的末尾添加下列代码。编辑结束后,保存关闭文件,按 Esc 键和":"键进入命令模式,在文件下方的冒号(:)后面输入"wq!",回车保存关闭该文件,如图 4.2 所示。

退出到 shell 命令行下,要输入"source~/.bashrc"命令编译该文件使其生效,如图 4.3 所示。

(4) 这样,JDK 就安装好了。现在,验证一下 JDK 是否安装成功。

在 shell 命令行中输入"java-version",显示有 Java 版本信息的字样,便安装成功了,如图 4.4 所示。

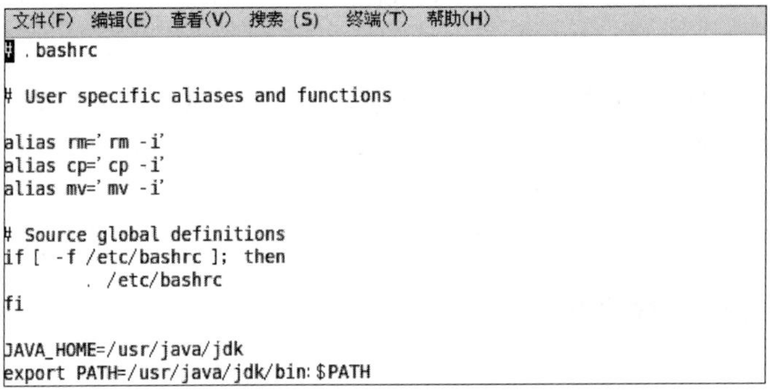

图 4.2　编辑配置文件

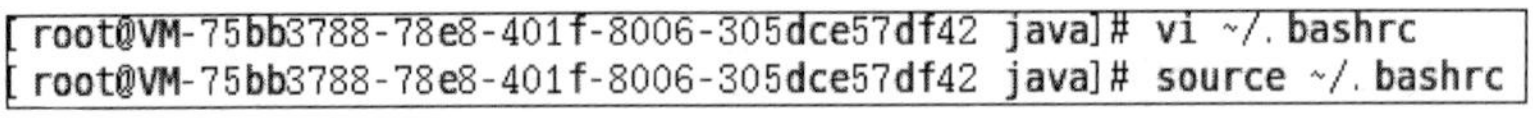

图 4.3　配置生效

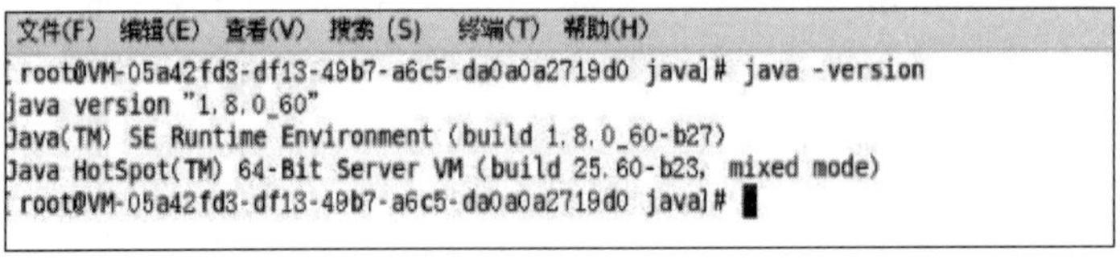

图 4.4　版本查看

2. 安装 Hadoop

（1）在 Apache 官网上下载 Hadoop 的安装包 hadoop-2.7.1.tar.gz。

（2）为了操作方便，先来配置免密码登录服务 SSH。在 shell 命令行下输入下列命令：

```
# ssh localhost
# ssh-keygen -t dsa -P '' -f ~/.ssh/id_dsa          ——此语句中''为两个单引号，在输入的
时候要注意
# cat ~/.ssh/id_dsa.pub >> ~/.ssh/authorized_keys
```

操作过程中，需要手动输入密码和"yes"继续工作。

（3）开始安装 Hadoop。

在 shell 命令行输入下列命令，在/usr 目录下新建一个 hadoop 文件夹，将 hadoop-2.7.1.tar.gz 复制到新建目录下，完成解压后对其重命名。

```
# mkdir /usr/hadoop
# cp /root/下载/hadoop-2.7.1.tar.gz /usr/hadoop
# cd /usr/hadoop
# tar -zxvf hadoop-2.7.1.tar.gz
# mv hadoop-2.7.1 hadoop
```

（4）开始配置环境变量。

在 shell 命令行中输入"vi ~/.bashrc"命令打开文件。这里配置环境变量同 JDK 的步骤相

同。我们需要在文件末尾添加下列代码。然后退出到 shell 命令行下，输入“source~/. bashrc”命令编译该文件使其生效。这里需要注意的是，“export PATH =…”这行无须重写，只需将 hadoop 路径加到 JDK 路径后面即可。

```
HADOOP_HOME = /usr/hadoop/hadoop
export PATH = /usr/java/jdk/bin:/usr/hadoop/hadoop/sbin:/usr/hadoop/hadoop/bin: $ PATH
                          -- 第二行不是重新写的，而是在 jdk 路径后面补充了 hadoop 的路径
```

完成之后如图 4.5 所示。

```
# .bashrc

# User specific aliases and functions

alias rm='rm -i'
alias cp='cp -i'
alias mv='mv -i'

# Source global definitions
if [ -f /etc/bashrc ]; then
        . /etc/bashrc
fi

JAVA_HOME=/usr/java/jdk
HADOOP_HOME=/usr/hadoop/hadoop

export PATH=/usr/java/jdk/bin:/usr/hadoop/hadoop/sbin:/usr/hadoop/hadoop/bin:$PA
TH
~
```

图 4.5 编辑配置文件

(5) 进入到/usr/hadoop/hadoop/etc/hadoop 目录下，修改配置文件。

在 shell 命令行输入“cd /usr/hadoop/hadoop/etc/hadoop”命令，然后使用 ls 命令，可以看到该目录下的所有文件，如图 4.6 所示。

```
文件(F) 编辑(E) 查看(V) 搜索(S) 终端(T) 帮助(H)
[root@VM-d69d317f-6b0b-4c49-99d8-8ac0283ff3ef ~]# cd /usr/hadoop/hadoop/etc/hado
op
[root@VM-d69d317f-6b0b-4c49-99d8-8ac0283ff3ef hadoop]# ls
capacity-scheduler.xml      httpfs-env.sh              mapred-env.sh
configuration.xsl           httpfs-log4j.properties    mapred-queues.xml.template
container-executor.cfg      httpfs-signature.secret    mapred-site.xml.template
core-site.xml               httpfs-site.xml            slaves
hadoop-env.cmd              kms-acls.xml               ssl-client.xml.example
hadoop-env.sh               kms-env.sh                 ssl-server.xml.example
hadoop-metrics2.properties  kms-log4j.properties       yarn-env.cmd
hadoop-metrics.properties   kms-site.xml               yarn-env.sh
hadoop-policy.xml           log4j.properties           yarn-site.xml
hdfs-site.xml               mapred-env.cmd
[root@VM-d69d317f-6b0b-4c49-99d8-8ac0283ff3ef hadoop]#
```

图 4.6 查看文件

(6) 在 shell 命令行下输入“vi hadoop-env. sh”命令打开 hadoop-env. sh 文件，并在该文件中添加 JAVA_HOME。

具体的做法是在该文件中找到“export JAVA_HOME= $ {JAVA_HOME}”这行代码，将其改成“export JAVA_HOME=/usr/java/jdk”，然后保存关闭该文件。如图 4.7 所示，用黑色底纹的命令行代替其上面的一行命令。

图 4.7　编辑配置文件

（7）在 shell 命令行下输入"vi core-site.xml"命令打开 core-site.xml 文件，修改该文件。

具体的做法是在该文件末尾处的＜configuration＞＜/configuration＞中添加代码，完成后保存并关闭该文件，如图 4.8 所示。

图 4.8　编辑配置文件

（8）在 shell 命令行下输入"vi hdfs-site.xml"命令打开 hdfs-site.xml 文件，并修改文件。

具体的做法是在该文件末尾处的＜configuration＞＜/configuration＞中添加下列代码，完成后保存并关闭文件。

```
< property >
  < name > dfs.replication </ name >
  < value > 1 </ value >
</ property >
< property >
  < name > dfs.data.dir </ name >
  < value > /usr/hadoop/hadoop/data </ value >
</ property >
```

(9) 将 mapred-site.xml.template 复制为 mapred-site.xml 文件。执行命令如下:

```
cp mapred-site.xml.template mapred-site.xml
```

在 shell 命令行下输入"vi mapred-site.xml"命令打开 mapred-site.xml 文件,修改该文件。

具体的做法是在该文件末尾处的<configuration></configuration>中添加下列代码,完成后保存并关闭文件。

```
<property>
    <name>mapred.job.tracker</name>
    <value>localhost:9001</value>
</property>
```

(10) 完成了上面的配置之后,在 shell 命令行输入下面的命令,进入到/usr/hadoop/hadoop/bin 目录,执行格式化操作(操作执行后 shell 界面会显示大量的日志文件)。

```
# cd /usr/hadoop/hadoop/bin
# ./hadoop namenode -format
```

(11) 再进入/usr/hadoop/hadoop/sbin 目录,启动 Hadoop。

```
# cd /usr/hadoop/hadoop/sbin
# ./start-all.sh
```

启动过程中,需要手动输入"yes"。

下面可以验证一下 Hadoop 是否安装成功。在 shell 命令行中输入"jps"(注意是小写的),查看是否出现如图 4.9 所示的界面。

```
[root@VM-75bb3788-78e8-401f-8006-305dce57df42 sbin]# jps
3088 SecondaryNameNode
3360 Jps
2897 DataNode
3233 ResourceManager
1924 VmServer.jar
3325 NodeManager
2799 NameNode
```

图 4.9 查看进程

(12) 打开浏览器,通过页面验证 Hadoop 是否安装成功。

在火狐浏览器地址栏中输入 localhost:50070,观察界面是否如图 4.10 所示;接着再在浏览器地址栏中输入 localhost:8088,观察界面是否如图 4.11 所示。

3. 安装 Hive

(1) 在 Apache 官网上下载 apache-hive-1.2.1-bin.tar.gz。

下载地址为 http://mirror.bit.edu.cn/apache/hive/hive-1.2.1/。

(2) 开始安装 Hive。

使用下列命令在/usr 目录下新建一个 Hive 的文件夹,将 apache-hive-1.2.1-bin.tar.gz

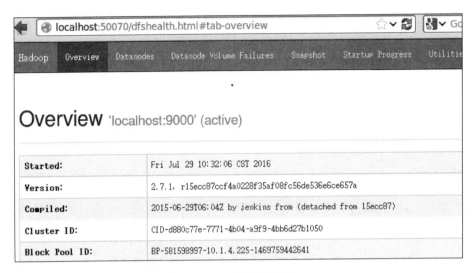

图 4.10　Web 界面访问（一）

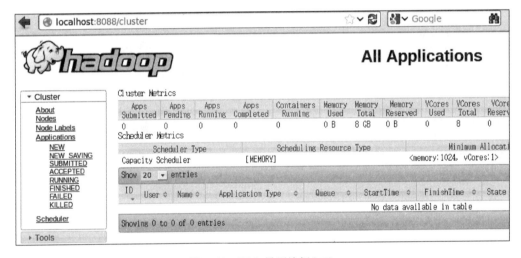

图 4.11　Web 界面访问（二）

复制到新建目录下,完成解压后对其重命名。

```
# mkdir /usr/hive
# cp /root/下载/apache-hive-1.2.1-bin.tar.gz /usr/hive/
# cd /usr/hive
# tar -zxvf apache-hive-1.2.1-bin.tar.gz
# mv apache-hive-1.2.1-bin hive
```

（3）开始配置环境变量,过程同安装 JDK 和 Hadoop 一样。

在 shell 命令行中输入"vi~/.bashrc"命令。打开文件后,在文件末尾添加下面的代码,如图 4.12 所示。然后,退出到 shell 命令行下,输入"source~/.bashrc"命令编译该文件使其生效。

```
HIVE_HOME = /usr/hive/hive
export PATH = /usr/java/jdk/bin:/usr/hadoop/hadoop/sbin:/usr/hadoop/hadoop/bin:
/usr/hive/hive/bin: $ PATH
-- 第二行代码不是重新写的,而是在 jdk、hadoop 路径后面补充了 hive 的路径
```

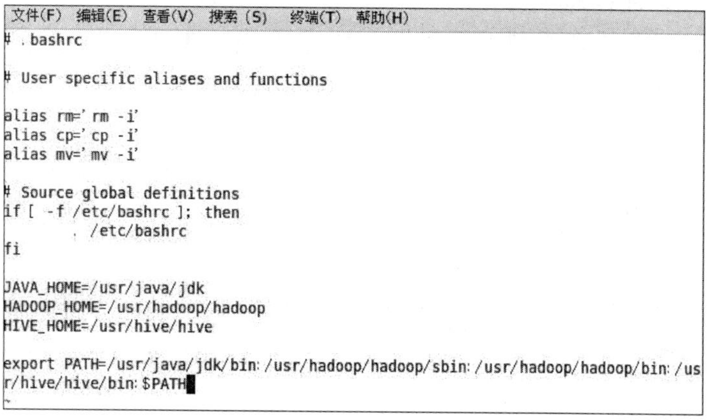

```
文件(F) 编辑(E) 查看(V) 搜索(S) 终端(T) 帮助(H)
# .bashrc

# User specific aliases and functions

alias rm='rm -i'
alias cp='cp -i'
alias mv='mv -i'

# Source global definitions
if [ -f /etc/bashrc ]; then
        . /etc/bashrc
fi

JAVA_HOME=/usr/java/jdk
HADOOP_HOME=/usr/hadoop/hadoop
HIVE_HOME=/usr/hive/hive

export PATH=/usr/java/jdk/bin:/usr/hadoop/hadoop/sbin:/usr/hadoop/hadoop/bin:/us
r/hive/hive/bin:$PATH
~
```

图 4.12　编辑配置文件

(4) 进入/usr/hive/hive/bin 目录下,修改 hive-config. sh 文件。

使用"vi hive-config. sh"命令打开 hive-config. sh 文件,在文件的末尾添加下列代码,如图 4.13 所示,然后保存并关闭文件。

```
export JAVA_HOME = /usr/java/jdk
export HADOOP_HOME = /usr/hadoop/hadoop
export HIVE_HOME = /usr/hive/hive
```

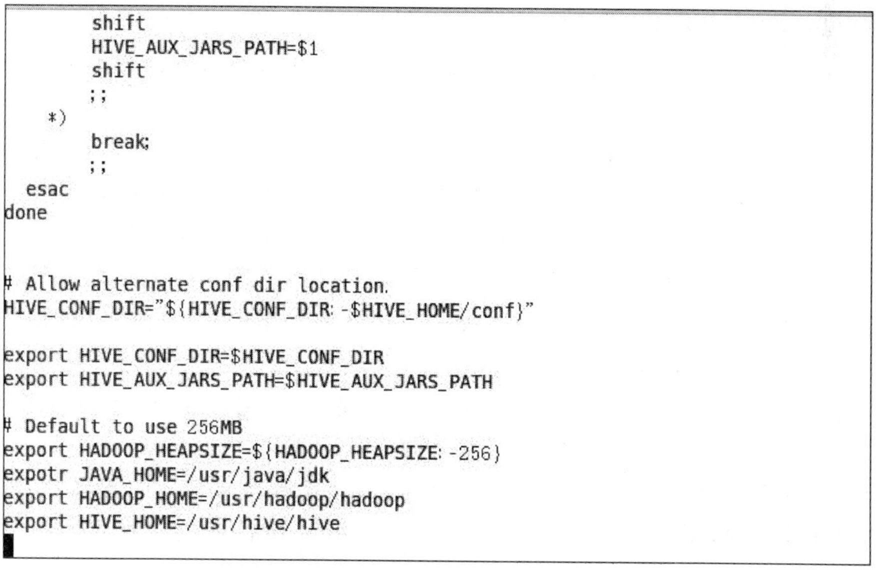

```
        shift
        HIVE_AUX_JARS_PATH=$1
        shift
        ;;
    *)
        break;
        ;;
  esac
done


# Allow alternate conf dir location.
HIVE_CONF_DIR="${HIVE_CONF_DIR:-$HIVE_HOME/conf}"

export HIVE_CONF_DIR=$HIVE_CONF_DIR
export HIVE_AUX_JARS_PATH=$HIVE_AUX_JARS_PATH

# Default to use 256MB
export HADOOP_HEAPSIZE=${HADOOP_HEAPSIZE:-256}
expotr JAVA_HOME=/usr/java/jdk
export HADOOP_HOME=/usr/hadoop/hadoop
export HIVE_HOME=/usr/hive/hive
```

图 4.13　编辑配置文件

(5) 现在就可以启动 Hive 了。

在 shell 命令行输入 hive，按回车键便进入了 Hive 的 CLI，退出 Hive 则只需输入"exit；"或者"quit；"后，按回车键即可。

在/usr/hive/hive/bin 目录下，可以看到有个 Hive 的文件。所以在 bin/目录下，输入 hive 命令，便启动了 Hive，如图 4.14 所示。

由于在安装 Hive 的时候，我们配置了环境变量，所以无论在什么目录下，都可以直接在 shell 命令行中输入 Hive 进入 Hive 的 CLI。需要注意的是，在 Hive 运行之前，必须保证你的 Hadoop 已经运行起来了。

图 4.14 进入 Hive Cli

4.1.2 Hive 内部是什么

Hive 二进制分支版本核心包含三个部分，主要部分是 Java 代码本身。在 ＄HIVE_HOME/lib 目录下可以发现有众多的 jar(Java 压缩包)文件，例如 hive-exec＊.jar 和 hive-metastore＊.jar。每个 jar 文件都实现了 Hive 功能中某个特定的部分。

＄HIVE_HOME/bin 目录下包含可以执行各种各样 Hive 服务的可执行文件，包括 Hive 命令行界面(也就是 CLI)。CLI 是使用 Hive 的最常用方式，除非有特别说明，否则都使用 hive(小写，固定宽度的字体)来代表 CLI。CLI 可用于提供交互式的界面供输入语句或者可以供用户执行含有 Hive 语句的"脚本"。

Hive 还有一些其他组件。Thrift 服务提供了可远程访问其他进程的功能，也提供使用 JDBC 和 ODBC 访问 Hive 的功能，这些都是基于 Thrift 服务实现的。

所有的 Hive 客户端都需要一个 metastoreservice(元数据服务)，Hive 使用这个服务来存储表模式信息和其他元数据信息。通常情况下会使用一个关系型数据库中的表来存储这些信息。默认情况下，Hive 会使用内置的 Derby SQL 服务器，其可以提供有限的、单进程的存储服务。例如，当使用 Derby 时，用户不可以执行两个并发的 Hive CLI 实例，然而，如果是在个人计算机上或者某些开发任务上使用，这样也是没问题的。对于集群来说，需要使用 MySQL 或者类似的关系型数据库。

最后，Hive 还提供了一个简单的网页界面，也就是 Hive 网页界面(HWI)，提供了远程访问 Hive 的服务。

conf 目录下存放了配置 Hive 的配置文件。Hive 具有非常多的配置属性，这些属性控制的功能包括元数据存储(如数据存放在哪里)、各种各样的优化和"安全控制"，等等。

4.2 数据定义

HiveQL 是 Hive 查询语言。与普遍使用的所有 SQL 语言一样，它不完全遵守任一种 ANSI SQL 标准的修订版。HiveQL 可能和 MySQL 的语言最接近，但是两者还是存在显著差异的。Hive 不支持行级插入操作、更新操作和删除操作。Hive 也不支持事务。Hive 增加了在 Hadoop 背景下的可以提供更高性能的扩展以及一些个性化的扩展，甚至还增加了一些外部程序。

接下来，要学习的是 HiveQL 所谓的数据定义语言部分，其可用于创建、修改和删除数据库、表、视图、函数和索引。

4.2.1 Hive 中的数据库

Hive 中数据库的概念本质上仅仅是表的一个目录或者命名空间。然而，对于具有很多组和用户的大集群来说，这是非常有用的，因为这样可以避免表命名冲突。通常会使用数据库来将生产表组织成逻辑组。

如果用户没有显式指定数据库，那么将会使用默认的数据库 default。

下面这个例子就展示了如何创建一个数据库：

打开终端后，进入/usr/hadoop/hadoop/sbin 目录，在 shell 命令行中运行“./start-all.sh”命令，启动 Hadoop。然后，退出到根目录下，运行 hive 命令，启动 Hive。

在 CLI 命令行下运行下面的代码，创建数据库。

```
hive > CREATE DATABASE financials;            -- 输入这行命令后按回车键，就创建了一个数据库
OK
Time taken:1.669 seconds
```

这时，已经创建了一个 financials 数据库。当再次运行刚刚这条代码的时候，就会出现错误，错误提示如图 4.15 所示。

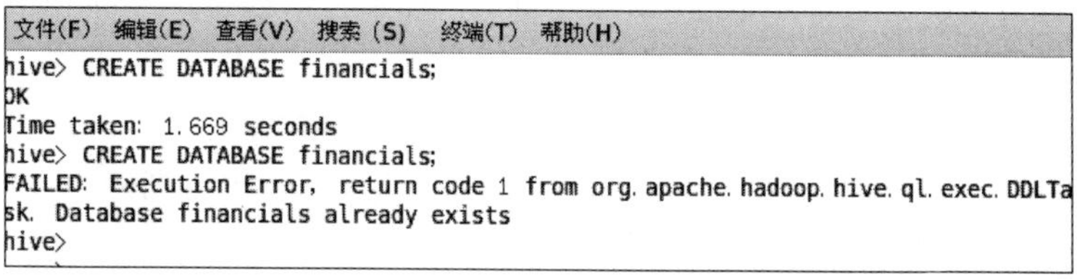
```
文件(F) 编辑(E) 查看(V) 搜索(S) 终端(T) 帮助(H)
hive> CREATE DATABASE financials;
OK
Time taken: 1.669 seconds
hive> CREATE DATABASE financials;
FAILED: Execution Error, return code 1 from org.apache.hadoop.hive.ql.exec.DDLTask. Database financials already exists
hive>
```

图 4.15 创建数据库错误提示

在刚刚创建失败后，使用如下语句创建数据库可以避免这种情况下抛出错误信息：

```
CREATE DATABASE IF NOT EXISTS financials;
```

重新在 CLI 命令行输入下面的代码，观察实验现象：

```
hive> CREATE DATABASE IF NOT EXISTS financials;
OK
Time taken:0.043 seconds
```

虽然通常情况下用户还是期望在同名数据库已经存在的情况下能够抛出警告信息的，但是 IF NOT EXISTS 这个子句对于那些在继续执行之前需要实时创建数据库的情况来说是非常有用的。

可以通过如下命令查看 Hive 中所包含的数据库：

```
hive> SHOW DATABASES;                        -- 显示数据库
OK
default
financials
Time taken: 0.035 second,Fetched:2 row(s)
hive> CREATE DATABASE human_resources;       -- 新建一个数据库 human_resources
OK
Time taken:0.17 seconds
hive> SHOW DATABASES;                        -- 显示数据库
OK
default
financials
human_resources
Time taken:0.052 secondes;FETCHED:3 row(s)
```

如果数据库非常多，那么可以使用正则表达式匹配来筛选出需要的数据库名，下面这个例子展示的是列举出所有以字母 h 开头，以其他字符结尾（即 .* 部分含义）的数据库名：

```
hive> SHOW DATABASES LIKE 'h.*';             -- 显示名称以 h 开头的数据库
OK
human_resources
Time taken:1.632 seconds,Fetched:1 row(s)
```

Hive 会为每个数据库创建一个目录。数据库中的表将会以这个数据库目录的子目录形式存储。有一个例外就是 default 数据库中的表，因为这个数据库本身没有自己的目录。数据库所在的目录位于属性 hive.metastore.warehouse.dir 所指定的顶层目录之后（这一项是在/usr/hive/hive/conf 目录下的 hive-site.xml 文件中配置的）。假设用户使用的是这个配置项默认的配置，也就是/user/hive/warehouse，那么当创建数据库 financials 时，Hive 将会对应地创建一个目录/user/hive/warehouse/financials.db。这里请注意，数据库的文件目录名是以 .db 结尾的。

用户可以通过如下命令来修改这个默认的位置：

```
hive> CREATE DATABASE mydatabase01           -- 将 mydatabase01 数据库存放到/my/preferred/
directory 下
    > LOCATION '/my/preferred/directory';
OK
Time taken: 1.602 seconds
```

用户也可以为这个数据库增加一个描述信息,这样通过 DESCRIBE DATABASE <database>命令就可以查看到该信息。

```
hive> CREATE DATABASE mydatabase02              -- 创建数据库 mydatabase02
    > COMMENT 'This is my second database.';    -- 为数据库 mydatabase02 增加描述信息
OK
Time taken: 0.1 seconds
hive> DESCRIBE DATABASE mydatabase02;           -- 显示数据库 mydatabase02 的描述信息
OK
mydatabase02  This is my second database.  hdfs://localhost:9000/user/hive/warehouse/
mydatabase02.db root USER
Time taken:0.09 seconds,Fetched:1 row(s)
```

从上面的例子中可以注意到,DESCRIEB DATABASE 语句也会显示出这个数据库所在的文件目录位置路径。

前面 DESCRIBE DATABASE 语句的输出中,使用了 maste-server 来代表 URI 权限,也就是说,应该是由文件系统的"主节点"(例如,HDFS 中运行 NameNode 服务的那台服务器)的服务器名加上一个可选的端口号构成的(例如,"服务器名:端口号"这样的格式)。如果用户执行的是伪分布式模式,那么主节点服务器名称就应该是 localhost。对于本地模式,这个路径应该是一个本地路径,例如,file:user/hive/warehouse/financials.db。

如果这部分信息省略了,那么 Hive 将会使用 Hadoop 配置文件中的配置项 fs.default.name 作为 maste-server 所对应的服务器名和端口号,这个配置文件可以在 /usr/hive/hive/conf 这个目录下找到。

需要明确的是,hdfs://user/hive/warehouse/financials.db 和 hdfs://master-server/user/hive/warehouse/financials.db 是等价的,其中 master-server 是主节点的 DNS 名和可选的端口号。

为了保持完整性,当用户指定一个相对路径(例如 some/relative/path)时,对于 HDFS 和 Hive,都会将这个相对路径放到分布式文件系统的指定根目录下(例如,hdfs://user/<username>)。然而,如果用户是在本地模式下执行,那么当前的本地工作目录将是 some/relative/path 的父目录。

为了脚本的可移植性,通常会省略掉那个服务器和端口号信息,而只有在涉及另一个分布式文件系统实例的时候才会指明该信息。

此外,用户还可以为数据库增加一些与其相关的键值对属性信息,尽管目前仅有的功能就是提供一种可以通过 DESCRIBE DATABASE EXTENDED<database>语句显示出这些信息的方式:

```
hive> CREATE DATABASE mydatabase03                              -- 创建数据库 mydatabase03
    > WITH DBPROPERTIES('creator' = 'root','date' = '2015-8-11');-- 为该数据库添加键值对属
                                                                   性信息
OK
Time taken: 0.116 seconds
hive> DESCRIBE DATABASE mydatabase03;                           -- 显示数据库 mydatabase03
                                                                   的描述信息
```

```
OK
mydatabase03        hdfs://localhost:9000/user/hive/warehouse/mydatabase03.db        root        USER
Time taken:0.049 seconds,Fetched:1 row(s)
hive > DESCRIBE DATABASE EXTENDED mydatabase03;    -- 显示数据库 mydatabase03 的相关键值对属性
                                                                                信息
OK
mydatabase03        hdfs://localhost:9000/user/hive/warehouse/mydatabase03.db        root
    USER
{date = 2015 - 8 - 11,creator = root}
Time taken: 0.058 seconds, Fetched: 1 row(s)
```

USE 命令用于将某个数据库设置为用户当前的工作数据库，和在文件系统中切换工作目录是一个概念：

```
hive > USE mydatabase01;        -- 将数据库 mydatabase01 设置为当前工作的数据库
OK
Time taken: 0.028 seconds
```

现在使用 SHOW TABLES 这样的命令就会显示当前这个数据库下所有的表。当然，我们的 mydatabase01 数据库下面还没有表。

不幸的是，并没有一个命令可以让用户查看当前所在的是哪个数据库。好在 Hive 中是可以重复使用 USE...命令的，这是因为在 Hive 中并没有嵌套数据库的概念。

可以通过设置一个属性值来在提示符里面显示当前所在的数据库（Hive V0.8.0 版本以及之后的版本才支持此功能）：

```
hive > set hive.cli.print.current.db = true;        -- 将"hive.cli.print.current.db"属性设
                                                        置为 true
hive (mydatabase01)> USE human_resources;
OK                                                    -- 将当前工作的数据库切换到 human
                                                        _resources
Time taken: 0.022 seconds
hive (human_resources)> set hive.cli.print.current.db = false;
-- 将"hive.cli.print.current.db"属性设置为 false
```

最后，用户可以使用"DROP DATABASE IF EXISTS tablename"来删除数据库：

```
hive > DROP DATABASE IF EXISTS mydatabase01;        -- 删除数据库 mydatabase01;
OK
Time taken: 0.793 seconds
```

IF EXISTS 子句是可选的，如果加了这个子句，就可以避免因数据库 financials 不存在而抛出警告信息。

默认情况下，Hive 是不允许用户删除一个包含有表的数据库的。用户可以先删除数据库中的表，然后再删除数据库，或者在删除命令的最后面加上关键字 CASCADE，这样可以使 Hive 自行删除数据库中的表。

下面在数据库 mydatabase02 中新建一个表，来观察一下现象：

```
hive> USE mydatabase02;                          -- 将数据库 mydatabase02 设置为当前工作
                                                    的数据库
OK
Time taken: 0.032 seconds
hive> CREATE TABLE table01 (name STRING);        -- 新建一个表 table01;
OK
Time taken: 1.51 seconds
hive> DROP DATABASE IF EXISTS mydatabase02;      -- 直接删除一个含有表的数据库,会报错
FAILED: Execution Error, return code 1 from org.apache.hadoop.hive.ql.exec.DDLTask.
InvalidOperationException (message:Database mydatabase02 is not empty. One or more tables
exist.)
hive> DROP DATABASE IF EXISTS mydatabase02 CASCADE;  -- 使用该命令可删除含有表的数据库
OK
Time taken: 2.398 seconds
hive>
```

如果使用的是 RESTRICT 而不是 CASCADE 这个关键字的话,那么就和默认情况一样,也就是说,如果想删除数据库,那么必须先要删除掉该数据库中的所有表。如果某个数据库被删除了,那么其对应的目录也同时会被删除。

4.2.2 修改数据库

用户可以使用 ALTER DATABASE 命令为某个数据库的 DBPROPERTIES 设置键值对属性值,来描述这个数据库的属性信息。数据库的其他元数据信息都是不可更改的,包括数据库名和数据库所在的目录位置。

下面的例子展示了如何修改数据库中的属性信息,首先在 shell 命令行中输入 hive 命令(前提是 Hadoop 已经运行起来了),进入到 CLI 命令行下;然后创建一个带有属性信息的数据库,再修改数据库的属性信息。

下面新建一个带有键值对属性信息的数据库 mydatabase04,然后为其修改属性信息:

```
hive> CREATE DATABASE mydatabase04               -- 创建一个带有日期键值对信息的数据库
    > WITH DBPROPERTIES('data' = '2012-01-01');  -- 日期为 2012-01-01
OK
Time taken: 1.643 seconds
hive> DESCRIBE DATABASE EXTENDED mydatabase04;   -- 显示数据库 mydatabase04 的键值对属性
                                                    信息
OK
mydatabase04      hdfs://localhost:9000/user/hive/warehouse/mydatabase04.dbrootUSER     {data
 = 2012-01-01}
    -- 显示日期为 2012-01-01
Time taken:0.643 seconds,Fetched:1 row(s)
hive> ALTER DATABASE mydatabase04 SET DBPROPERTIES('data' = '2015-08-12');
    -- 修改日期信息为 2015-08-12
OK
Time taken:0.167 seconds
hive> DESCRIBE DATABASE EXTENDED mydatabase04;   -- 再次显示键值对属性信息
OK
mydatabase04      hdfs://localhost:9000/user/hive/warehouse/mydatabase04.db     rootUSER{data
 = 2015-08-12}                                    -- 这里的日期信息已经变了
Time taken:0.057 seconds,Fetched:1 row(s)
```

没有办法可以删除或者“重置”数据库属性。

4.2.3 创建表

CREATE TABLE 语句遵从 SQL 语法惯例，但是 Hive 的这个语句中具有显著的功能扩展，使其可以具有更广泛的灵活性。例如，可以定义表的数据文件存储在什么位置、使用什么样的存储格式等。本节会讨论其他一些在 CREATE TABLE 语句中可以使用到的选项。

在 4.2.2 节的基础上，在 shell 命令行使用 hadoop dfs-ls /user/hive/warehouse/命令可以查看 HDFS 中有哪些文件，如图 4.16 所示。其中，带有后缀.db 的为数据库，没有后缀的为表。

```
[root@VM-75bb3788-78e8-401f-8006-305dce57df42 ~]# hadoop dfs -ls /user/
hive/warehouse/
DEPRECATED: Use of this script to execute hdfs command is deprecated.
Instead use the hdfs command for it.

16/08/01 16:35:33 WARN util.NativeCodeLoader: Unable to load native-had
oop library for your platform... using builtin-java classes where appli
cable
Found 6 items
drwxr-xr-x   - root supergroup          0 2016-08-01 15:22 /user/hive/w
arehouse/financials.db
drwxr-xr-x   - root supergroup          0 2016-08-01 15:24 /user/hive/w
arehouse/human_resources.db
drwxr-xr-x   - root supergroup          0 2016-08-01 14:03 /user/hive/w
arehouse/my_table
drwxr-xr-x   - root supergroup          0 2016-08-01 15:52 /user/hive/w
arehouse/mydatabase04.db
drwxr-xr-x   - root supergroup          0 2016-08-01 16:03 /user/hive/w
arehouse/mydatabase05.db
drwxr-xr-x   - root supergroup          0 2016-07-29 15:36 /user/hive/w
arehouse/whatsit
```

图 4.16 查看文件

另外，也可以使用 Hadoop 的 Web 页面工具来查看文件。

在浏览器地址栏中输入“localhost：50070”，进入到如图 4.17 所示界面，单击 Utilities→Browse the files system 命令，便可以查看到/user/hive/warehouse 目录下的文件了。

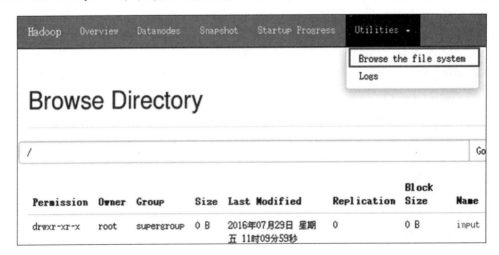

图 4.17 浏览器查看

下面开始学习在数据库中创建表。首先,在 shell 命令行的根目录下输入 hive 命令,进入 CLI 命令行界面。创建一个数据库 mydatabase,然后在该表下创建表 employees(前提是 Hadoop 环境已经在运行)。

接着在 CLI 命令行输入下面的代码,创建表 employees_01。需要注意的是,在输入代码的时候,为了看起来整齐直观,可以借助回车键多行输入,";"是终止符。

```
hive> CREATE DATABASE mydatabase;                        -- 创建数据库 mydatabase
OK
Time taken: 1.133 seconds
hive> CREATE TABLE IF NOT EXISTS mydatabase.employees_01(  -- 开始创建表 employees_01
    > name STRING COMMENT 'Employee name',                 -- COMMENT 后面为该项的描述信息
    > salary FLOAT COMMENT 'Employee salary',
    > subordinates ARRAY<STRING> COMMENT 'Names of subordinates',
    > deductions MAP<STRING,FLOAT> COMMENT 'Keys are deductions are percentges',
    > address STRUCT<street:STRING,city:STRING,state:STRING,zip:INT> COMMENT 'Home address')
    > COMMENT ' Description of the table '
    > LOCATION '/usr/hive/warehouse/mydatabase.db/employees_01';
OK
Time taken: 0.557 seconds
```

可以注意到,如果用户当前所处的数据库不一定就是目标数据库,那么用户是可以在表名前增加一个数据库名来进行指定的,也就是例子中的 mydatebase。

如果用户增加了可选项 IF NOT EXITS,而且表已经存在了,Hive 就会忽略掉后面的执行语句,而且不会有任何提示。在那些第一次执行时需要创建表的脚本中,这么写是非常有用的。其实,如果用户使用了 IF NOT EXISTS,而且这个已经存在的表和 CREATE TABLE 语句后指定的模式是不同的,但 Hive 会忽略掉这个差异。所以,在这里就要注意,如果所指定的表的模式和已经存在的这个表的模式不同,Hive 不会为此做出提示。如果用户的意图是使这个表具有重新指定的那个新的模式,那么需要先删除这个表,也就是丢弃之前的数据,然后再重建这张表。当然,也可以考虑使用一个或多个 ALTER TABLE 语句来修改已经存在的表的结构。

用户可以在字段类型后为每个字段增加一个注释。与数据库一样,用户也可以为这个表本身添加一个注释,还可以自定义一个或多个表属性。大多数情况下,TBLPROPERTIES 的主要作用是按键值对的格式为表增加额外的文档说明。但是,当检查 Hive 和像 DymamoDB 这样的数据库间的集成时,可以发现 TBLPROPERTIES 还可用作表示关于数据库连接的必要的元数据信息。

Hive 会自动增加两个表属性:一个是 last_modified_by,其保存着最后修改这个表的用户的用户名;另一个是 last_modified_time,其保存着最后一次修改的时间。

最后,可以看到,可以根据情况为表中的数据指定一个存储路径(和元数据截然不同的是,元数据总是会保存这个路径)。在这个例子中,Hive 将会使用默认的路径 usr/hive/warehouse/mydatabase.db/employees_01。其中,/usr/hive/warehouse 是默认的"数据仓库"路径地址,mydatabase.db 是数据库目录,employees_01 是表目录。

默认情况下,Hive 总是将创建的表的目录放置在这个表所属的数据库目录之后。不过,default 数据库是个例外,其在/usr/hive/warehouse 下并没有对应一个数据库目录。因

此 default 数据库中的表目录会直接位于/usr/hive/warehouse 目录之后（除了用户明确指定为其他路径的情况，这里并没有为其指定路径）。

用户还可以复制一张已经存在的表的表模式（而无须复制数据）。

可以在 CLI 命令行中输入下列代码：

```
hive> CREATE TABLE IF NOT EXISTS mydatabase.employees_02      -- 创建一张表 employees_02
    > LIKE mydatabase.employees_01;                            -- 表 employees_02 复制表
employees_01 的表模式
OK
Time taken: 0.629 seconds
hive>
```

SHOW TABLES 命令可以列举出所有的表。如果不增加其他参数，那么只会显示当前工作数据库下的表。在上面的介绍中，工作数据库是 mydatabase.db，我们已经创建了表 employees_01 和表 employees_02。

接下来，可以在 CLI 命令行中输入下列代码来显示表：

```
hive> USE mydatabase;                 -- 将当前工作数据库切换到 mydatabase
OK
Time taken: 0.03 seconds
hive> SHOW TABLES;                    -- 显示当前数据库下的表
OK
employees_01
employees_02
Time taken: 0.119 seconds, Fetched: 2 row(s)
hive>
```

即使不在那个数据库下，也可以列举指定数据库下的表。

在 CLI 命令行下输入下列代码：

```
hive> USE default;                         -- 将当前工作数据库切换到 dafault 数据库
OK
Time taken: 0.022 seconds
hive> SHOW TABLES IN mydatabase;           -- 显示另一数据库 mydatabase 下的表
OK
employees_01
employees_02
Time taken: 0.06 seconds, Fetched: 2 row(s)
```

如果有很多的表，那么可以使用正则表达式来过滤出所需要的表名。

在 mydatabase 数据库下再新建一个表 mytable，然后使用"SHOW TABLES 'empl.*';"命令，观察能显示出哪些表。

下面在 CLI 命令行输入下列的命令：

```
hive> USE mydatabase;                              -- 将当前工作数据库切换到 mydatabase 数据库
OK
Time taken: 0.025 seconds
hive> CREATE TABLE mytable(name STRING);           -- 创建一个简单的表 mytable
```

```
OK
Time taken: 0.14 seconds
hive > SHOW TABLES 'empl. * ';  -- 使用正则表达式来过滤出所需要的表名
OK  -- 下面只显示出来表 employees_01、表 employees_02
employees_01
employees_02
Time taken: 0.235 seconds, Fetched: 2 row(s)
hive > SHOW TABLES;  -- 显示该数据库下的所有的表
OK  -- 共显示出 3 张表,有 employees_01、employees_02 和 mytable
employees_01
employees_02
mytable
Time taken: 0.058 seconds, Fetched: 3 row(s)
hive >
```

Hive 并非支持所有的正则表达式功能。如果用户了解正则表达式,最好事先测试一下备选的正则表达式是否真正奏效。

单引号中的正则表达式表示的是所有以 empl 开头并以其他任意字符(也就是 . * 部分)结尾的表名。

也可以使用"DESCRIBE EXTENDED mydatabase. employees_01"命令来查看这个表的详细表结构信息。

下面在 CLI 命令行输入下列代码(如果当前的工作数据库是 mydatabase,就不必加前缀 mydatabase.),观察实验现象:

```
hive > DESCRIBE EXTENDED mydatabase.employees_01;      -- 查看表 employees_01 的详细信息
OK
name                     string                   Employee name
salary                   float                    Employee salary
subordinates             array < string >         Names of subordinates
deductions               map < string,float >     Keys are deductions are percentges
address                  struct < street:string,city:string,state:string,zip:intHome address
    Detailed Table Information. . . . . .                -- 后面还有很多内容,此处省略
hive >
```

若使用 FORMATTED 关键字替代 EXTENDED 关键字,则可以提供更加可读的和冗长的输出信息。在实际使用中,使用 FORMATTED 要更多些,因为其输出内容更加详细而且可读性强。不论是使用关键字 FORMATTED 还是 EXTENDED,上面输出信息的第一段和没有使用关键字 EXTENDED 或者 FROMATTED 时输出的结果是一样的,都包含有列描述信息的表结构信息。如果用户只想查看某一个列的信息,那么只要在表名后增加这个字段的名称即可。这种情况下,使用 EXTENDED 关键字也不会增加更多的输出信息。

下面在 CLI 命令行下输入下列代码,观察现象:

```
hive > USE mydatabase;                        -- 将工作目录切换到 mydatabase 下
OK
Time taken: 0.018 seconds
hive > DESCRIBE employees_01.salary;          -- 查询 employees_01 的 salary 这一列的信息
```

```
OK
salary            float             from deserializer
Time taken: 0.307 seconds, Fetched: 1 row(s)
hive> DESCRIBE EXTENDED employees_01.salary; --增加关键字 EXTENDED 也不会输出很多的信息
OK
salary            float             from deserializer
Time taken: 0.395 seconds, Fetched: 1 row(s)
```

1. 管理表

目前所创建的表都是所谓的管理表,有时也被称为内部表。因为这种表,Hive 会(或多或少地)控制着数据的生命周期。正如我们所看见的,Hive 默认情况下会将这些表的数据存储在由配置项 hive.metastore.warehouse.dir(例如,/usr/hive/warehouse)所定义的目录的子目录下。

当删除一个管理表时,Hive 也会删除这个表中的数据。但是,管理表不方便和其他工作共享数据。例如,假设有一份由 Pig 或者其他工具创建并且主要由这一工具使用的数据,同时还想使用 Hive 在这份数据上执行一些查询,可是并没有给予 Hive 对数据的所有权,此时可以创建一个外部表指向这份数据,而并不需要对其具有所有权。

2. 外部表

外部表是指向已经在 HDFS 中存在数据的表,它和内部表在元数据的组织上是相同的,但是它的实际数据的存储却有较大的差异。通俗地讲,外部表只是一个过程,它在建表的过程中,加载数据和创建表同时完成,并不会移动到数据仓库目录中,只是与外部数据建立一个链接,当删除一个外部表时,仅删除该链接。

下面学习如何创建一个外部表。先将数据存到 HDFS 上,才能建立一个外部表。

首先,进入到/usr/tmp 目录下,使用 vi 命令新建一个文件 stocks.txt:

```
vi stocks.txt
```

并在其中写入下面的信息:

```
NASDAQ♯AAPL♯2010-01-01♯195.69♯197.88♯194.00♯194.12♯17036300♯194.12
NYSE♯IBM♯2010-01-01♯129.46♯129.86♯127.94♯128.49♯78626000♯127.92
NYSE♯GE♯2010-01-01♯37.00♯37.00♯36.50♯36.60♯27717300♯32.32
```

其次,要在 HDFS 上新建一个/stocks 文件目录,然后使用"hadoop dfs - put 本地文件路径 /hdfs 的存储目录"命令将本地文件 stocks.txt 上传到 hdfs 的/stocks 的目录下,操作如图 4.18 所示。

在 shell 命令行输入下面的命令:

```
♯ hadoop dfs - mkdir /stocks
♯ hadoop dfs - put /usr/tmp/stocks.txt /stocks
```

接着可以使用 Hadoop 的 Web 页面工具来查看表是否上传成功。打开火狐浏览器,在地址栏输入"localhost:50070",进入 Hadoop 的 Web 页面工具来查看文件。进入 stocks 文

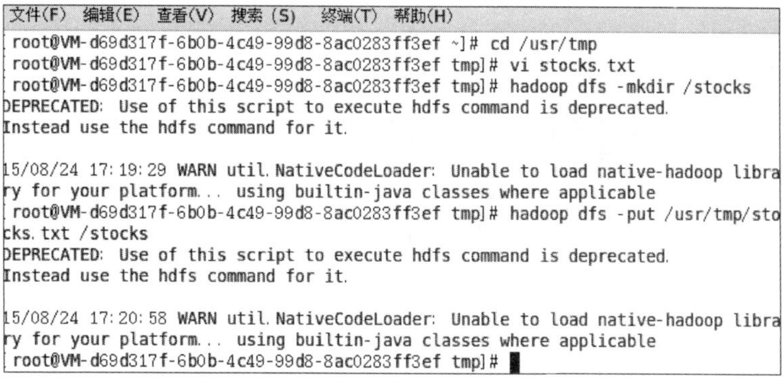

件夹后,将会看到 stocks.txt 文件,单击该文件后将会进入到文件下载页面,将文件下载到本地,如图 4.19 所示。

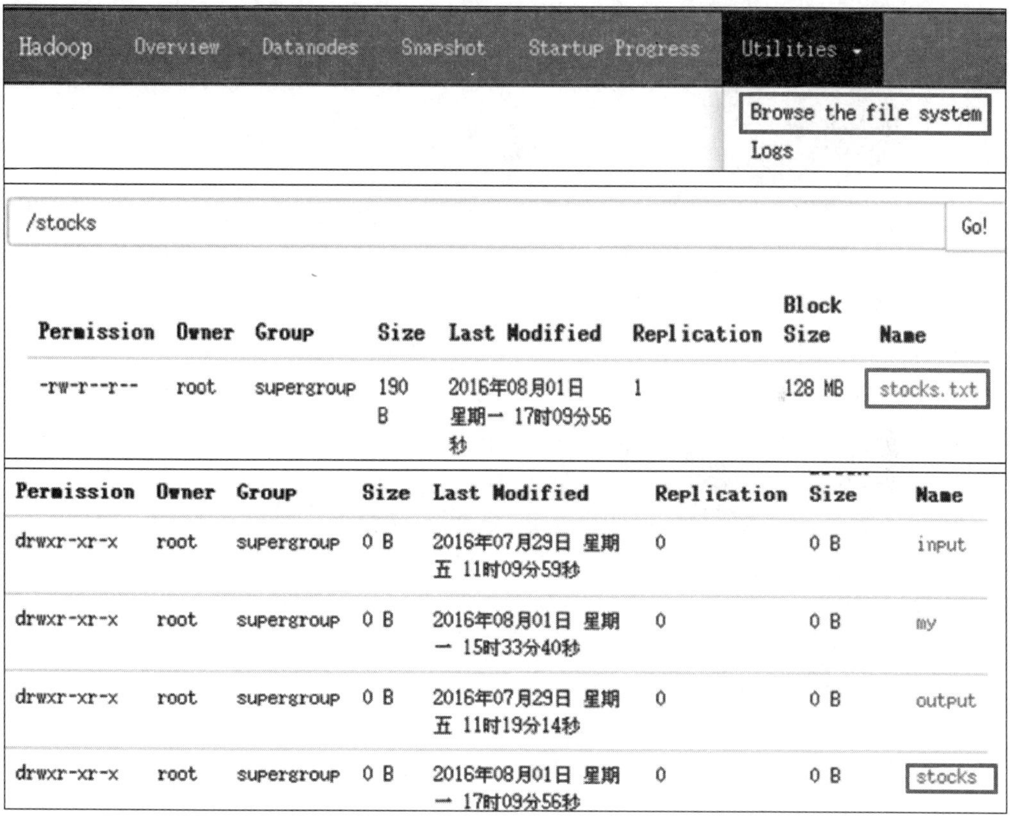

图 4.18 上传文件

图 4.19 在浏览器中查看

将文件下载到本地后,打开本地文件夹"下载",找到 stocks.txt 文件,右击此文件,单击"使用其他程序打开",弹出"打开方式"窗口,选择"Firefox Web Browser"打开文件,就可以看到其内容正是我们所写的内容,在本地/usr/tmp/目录下新建的文件 stocks.txt 中的内容相同,如图 4.20 所示。

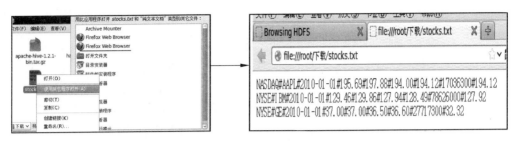

图 4.20 查看文件内容

最后，退出到根目录下，在 shell 命令行输入 hive，进入 Hive 的 CLI。在 mydatabase 数据库下，创建外部表 stocks。其中，location 后面所指的位置就是刚刚上传到 HDFS 上的 stock.txt 文件的目录。

```
hive> USE mydatabase;                          -- 将数据库 mydatabase 切换成当前数据库
OK
Time taken: 0.042 seconds
hive> CREATE EXTERNAL TABLE IF NOT EXISTS stocks (  -- 创建一个外部表,有关键字 EXTERNAL
    > bourse STRING,
    > symbol STRING,
    > ymd STRING,
    > price_open FLOAT,
    > price_high FLOAT,
    > price_low FLOAT,
    > price_close FLOAT,
    > volume INT,
    > price_adj_close FLOAT)
    > ROW FORMAT DELIMITED FIELDS TERMINATED BY '#'
    > LOCATION '/stocks';                      -- 该表的数据来源于 hdfs 的
/stocks 目录下的数据
OK
Time taken: 1.311 seconds
hive> SELECT * FROM stocks;                    -- 查询表 stocks 中的信息

OK
NASDAQ   AAPL   2010-01-01   195.69   197.88   194.0   194.12   17036300   194.12
NYSE     IBM    2010-01-01   129.46   129.86   127.94  128.49   78626000   127.92
NYSE     GE     2010-01-01   37.0     37.0     36.5    36.6     27717300   32.32
Time taken: 0.292 seconds, Fetched: 3 row(s)   -- 以上信息就是文件 stocks.txt 中的信息
```

再来总结一下创建外部表的知识点。关键字 EXTENAL 告诉 Hive 这个表是外部的，而后面的 LOCATION…子句则用于告诉 Hive 数据位于哪个路径下。因为表是外部的，所以 Hive 并非认为其完全拥有这份数据。因此，删除该表并不会删除这份数据，不过描述表的元数据信息会被删除。

管理表和外部表有一些小小的区别，有些 HiveQL 语法结构并不适用于外部表。

然而，需要弄清楚的重要一点是管理表和外部表之间的差异要比刚开始所看到的小得多。即使对于管理表，用户也是可以知道数据是位于哪个路径下的，因此用户也可以使用其他工具（例如 Hadoop 的 dfs 命令等）来修改甚至删除管理表所在的路径目录下的数据。可

能从严格意义上来说,Hive 虽管理着这些目录和文件,但是其并非具有对它们的完全控制权限。尽管如此,好的软件设计的一般原则是表达意图。如果数据被多个工具共享,那么可以创建一个外部表,来明确对数据的所有权。

用户可以使用 DESCRIBE EXTENDED tablename 命令来观察输出中的 tableType 信息,就刚刚创建的外部表 stocks,在 CLI 命令行中输入下列命令,观察实验现象:

```
hive > DESCRIBE EXTENDED stocks;                 -- 显示表结构信息
OK
bourse                    string
symbol                    string
ymd                       string
price_open                float
price_high                float
price_low                 float
price_close               float
volume                    int
price_adj_close           float
Detailed Table Information    Table (tableName: stocks, dbName: mydatabase, owner: root,
createTime: 1440409867, lastAccessTime: 0, retention: 0, sd: StorageDescriptor ( cols:
[FieldSchema(name:bourse, type:string, comment:null), FieldSchema(name:symbol,type:string,
comment:null), FieldSchema(name:ymd, type:string, comment:null), FieldSchema(name:price_
open, type:float, comment:null), FieldSchema(name:price_high, type:float, comment:null),
FieldSchema(name:price_low, type:float, comment:null), FieldSchema(name:price_close, type:
float, comment:null), FieldSchema(name:volume, type:int, comment:null), FieldSchema(name:
price_ adj _ close, type: float, comment: null)], location: hdfs://localhost: 9000/stocks,
inputFormat:org. apache. hadoop. mapred. TextInputFormat, outputFormat:org. apache. hadoop. hive.
ql. io. HiveIgnoreKeyTextOutputFormat, compressed:false, numBuckets: - 1, serdeInfo:SerDeInfo
(name: null, serializationLib: org. apache. hadoop. hive. serde2. lazy. LazySimpleSerDe,
parameters:{field. delim = #, serialization. format = #}), bucketCols:[], sortCols:[],
parameters: { }, skewedInfo: SkewedInfo ( skewedColNames: [ ], skewedColValues: [ ],
skewedColValueLocationMaps: { }), storedAsSubDirectories: false ), partitionKeys: [ ],
parameters:{totalSize = 0, EXTERNAL = TRUE, numRows = - 1, rawDataSize = - 1, COLUMN_STATS_
ACCURATE = false, numFiles = 0, transient_lastDdlTime = 1440409867}, viewOriginalText:null,
viewExpandedText:null, tableType:EXTERNAL_TABLE)           -- 这里指明了表类型
Time taken: 0.248 seconds, Fetched: 11 row(s)
```

对于管理表,用户可以查看到如下信息:

```
tableType:MANAGED_TABLE)
```

4.2.4 分区表

数据分区的一般概念存在已久。其可以有多种形式,但是通常使用分区来水平分散压力,将数据从物理上转移到离使用最频繁的用户更近的地方,或实现其他目的。

Hive 中有分区表的概念。可以看到分区表具有重要的性能优势,而且分区表还可以将数据以一种符合逻辑的方式进行组织,比如分层存储。

首先讨论分区管理表。重新来看之前那张 employees_01 表,并假设我们在一个非常大的跨国公司工作。我们的 HR 人员经常会执行一些带 WHERE 语句的查询,这样可以将结

果限制在某个特定条件下的第一级细分(例如美国的州或者加拿大的省)。那么,让我们按照country(国家)和state(州)作为分区字段来进行分区。

下面在mydatabase数据库中创建一个分区表employees,并设置其分区字段为country和state。

(1) 利用vi命令在/usr/tmp/目录下新建文档 employees01.txt、employees02.txt、employees03.txt和employees04.txt,并在其中写入下列数据。

vi employees01.txt:

```
John Doe#100000#Mary Smith*Todd Jones#Federal Taxes:0.2*State Taxes:0.05*Insurance:0.1#1 Michigan Ave.*Hilo*HI*60500
Boss Man#200000#Fred Finance#Federal Taxes:0.2*State Taxes:0.05*Insurance:0.1#1 Pretentious Drive.*Maui*HI*60500
Fred Finance#150000#Stacy Accountant#Federal Taxes:0.2*State Taxes:0.05*Insurance:0.1#2 Pretentious Drive*Kona*HI*60500
```

vi employees02.txt:

```
Mary Smith#80000#Bill King#Federal Taxes:0.2*State Taxes:0.05*Insurance:0.1#100 Ontario St.*Miami*FI*60601
```

vi employees03.txt:

```
Todd Jones#70000##Federal Taxes:0.15*State Taxes:0.03*Insurance:0.1#200 Chicago AVE.*Whitehouse*YT*60700
```

vi employees04.txt:

```
Bill King#60000##Federal Taxes:0.3*State Taxes:0.03*Insurance:0.1#300 Obscure Dr.*Winninpeg*MB*60100
```

(2) 退出到根目录下,在shell命令行输入命令进入Hive,在CLI命令行中输入下列代码创建一个分区表employees。

```
hive>USE mydatabase;
OK
Time taken: 0.01 seconds
hive>CREATE TABLE employees (                    --创建表employees
    >name STRING,
    >salary INT,
    >subordinates ARRAY<STRING>,
    >deductions MAP<STRING, FLOAT>,
    >address STRUCT<street:STRING,city:STRING,state:STRING,zip:INT>
    >)
    >PARTITIONED BY (country STRING, state STRING)    --分区字段为country和state
    >ROW FORMAT DELIMITED FIELDS TERMINATED BY '#'
    >COLLECTION ITEMS TERMINATED BY '*'
    >MAP KEYS TERMINATED BY ':'
```

```
   > LINES TERMINATED BY '\n';
OK
Time taken: 1.694 seconds
hive>
```

（3）将本地文件 employees01.txt、employees02.txt、employees03.txt 和 employees04.txt 中的数据装载到新建的分区表 employees 中。

接下来，在 CLI 命令行中输入下面的代码：

```
hive> LOAD DATA LOCAL INPATH '/usr/tmp/employees01.txt' INTO TABLE employees
   > PARTITION (country = 'US', state = 'HI');         -- 设置分区为 country = 'US', state = 'HI'
Loading data to table mydatabase.employees partition (country = US, state = HI)
Partition mydatabase.employees{country = US, state = HI} stats: [numFiles = 1, numRows = 0,
totalSize = 350, rawDataSize = 0]
OK
Time taken: 2.898 seconds
hive> LOAD DATA LOCAL INPATH '/usr/tmp/employees02.txt' INTO TABLE employees
   > PARTITION (country = 'US', state = 'FL');         -- 设置分区为 country = 'US', state = 'FL'
Loading data to table mydatabase.employees partition (country = US, state = FL)
Partition mydatabase.employees{country = US, state = FL} stats: [numFiles = 1, numRows = 0,
totalSize = 107, rawDataSize = 0]
OK
Time taken: 0.952 seconds
hive> LOAD DATA LOCAL INPATH '/usr/tmp/employees03.txt' INTO TABLE employees
   > PARTITION (country = 'CA', state = 'YT');         -- 设置分区为 country = 'CA', state = 'YT'
Loading data to table mydatabase.employees partition (country = CA, state = YT)
Partition mydatabase.employees{country = CA, state = YT} stats: [numFiles = 1, numRows = 0,
totalSize = 105, rawDataSize = 0]
OK
Time taken: 0.783 seconds
hive> LOAD DATA LOCAL INPATH '/usr/tmp/employees04.txt' INTO TABLE employees
   > PARTITION (country = 'CA', state = 'MB');         -- 设置分区为 country = 'CA', state = 'MB'
Loading data to table mydatabase.employees partition (country = CA, state = MB)
Partition mydatabase.employees{country = CA, state = MB} stats: [numFiles = 1, numRows = 0,
totalSize = 101, rawDataSize = 0]
OK
Time taken: 0.677 seconds
hive>
```

（4）查询刚刚新建的表 employees。

```
hive> SELECT * FROM employees;
OK
Bill King    60000    []    {"Federal Taxes":0.3,"State Taxes":0.03,"Insurance":0.1}    {"street":"300 Obscure Dr.","city":"Winninpeg","state":"MB","zip":60100}    CA    MB
Todd Jones    70000    []    {"Federal Taxes":0.15,"State Taxes":0.03,"Insurance":0.1}    {"street":"200 Chicago AVE.","city":"Whitehouse","state":"YT","zip":60700}    CA    YT
Mary Smith    80000    ["Bill King"]    {"Federal Taxes":0.2,"State Taxes":0.05,"Insurance":0.1}    {"street":"100 Ontario St.","city":"Miami","state":"FI","zip":60601}    US    FL
John Doe    100000    ["Mary Smith","Todd Jones"]    {"Federal Taxes":0.2,"State Taxes":0.05,"Insurance":0.1}    {"street":"1 Michigan Ave.","city":"Hilo","state":"HI","zip":60500}    US    HI
```

```
Boss Man    200000    ["Fred Finance"]    {"Federal Taxes":0.2,"State Taxes":0.05,"Insurance":0.
1}    {"street":"1Pretentious Drive.","city":"Maui","state":"HI","zip":60500}    US    HI
Fred Finance    150000    ["Stacy Accountant"]    {"Federal Taxes":0.2,"State Taxes":0.05,"
Insurance":0.1}    {"street":"2 Pretentious Drive","city":"Kona","state":"HI","zip":60500}
    US    HI
Time taken: 0.177 seconds, Fetched: 6 row(s)
hive >
```

分区表改变了 Hive 对数据存储的组织方式。我们是在 mydatabase 数据库中创建的这个表，那么对于这个表只会有一个 employees 目录与之对应：hdfs://localhost:9000/user/hive/warehouse/mydatabase.db/employees，但是，Hive 现在将会创建好可以反映分区结构的子目录。例如：

```
…/employees/country = US/state = HI
…/employees/country = US/state = FL
…/employees/country = CA/state = YT
…/employees/country = CA/state = MB
```

这些就是实际的目录名称。州目录下将会包含有文件，这些文件中存放着对应州的雇员信息。

分区字段(这个例子中就是 country 和 state)一旦创建好，表现得就和普通的字段一样。事实上，除非需要优化查询性能，否则使用这些表的用户不需要关心这些"字段"是否是分区字段。

下面的这个查询将会查询出 country＝'US', state＝'HI'的雇员信息：

```
hive > SELECT * FROM employees
    > WHERE country = 'US' AND state = 'HI';
-- 查询表 employees 中 country = 'US' AND state = 'HI'的雇员信息
OK
John Doe    100000    ["Mary Smith","Todd Jones"]    {"Federal Taxes":0.2,"State Taxes":0.05,"
Insurance":0.1}    {"street":"1 Michigan Ave.","city":"Hilo","state":"HI","zip":60500}    US
    HI
Boss Man    200000    ["Fred Finance"]    {"Federal Taxes":0.2,"State Taxes":0.05,"Insurance":0.
1}    {"street":"1Pretentious Drive.","city":"Maui","state":"HI","zip":60500}    US    HI
Fred Finance    150000    ["Stacy Accountant"]    {"Federal Taxes":0.2,"State Taxes":0.05,"
Insurance":0.1}    {"street":"2 Pretentious Drive","city":"Kona","state":"HI","zip":60500}
    US    ZHI
Time taken: 1.159 seconds, Fetched: 3 row(s)
hive >
```

需要注意的是，因为 country 和 state 的值已经包含在文件目录名称中了，所以也就没有必要将这些值存放到它的目录下的文件中。事实上，数据只能从这些文件中获得，因此用户需要在表模式中说明这一点，从而避免浪费空间。

对数据进行分区，最重要的原因就是为了更快地查询。在前面的例子中，可以将查询的结果限制在对美国夏威夷州的雇员中，仅仅需要扫描一个目录下的内容即可。即使有成千上万条的员工信息，除了一个目录其他的都可以忽略不计。对于非常大的数据集，分区可以显著提高查询性能,除非对分区进行常见的范围筛选(例如,按照地理位置范围或按照时间范围等)。

当在 WHERE 子句中增加谓词来按照分区值进行过滤时,这些谓词被称为分区过滤器。

当然,如果用户需要做一个查询,查询对象是公司的所有员工,那么这也是可以做到的。Hive 会读取每个文件目录,但这种宽范围的磁盘扫描还是比较少见的。

但是,如果表中的数据以及分区个数都非常大,那么执行这样一个包含所有分区的查询可能会触发一个巨大的 MapReduce 任务。一个强烈推荐的安全措施就是将 Hive 设置为 strict(严格)模式,这样如果对分区表进行查询而 WHERE 子句没有加分区过滤的话,将会禁止提交这个任务。用户也可以按照下面的语句将属性值设置为 nostrict(非严格)。

在 CLI 命令行下输入下面的代码:

```
hive > set hive.mapred.mode = strict;          -- 将 hive 设置为 strict(严格)模式
hive > SELECT name,salary FROM employees;      -- 查询表 employees 里面的员工信息
FAILED: SemanticException [Error 10041]: No partition predicate found for Alias "employees"
Table "employees"
hive > set hive.mapred.mode = nonstrict;       -- 将 hive 设置为 nonstrict(非严格)模式
hive > SELECT name,salary FROM employees;
OK
Bill King           60000
Todd Jones          70000
Mary Smith          80000
John Doe            100000
Boss Man            200000
Fred Finance        150000
Time taken: 0.174 seconds, Fetched: 6 row(s)
hive >
```

可以通过"SHOW PARTITIONS 表名"命令查看表中存在的所有分区:

```
hive > SHOW PARTITIONS employees;
OK
country = CA/state = MB
country = CA/state = YT
country = US/state = FL
country = US/state = HI
Time taken: 0.187 seconds, Fetched: 4 row(s)
hive >
```

如果表中现在存在很多分区,而用户只想查看是否存储某个特定分区键的分区,那么用户还可以在这个命令上增加一个指定了一个或者多个特定分区字段值的 PARTITION 子句,进行过滤查询:

```
hive > SHOW PARTITIONS employees PARTITION(country = 'US');
OK
country = US/state = HI
```

```
country = US/state = FL
Time taken: 0.211 seconds, Fetched: 2 row(s)
hive >
```

DESCRIBE EXTENDED employees 命令也会显示出分区键：

```
hive > DESCRIBE EXTENDED employees;
OK
name               string
salary             int
subordinates       array < string >
deductions         map < string, float >
address            struct < street:string, city:string, state:string, zip:int >
country            string
state              string

# Partition Information
# col_name         data_type         comment

country            string
state              string

Detailed Table Information    Table ( tableName: employees, dbName: mydatabase, owner: root,
createTime: 1440412809, lastAccessTime: 0, retention: 0, sd: StorageDescriptor ( cols:
[FieldSchema ( name: name, type: string, comment: null ), FieldSchema ( name: salary, type: int,
comment: null ), FieldSchema ( name: subordinates, type: array < string >, comment: null ),
FieldSchema ( name: deductions, type: map < string, float >, comment: null ), FieldSchema ( name:
address, type: struct < street: string, city: string, state: string, zip: int >, comment: null ),
FieldSchema ( name: country, type: string, comment: null ), FieldSchema ( name: state, type: string,
comment: null )], location: hdfs://localhost: 9000/user/hive/warehouse/mydatabase. db/
employees, inputFormat: org. apache. hadoop. mapred. TextInputFormat, outputFormat: org. apache.
hadoop. hive. ql. io. HiveIgnoreKeyTextOutputFormat, compressed: false, numBuckets: - 1,
serdeInfo: SerDeInfo ( name: null, serializationLib: org. apache. hadoop. hive. serde2. lazy.
LazySimpleSerDe, parameters:{line. delim =
, field. delim = #, colelction. delim = *, mapkey. delim = :, serialization. format = # }),
bucketCols:[ ], sortCols:[ ], parameters: { }, skewedInfo: SkewedInfo ( skewedColNames: [ ],
skewedColValues: [ ], skewedColValueLocationMaps: { }), storedAsSubDirectories: false ),
partitionKeys:[ FieldSchema ( name: country, type: string, comment: null ), FieldSchema ( name:
state, type: string, comment: null )], parameters: { transient _ lastDdlTime = 1440412809},
viewOriginalText:null, viewExpandedText:null, tableType:MANAGED_TABLE)
Time taken: 0.127 seconds, Fetched: 16 row(s)
```

输出信息中的模式信息部分会将 country 和 state 以及其他字段列在一起，因为就查询而言，它们就是字段。Detailed Table Information(详细表信息)将 country 和 state 作为分区键处理。

外部分区表(简称外部表)

外部表同样可以使用分区。事实上，用户可能会发现，这是管理大型生产数据集最常见的情况。这种结合给用户提供了一个可以和其他工具共享数据的方式，同时也可以优化查询性能。

因为用户可以自己定义目录结构，所以用户对于目录结构的使用具有更多的灵活性。下面举一个新例子。前面介绍过一个外部表 stocks，用来记录公司的股票市场的数据，现在建立一个 dividends 表，用来记录股息情况。对于股息信息，其中记录有交易所名称、公司名称、时间和股息。

下面创建外部分区表 dividends。

(1) 需要在本地的/usr/tmp 的目录下新建三份文档：dividends_2008.txt、dividends_2009.txt 和 dividends_2010.txt，分别在其中输入下面的数据：

vi dividends_2008.txt。

```
NASDAQ#AAPL#0.03
NYSE#IBM#0.5
NYSE#GE#0.31
```

vi dividends_2009.txt。

```
NASDAQ#AAPL#0.02
NYSE#IBM#0.4
NYSE#GE#0.1
```

vi dividends_2010.txt。

```
NASDAQ#AAPL#0.03
NYSE#IBM#0.3
NYSE#GE#0.31
```

(2) 需要在 HDFS 上创建/dividends/目录，并将刚刚创建的三个文档上传上去。

(3) 在 shell 命令行下输入下面的命令：

```
# hadoop dfs -mkdir /dividends/
# hadoop dfs -mkdir /dividends/2008
# hadoop dfs -mkdir /dividends/2009
# hadoop dfs -mkdir /dividends/2010                          -- 以上步骤是在 hdfs 上面创
                                                                建三个文件夹
# hadoop dfs -put /usr/tmp/dividends_2008.txt /dividends/2008
# hadoop dfs -put /usr/tmp/dividends_2009.txt /dividends/2009
# hadoop dfs -put /usr/tmp/dividends_2010.txt /dividends/2010 -- 以上三步完成了文件的
                                                                上传
```

(4) 退出到根目录下，在 shell 命令行输入 hive 命令进入 Hive，创建一个外部分区表 dividends 来存储一些信息。为了分析简便，在 Hive 的 CLI 命令行中输入下面的代码创建一个外部分区表 dividends：

```
hive> USE mydatabase;
OK
Time taken: 1.264 seconds
hive> CREATE EXTERNAL TABLE dividends (       -- 创建外部分区表 dividends
    > bourse STRING,                          -- 交易所的信息
    > symbol STRING,                          -- 公司的信息
    > dividend FLOAT)                         -- 股息
    > PARTITIONED BY (year INT)               -- 按照年份分区
```

```
        > ROW FORMAT DELIMITED FIELDS TERMINATED BY '#';
OK
Time taken: 1.134 seconds
hive>
```

之前，我们创建过一个非分区外部表 stocks，那时要求使用一个 LOCATION 子句，对于外部分区表则没有这样的要求。有一个 ALTER TABLE 语句可以单独实现增加分区，这个语句需要为每一个分区键指定一个值，本例中，也就是要为 year 这个分区键指定值。

下面是一个例子，演示如何增加一个 2008 年分区：

```
hive> ALTER TABLE dividends ADD PARTITION (year = '2008')    -- 为表 dividends 增加一个分区
    > LOCATION 'hdfs://localhost:9000/dividends/2008';       -- 数据存放的地址
OK
Time taken: 0.962 seconds
hive> SELECT * FROM dividends;                               -- 查询表中的信息
OK
NASDAQ      AAPL 0.03 2008
NYSE        IBM 0.5 2008
NYSE        GE 0.31 2008
Time taken: 0.874 seconds, Fetched: 3 row(s)
hive>
```

我们使用的目录组织习惯完全由自己定义。这里按照分层目录结构组织，因为这是一个合乎逻辑的数据组织方式，但是并非要求一定如此。我们可以遵从 Hive 的目录命名习惯（例如，…bourse=NASDAD/symbol=AAPL），但是并不一定要求如此。

接下来继续为外部分区表增加分区：

```
hive> ALTER TABLE dividends ADD PARTITION(year = '2009')
    > LOCATION 'hdfs://localhost:9000/dividends/2009';
OK
Time taken: 0.167 seconds
hive> ALTER TABLE dividends ADD PARTITION(year = '2010')
    > LOCATION 'hdfs://localhost:9000/dividends/2010';
OK
Time taken: 0.173 seconds
hive>
```

这样，就为表 dividends 增加了三个分区。

顺便说一下，Hive 不关心一个分区对应的分区目录是否存在或者分区目录下是否有文件。如果分区目录不存在或分区目录下没有文件，则对于这个过滤分区的查询将没有返回结果。当用户想在另外一个进程开始向分区中写数据之前创建好分区时，这样做是很方便的。数据一旦存在，对于这份数据的查询就会有返回结果。

这个功能所具有的另一个好处是：可以将新数据写入到一个专用的目录中，并与位于其他目录中的数据存在明显的区别。同时，不管用户是将旧数据转移到一个"存档"位置还是直接删除掉，新数据被篡改的风险都被降低了，因为新数据的数据子集位于不同的目录下。

与非分区外部表一样，Hive 并不控制这些数据。即使表被删除，数据也不会被删除。与分区管理表一样，通过 SHOW PARTITIONS 命令可以查看一个外部表的分区：

```
hive> SHOW PARTITIONS dividends;              -- 查看表 dividends 的分区信息
OK
year = 2008
year = 2009
year = 2010
Time taken: 0.28 seconds, Fetched: 3 row(s)
hive>
```

同样，DESCRIBE EXTENDED dividends 语句会将分区键作为表的模式的一部分，partitionKeys 列表的内容也同时进行显示：

```
hive> DESCRIBE EXTENDED dividends;
OK
bourse                  string
symbol                  string
dividend                float
year                    int

# Partition Information
# col_name              data_type               comment

year                    int

Detailed Table Information    Table(tableName:dividends, dbName:mydatabase, owner:root, createTime:1440468530, lastAccessTime:0, retention:0, sd:StorageDescriptor(cols:[FieldSchema(name:bourse, type:string, comment:null), FieldSchema(name:symbol, type:string, comment:null), FieldSchema(name:dividend, type:float, comment:null), FieldSchema(name:year, type:int, comment:null)], location:hdfs://localhost:9000/user/hive/warehouse/mydatabase.db/dividends, inputFormat:org.apache.hadoop.mapred.TextInputFormat, outputFormat:org.apache.hadoop.hive.ql.io.HiveIgnoreKeyTextOutputFormat, compressed:false, numBuckets:-1, serdeInfo:SerDeInfo(name:null, serializationLib:org.apache.hadoop.hive.serde2.lazy.LazySimpleSerDe, parameters:{field.delim=#, serialization.format=#}), bucketCols:[], sortCols:[], parameters:{}, skewedInfo:SkewedInfo(skewedColNames:[], skewedColValues:[], skewedColValueLocationMaps:{}), storedAsSubDirectories:false), partitionKeys:[FieldSchema(name:year, type:int, comment:null)], parameters:{EXTERNAL=TRUE, transient_lastDdlTime=1440468530}, viewOriginalText:null, viewExpandedText:null, tableType:EXTERNAL_TABLE)
Time taken: 0.23 seconds, Fetched: 11 row(s)
```

这个输出缺少了一个非常重要的信息，那就是分区数据实际存在的路径。这里有一个路径字段，但是该字段仅仅表示如果表是管理表，那么其会使用到 Hive 默认的目录。不过，可以通过如下方式查看分区数据所在的路径：

```
hive> DESCRIBE EXTENDED dividends PARTITION (year = '2010');
OK
…
location:hdfs://localhost:9000/dividends/2010,              -- 分区路径
…
```

```
Time taken: 0.501 seconds, Fetched: 11 row(s)
hive>
```

通常会使用分区外部表,因为它具有非常多的优点,例如逻辑数据管理、高性能的查询等。ALTER TABLE … ADD PARTITION 语句并非只有对外部表才能够使用。对于管理表,当有分区数据不是由我们之前讨论过的 LOAD 和 INSERT 语句产生时,用户同样可以使用这个命令指定分区路径。用户需要记住并非所有的表数据都是放在通常的 Hive 的 ware-house 目录下的,同时,当删除管理表时,这些数据不会连带被删除。

4.2.5 删除表

Hive 支持与 SQL 中 DROP TABLE 命令类似的操作。在之前的实验中,我们创建了 my-database 数据库,并且在该数据库下建立了 mytable、employees_01 和 employees_02 等多张表。下面就尝试删除表 mytable。从根目录下进入 CLI 命令行,在 CLI 命令行执行下列删除表的命令操作:

```
hive> USE mydatabase;                    -- 将工作目录切换到 mydatabase 数据库
OK
Time taken: 0.023 seconds
hive> SHOW TABLES;                       -- 显示该数据库下的所有表
OK
binary_table
dividends
employees
employees_01
employees_02
mytable
stocks
Time taken: 0.044 seconds, Fetched: 6 row(s)
hive> DROP TABLE IF EXISTS mytable;      -- 删除表 mytable
OK
Time taken: 2.538 seconds
hive> SHOW TABLES;                       -- 再次显示该数据库下的表,就没有表 mytable 了
OK
dividends
employees
employees_01
employees_02
stocks
Time taken: 0.066 seconds, Fetched: 5 row(s)
```

还可以在浏览器地址栏中输入"localhost:50070",进入 Hadoop 的 Web 管理界面查看文件情况。

在执行删除表的时候,可以选择是否支持使用 IF EXITST 关键字,如果没有使用这个关键字而且表并不存在,那么将会抛出一个错误信息。对于管理表,表的元数据信息和表内的数据都会被删除。对于外部表,表的元数据信息会被删除,但是表中的数据不会被删除。

4.2.6 修改表

大多数的表属性可以通过 ALTER TABLE 语句来进行修改。这种操作会修改元数据，但不会修改数据本身。这些语句可用于修改表模式中出现的错误、改变分区路径以及其他一些操作。

1. 表重命名

使用"ALTER TABLE tablename1 RENAME TO tablename2；"就可以将 tablename1 的表名改为 tablename2。

在前面创建过一个外部分区表 dividends，现在使用表重命名的命令将其改名为 my_dividends：

```
hive> USE mydatabase;                                  --将工作目录切换到 mydatabase 数据库
OK
Time taken: 0.028 seconds
hive> ALTER TABLE dividends RENAME TO my_dividends;  --对表 dividends 进行重命名
OK
Time taken: 0.252 seconds
hive> SHOW TABLES;                                     --显示该数据库下的表，这时就没有表
                                                         dividends 了，而是 my_dividends
OK
employees
employees_01
employees_02
my_dividends
stocks
Time taken: 0.061 seconds, Fetched: 5 row(s)
hive>
```

2. 增加、修改和删除表分区

如前所述，ALTER TABLE tablename ADD PARTITION…语句用于为表（通常是外部表）增加一个新的分区。这里增加可提供的可选项，然后多次重复前面的分区路径语句，在前面我们就说过，Hive 不关心一个分区对应的分区目录下是否存在文件，所以 location 子句后面的路径为空时，也能为表增加分区：

```
hive> ALTER TABLE my_dividends ADD IF NOT EXISTS
    > PARTITION (year = '2011') LOCATION 'hdfs://localhost:9000/dividends/2011'
    > PARTITION (year = '2012') LOCATION 'hdfs://localhost:9000/dividends/2012';
OK
Time taken: 0.409 seconds
```

当使用 Hive v0.80 或其后的版本时，在同一个查询中可以同时增加多个分区。一如既往，IF NOT EXISTS 也是可选的，而且含义作用不变。

同时，用户还可以通过高效地移动位置来修改某个分区的路径：

```
hive> SELECT * FROM my_dividends            --查看表 my_dividends 的 year = 2011 分区下的内
                                              容，里面为空
```

```
    > WHERE year = '2011';
OK
Time taken: 2.183 seconds
hive > ALTER TABLE my_dividends PARTITION (year = 2011)
 -- 修改 year = 2011 分区的路径为 hdfs://localhost:9000/dividends/2008'
    > SET LOCATION 'hdfs://localhost:9000/dividends/2008';
OK
Time taken: 0.422 seconds
hive > SELECT * FROM my_dividends  -- 再次查看 my_dividends 下 year = 2011 分区的内容
    > WHERE year = 2011;
OK
NASDAQ     AAPL 0.03 2011
NYSE       IBM    0.5  2011
NYSE       GE     0.31 2011
Time taken: 0.276 seconds, Fetched: 3 row(s)
hive >
```

这个命令不会将数据从旧的路径转移走，也不会删除旧的数据。

最后，用户可以通过 ALTER TABLE tablename DROP PARTITION(…) 删除某个分区：

```
hive > SHOW PARTITIONS my_dividends;
OK
year = 2008
year = 2009
year = 2010
year = 2011
year = 2012
Time taken: 0.137 seconds, Fetched: 5 row(s)
hive > ALTER TABLE my_dividends DROP PARTITION(year = 2012);  -- 删除分区 year = 2012
Dropped the partition year = 2012
OK
Time taken: 0.189 seconds
hive > SHOW PARTITIONS my_dividends;                           -- 表 my_dividends 下就没有分区
                                                                  year = 2012 了
OK
year = 2008
year = 2009
year = 2010
year = 2011
Time taken: 0.098 seconds, Fetched: 4 row(s)
hive >
```

按照常规，上面语句中的 IF EXISTS 子句是可选的。对于管理表，即使是使用 ALTER TABLE…ADD PARTITION 语句增加的分区，分区内的数据也是会同时和元数据信息一起被删除的。但是对于外部表，分区内数据不会被删除。

3. 修改列信息

用户可以对某个字段进行重命名，并修改其位置、类型或者注释。

```
hive > DESCRIBE my_dividends;              -- 先查看一下表 my_dividends 的表结构信息
OK
bourse                  string
symbol                  string
dividend                float              -- 注意这里的 dividend 列是 float 类型的 year int

# Partition Information
# col_name              data_type          comment

year                    int
Time taken: 0.216 seconds, Fetched: 9 row(s)
hive > ALTER TABLE my_dividends
    > CHANGE COLUMN dividend my_dividend INT
-- 修改列 dividend 的名字为 my_dividend,类型为 int
    > COMMENT 'change the column_name and column_type. ';          -- 增加注释
OK
Time taken: 0.498 seconds
hive > DESCRIBE my_dividends ;
-- 再查看表信息,发现列 dividend 的名称、类型发生了变化,而且还多了说明部分
OK
bourse                  string
symbol                  string
my_dividend             int                change the column_name and column_type.
year                    int

# Partition Information
# col_name              data_type          comment

year                    int
Time taken: 0.157 seconds, Fetched: 9 row(s)
hive >
```

下面修改列的位置信息:

```
hive > ALTER TABLE my_dividends
    > CHANGE COLUMN bourse bourse STRING   -- 即使该列的名称、类型不变,也要写出来
    > AFTER symbol;                        -- 将 bourse 列置于 symbol 列之后
OK
Time taken: 0.247 seconds
hive > DESCRIBE my_dividends;
-- 查看表 my_dividends 的表结构信息,可以看出列 bourse 在列 symbol 的后面
OK
symbol                  string
bourse                  string
my_dividend             int                change the column_name and column_type.
year                    int

# Partition Information
# col_name              data_type          comment

year                    int
Time taken: 0.132 seconds, Fetched: 9 row(s)
```

即使字段名或者字段类型没有改变,用户也需要完全指定旧的字段名,并给出新的字段名及新的字段类型。关键字 COLUMN 和 COMMENT 子句都是可选的。在前面所演示的例子中,我们将字段 bourse 转移到字段 symbol 之后。如果用户想将这个字段移动到第一个位置,那么只需要使用 FIRST 关键字替代 AFTER other_column 子句即可。

和通常一样,这个命令只会修改元数据信息。如果用户移动的是字段,那么数据也应当和新的模式匹配或者通过其他某些方法修改数据以使其能够和模式匹配。

4. 增加列

用户可以在分区字段之前、已有的字段之后增加新的字段。

```
hive> ALTER TABLE my_dividends
    > ADD COLUMNS (month INT COMMENT 'month',    -- 增加字段 month
    > day INT COMMENT 'day');                    -- 增加字段 day
OK
Time taken: 0.142 seconds
hive> DESCRIBE my_dividends;                     -- 查看表结构
OK                                               -- 在后面增加了 month、day 字段,并且还有
                                                    注释
symbol                  string
bourse                  string
my_dividend             int
month                   int            month
day                     int            day
year                    int

# Partition Information
# col_name              data_type      comment

year                    int
Time taken: 0.095 seconds, Fetched: 11 row(s)
```

COMMENT 子句和通常一样,是可选的。如果新增的字段中有某个或多个字段位置是错误的,那么需要使用"ALTER COULME 表名 CHANGE COLUMN"语句逐一将字段调整到正确的位置。

5. 删除或者替换列

下面这个例子移除了之前所有的字段并重新指定了新的字段:

```
hive> ALTER TABLE my_dividends
    > REPLACE COLUMNS (bourse STRING,   -- 重新指定该表的字段为 bourse、symbol、dividend
    > symbol STRING,
    > dividend FLOAT);
OK
Time taken: 0.133 seconds
hive> DESCRIBE my_dividends;            -- 查看表结构,是按照我们的指定排序的,分区字段不变还
                                           在最后
OK
bourse                  string
symbol                  string
```

```
dividend                float
year                    int
# Partition Information
# col_name              data_type            comment
year                    int
Time taken: 0.088 seconds, Fetched: 9 row(s)
hive>
```

这个语句实际上重新指定了表 my_dividends 中的所有字段。因为是 ALTER 语句,所以只有表的元数据信息改变了,表中的数据信息需要重新装载。

6. 修改表属性

用户可以增加附加的表属性或者修改已经存在的属性,但是无法删除属性。

```
hive> ALTER TABLE my_dividends SET TBLPROPERTIES (   -- 增加表属性
    > 'notes' = 'SET TBLPROPERTIES : created at 2015 - 08 - 25');
OK
Time taken: 0.076 seconds
hive> DESC FORMATTED my_dividends;                    -- 查看表结构信息
OK
# col_name              data_type            comment

bourse                  string
symbol                  string
dividend                float

# Partition Information
# col_name              data_type            comment

year                    int

# Detailed Table Information
  ...
    notes               SET TBLPROPERTIES:created at 2015 - 08 - 25
-- 可以注意到,相比之前的信息,这里多了一条属性
    transient_lastDdlTime1440486395
...
Time taken: 0.082 seconds, Fetched: 38 row(s)
hive>
```

7. 修改存储属性

有几个 ALTER TABLE 语句用于修改存储格式。

下面这个语句将一个分区的存储格式修改成了 SEQUENCEFILE。

```
hive> ALTER TABLE my_dividends
    > PARTITION (year = 2008)
    > SET FILEFORMAT SEQUENCEFILE;
OK
Time taken: 0.836 seconds
```

如果表是分区表,那么需要使用 PARTITION 子句,这里的表 my_dividends 是分区表。而且,我们可以修改表的存储属性:

```
hive> ALTER TABLE stocks
    > CLUSTERED BY (bourse,symbol)
    > SORTED BY (symbol)
    > INTO 3 BUCKETS;
OK
Time taken: 0.574 seconds
```

SORTED BY 子句是可选的,但是 CLUSTER BY 和 INTO…BUCKETS 子句是必选的。

第 5 章

Hive数据操作与查询

5.1 数据操作

接下来将继续讨论 HiveQL，也就是 Hive 查询语言，并关注向表中装载数据和从表中抽取数据到文件系统的数据操作语言部分。前面讨论了如何创建表，现在希望解决随之而来的下一个问题，即如何装载数据到这些表中，然后才能有供用户查询的内容。

5.1.1 向管理表中装载数据

Hive 没有行级别的数据插入、数据更新和删除操作，那么向表中装载数据的唯一途径就是使用一种"大量"的数据装载操作，或者通过其他方式仅仅将文件写入到正确的目录下。

前面已经简单地讲过如何向表中装载数据。这里再详细介绍一下。

首先，需要在本地新建文本文件，在该文件中写入数据，然后再使用"LOAD DATA LOCAL INPATH '/本地文件目录'INTO TABLE tablename"命令将文件导入到 HDFS 下的表中。下面完成一个例子。

打开终端，在/usr/tmp 目录下新建三个文本文件：stu_table.txt、Mstu_table.txt 和 Fstu_table.txt，并在其中写入相关数据，如图 5.1 所示。

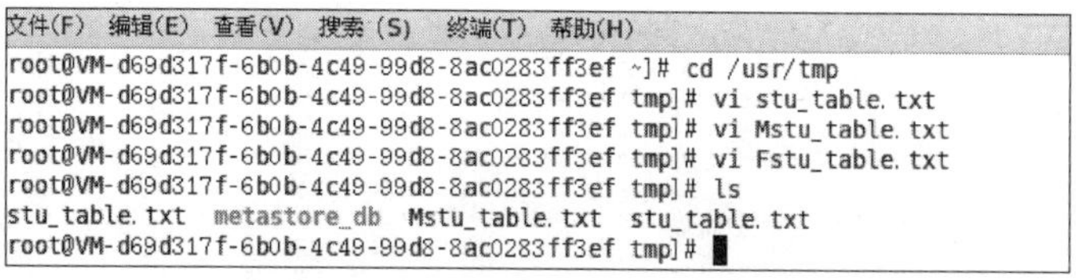

```
文件(F) 编辑(E) 查看(V) 搜索(S) 终端(T) 帮助(H)
root@VM-d69d317f-6b0b-4c49-99d8-8ac0283ff3ef ~]# cd /usr/tmp
root@VM-d69d317f-6b0b-4c49-99d8-8ac0283ff3ef tmp]# vi stu_table.txt
root@VM-d69d317f-6b0b-4c49-99d8-8ac0283ff3ef tmp]# vi Mstu_table.txt
root@VM-d69d317f-6b0b-4c49-99d8-8ac0283ff3ef tmp]# vi Fstu_table.txt
root@VM-d69d317f-6b0b-4c49-99d8-8ac0283ff3ef tmp]# ls
stu_table.txt  metastore_db  Mstu_table.txt  stu_table.txt
root@VM-d69d317f-6b0b-4c49-99d8-8ac0283ff3ef tmp]#
```

图 5.1 创建文件

向 stu_table.txt、Mstu_table.txt 和 Fstu_table.txt 中写入数据：

vi stu_table.txt。

```
12001#Mary#21
12002#David#23
12003#Tom#22
12004#Ella#20
```

vi Mstu_table.txt。

```
12005#Frank#21
12006#Tony#23
12007#Pete#21
```

vi Fstu_table.txt。

```
12008#Angela#21
12009#Alice#20
12010#Amy#22
```

(1) 退出到根目录下，在 shell 命令行下输入 hive 命令，进入 CLI 命令行，新建一个名为 school 的数据库，并新建一个名为 student 的表。

```
hive> CREATE DATABASE school;                     -- 新建一个数据库 school
OK
Time taken: 1.742 seconds
hive> USE school;
OK
Time taken: 0.053 seconds
hive> CREATE TABLE stu_table (id INT,name STRING, age INT)   -- 新建一个以"#"为分隔符的表
    > ROW FORMAT DELIMITED FIELDS TERMINATED BY '#';
OK
Time taken: 1.083 seconds
```

(2) 使用"LOAD DATA LOCAL INPATH '本地数据目录' INTO TABLE tablename;"命令将 Mstu_table.txt 和 Fstu_table.txt 中的数据装载到表 student 中，注意这里没有关键字 OVERWRITE。

```
hive> LOAD DATA LOCAL INPATH '/usr/tmp/Mstu_table.txt' INTO TABLE stu_table;
-- 将数据装载到表 stu_table 中,没有关键字 OVERWRITE
Loading data to table school.stu_table
Table school.stu_table stats: [numFiles = 1, totalSize = 32]
OK
Time taken: 1.085 seconds
hive> SELECT * FROM stu_table;              -- 查询表 stu_table 里面的内容
OK
12005      Frank      21
12006      Tony       23
12007      Pete       21                    -- 包含有 3 个人的信息
```

```
Time taken: 1.006 seconds, Fetched: 3 row(s)
hive> LOAD DATA LOCAL INPATH '/usr/tmp/Fstu_table.txt' INTO TABLE stu_table;
 -- 将数据装载到表 stu_table 中,没有关键字 OVERWRITE
Loading data to table school.stu_table
Table school.stu_table stats: [numFiles=2, totalSize=64]
OK
Time taken: 0.497 seconds
hive> SELECT * FROM stu_table;                    -- 查询表 stu_table 里面的内容
OK
12008  Angela  21
12009  Alice   20
12010  Amy     22
12005  Frank   21
12006  Tony    23
12007  Pete    21                                  -- 包含 6 个人的信息
Time taken: 0.182 seconds, Fetched: 6 row(s)
```

（3）在上面命令的基础上,增加一个关键字 OVERWRITE,将 stu_table.txt 中的数据装载到表 stu_table 中。

```
hive> LOAD DATA LOCAL INPATH '/usr/tmp/stu_table.txt' OVERWRITE INTO TABLE stu_table;
 -- 将 stu_table.txt 中的数据装载到表 stu_table 中,这里有关键字 OVERWRITE
Loading data to table school.stu_table
Table school.stu_table stats: [numFiles=1, numRows=0, totalSize=40, rawDataSize=0]
OK
Time taken: 0.573 seconds
hive> SELECT * FROM stu_table;                    -- 查询表 stu_table 里面的内容
OK
12001   Mary    21                                 -- 之前的 6 条的信息被覆盖掉了
12002   David   23
12003   Tom     22
12004   Ella    20
Time taken: 0.122 seconds, Fetched: 4 row(s)
hive>
```

那么如何向分区表中装载数据呢？让我们继续往下学习。

（4）要创建一个分区表,在 CLI 命令行输入下面的命令,创建一个分区表：

```
hive> USE school;
OK
Time taken: 0.049 seconds
hive> CREATE TABLE students ( id INT, name STRING, age INT)    -- 创建一个分区表 students
    > PARTITIONED BY (gender STRING)
    > ROW FORMAT DELIMITED FIELDS TERMINATED BY '#';
OK
Time taken: 0.142 seconds
hive>
```

（5）将 Mstu_table.txt 中的数据装载到表 students 中：

```
hive> LOAD DATA LOCAL INPATH '/usr/tmp/Mstu_table.txt'
```

```
    > INTO TABLE students
    > PARTITION (gender = 'M');
Loading data to table school.students partition (gender = M)
Partition school.students{gender = M} stats: [numFiles = 1, numRows = 0, totalSize = 44,
rawDataSize = 0]
OK
Time taken: 0.737 seconds
```

(6) 将 Fstu_table.txt 中的数据装载到表 students 中：

```
hive> LOAD DATA LOCAL INPATH '/usr/tmp/Fstu_table.txt'
    > INTO TABLE students
    > PARTITION (gender = 'F');
Loading data to table school.students partition (gender = F)
Partition school.students{gender = F} stats: [numFiles = 1, numRows = 0, totalSize = 44,
rawDataSize = 0]
OK
Time taken: 1.097 seconds
```

如果分区目录不存在，那么这个命令会先创建分区目录，然后再将数据复制到该目录下。

通常情况下指定的路径应该是一个目录，而不是单个独立的文件。Hive 会将所有文件都复制到这个目录中。这使得用户将更方便地将数据组织到多文件中，同时，在不修改 Hive 脚本的前提下修改文件命名规则。不管怎么样，文件都会被复制到目标表路径下而且文件名保持不变。

使用了 LOCAL 这个关键字，表明这个路径应该为本地文件系统路径。数据将会被复制到目标位置。如果省略 LOCAL 关键字，那么这个路径应该是分布式文件系统中的路径。这种情况下，数据是从这个路径转移到目标位置的。

之所以会存在这种差异，是因为用户在分布式文件系统中可能并不需要重复的多份数据文件副本。

同时，因为文件是以这种方式移动的，Hive 要求源文件和目标文件以及目录应该在同一个文件系统中。例如，用户不可以使用 LOAD DATA 语句将数据从一个集群的 HDFS 中转载（转移）到另一个集群的 HDFS 中。

指定全路径会具有更好的鲁棒性，但也同样支持相对路径。当使用本地模式执行时，相对路径相对的是当 Hive CLI 启动时用户的工作目录。对于分布式或者伪分布式模式，这个路径解读为相对于分布式文件系统中用户的根目录，该目录在 HDFS 和 MapRFS 中默认为 /user/$USER。

如果用户指定了 OVERWRITE 关键字，那么目标文件夹中之前存在的数据将会被先删除掉。如果没有这个关键字，仅仅会把新增的文件增加到目标文件夹中而不会删除之前的数据。然而，如果目标文件夹中已经存在和装载的文件同名的文件，那么旧的同名文件将会被覆盖重写。

如果目标表是分区表，那么需要使用 PARTITION 子句，而且用户还必须为每个分区的键指定一个值。

按照之前所说的那个例子，我们创建的表 stu_table 中的数据现在将会存放到如下这个文件夹中：hdfs://localhost：9000/user/hive/warehouse/school.db/stu_table。而我们创建的分区表 students 中的数据将存放在 hdfs://localhost：9000/user/hive/warehouse/school.db/students/gender＝M 目录下和 hdfs://localhost：9000/user/hive/warehouse/school.db/students/gender＝F 目录下。

对于 INPATH 子句中使用的文件路径还有一个限制，那就是这个路径下不可以包含任何文件夹。

Hive 并不会验证用户装载的数据和表的模式是否匹配。然而，Hive 会验证文件格式是否和表结构定义的一致。例如，如果表在创建时定义的存储格式是 SEQUENCEFILE，那么装载进去的文件也应该是 SEQUENCEFILE 格式。

5.1.2 通过查询语句向表中插入数据

INSERT 语句允许用户通过查询语句向目标表中插入数据。这里通过新建表，向其中插入表 stu_table 和表 students 中的数据。

（1）下面新建一个表 students_01：

```
hive> USE school;
OK
Time taken: 0.015 seconds
hive> CREATE TABLE students_01( id INT , name STRING , age INT )    --创建一个表 students_01
    > ROW FORMAT DELIMITED FIELDS TERMINATED BY ' # ';
OK
Time taken: 0.097 seconds
```

（2）向表 students01 中插入数据：

```
hive> INSERT INTO TABLE students_01              -- 使用 INSERT INTO…向表 students_01 中插入数据
    > SELECT * FROM stu_table                    -- 插入的数据来自表 stu_table
    > WHERE id = '12001';                        -- 可以使用 where 语句来选择插入的数据
...
OK
Time Taken: 6.665 seconds
hive> SELECT * FROM students_01;                 -- 查询表 students_01 中的数据
12001    Mary     21
Time taken:0.13 seconds, Fetched: 1 row(s)
hive> INSERT OVERWRITE TABLE students_01   -- 使用 INSERT OVERWRITE…向表 students_01 中插入
                                                    数据
    > SELECT * FROM stu_table                    -- 插入的数据来自表 stu_table
        > WHERE id = '12002';
...
OK
Time taken: 2.137 seconds
hive> SELECT * FROM students_01;                 -- 查询表 students_01 中的数据,原来的数据已经被覆
                                                    盖了
OK
12002    David     23
Time taken: 0.138 seconds, Fetched: 1 row(s)
hive>
```

这里使用了 OVERWRITE 关键字，因此之前分区中的内容将会被覆盖掉。如果使用 INTO 关键字替换掉它，那么 Hive 将会以追加的方式写入数据而不会覆盖掉之前已经存在的内容。

接下来，再来看看分区表中怎么使用 INSERT 来插入数据。

(1) 如上面的例子一样，首先要创建一个分区表 students_02。

```
hive> CREATE TABLE students_02 (id INT , name STRING , age INT )    -- 创建分区表 students_02
    > PARTITIONED BY (gender STRING)                                 -- 以性别为分区字段
    > ROW FORMAT DELIMITED FIELDS TERMINATED BY '#';
OK
Time taken: 0.105 seconds
hive>
```

(2) 向表 students_02 中插入数据。

```
hive> INSERT INTO TABLE students_02                     -- 向分区表中插入数据，分区为 gender = M
    > PARTITION( gender = 'M' )
    > SELECT * FROM stu_table
    > WHERE id = '12002';
...
OK
Time taken: 2.398 seconds
hive> SELECT * FROM students_02;
OK
12002       David       23      M
Time taken: 0.093 seconds,Fetched: 1 row(s)
```

同非分区表一样，使用 OVERWRITE 会覆盖原来的数据，不使用 OVERWRITE 便会在表中原来数据上叠加。

这里需要注意的是，向分区表中插入数据时，要注意两个表中的字段数是一样的（分区字段除外）。例如，该例子中，表 students_02 中有三个字段：id、name、age（不包括 gender），那么表 stu_table 中也有同样的三个字段：id、name、age。

这个例子展示了这个功能非常有用的一个常见场景，即：数据已经存在于某个目录下，对于 Hive 来说其为一个外部表，而现在想将其导入到最终的分区表中。

然而，如果表中数据特别多，这样分批插入就会很麻烦。Hive 提供了另一种 INSERT 语法，可以只扫描一次输入数据，然后按多种方式进行划分。如下例子显示了如何为两个分区表一次性插入数据。

上面已经建立了一个分区字段为 gender 的分区表 students_02，接下来创建一个分区字段为 gender 的分区表 students_03。

```
hive> CREATE TABLE students_03 (id INT, name STRING, age INT)
    > PARTITIONED BY (gender STRING)
    > ROW FORMAT DELIMITED FIELDS TERMINATED BY '#';
OK
Time taken: 1.277 seconds
```

现在数据库 school 中有两个分区字段为 gender 的分区表。接下来,向两个分区表中插入数据,分别存放 gender='M'和 gender='F'的数据,数据来源于分区表 students。

```
hive> FROM students                             -- 从表 students 中获取数据
    > INSERT OVERWRITE TABLE students_02        -- 向表 students_02 中插入数据
    > PARTITION (gender = 'M')                  -- 分区为 gender = 'M'
    > SELECT id,name,age WHERE gender = 'M'     -- 从表 students 中找满足 gender = 'M'的数据
    > INSERT OVERWRITE TABLE students_03        -- 向表 students_03 中插入数据
    > PARTITION(gender = 'F')                   -- 分区为 gender = 'F'
    > SELECT id,name,age WHERE gender = 'F';    -- 从表 students 中找满足 gender = 'F'的数据
...
OK
Time taken: 11.362 seconds
hive> SELECT * FROM students_02;
OK
12005     Frank      21      M
12006     Tony       23      M
12007     Peter      21      M
Time taken: 0.438 seconds,Fetched:3 row(s)
hive> SELECT * FROM students_03;
OK
12008     Angela     21      F
12009     Alice      20      F
12010     Amy        22      F
Time taken: 0.29 seconds,Fetched:3 row(s)
```

从 students 表中读取的每条记录都会经过一条 SELECT…WHERE…句子进行判断。这些句子都是独立进行判断的,并不是 IF…THEN…ELSE…结构。

事实上,通过使用这个结构,源表中的某些数据可以被写入目标表的多个分区中或者不被写入任一个分区中。

如果某条记录满足某个 SELECT…WHERE…语句,那么这条记录就会被写入到指定的表和分区中。简单地说,只要有需要,每个 INSERT 子句都可以插入到不同的表中,而那些目标表可以是分区表也可以是非分区表。

因此,输入的某些数据可能输出到多个输出位置而其他一些数据可能就被删除了!

当然,这里可以混合使用 INSERT OVERWRITE 语句和 INSERTINTO 语句。

动态分区插入

前面所说的语法中还是有一个问题,即:如果需要创建非常多的分区,那么用户就需要写非常多的 SQL! 不过幸运的是,Hive 提供了一个动态分区功能,其可以基于查询参数推断出需要创建的分区名称。到目前为止,我们所看到的都是静态分区。

下面来看看如何动态插入数据。

新建一个分区表 students_04,其模式同表 students_02 相同,所以,这里可以使用表复制语句"CREATE TABLE 新建表名 LIKE 源表名"。

```
hive> CREATE TABLE students_04 LIKE students_02 ;
    -- 只会复制表 students_02 的表结构,不会复制其数据
OK
```

```
Time taken: 0.303 seconds
hive> INSERT OVERWRITE TABLE students_04      -- 向表 students_04 中插入数据
    > PARTITION ( gender )                    -- 分区字段为 gender
    > SELECT id,name,age,gender               -- 最后一个字段为 gender,是用来确定分区字段的
                                                 值
    > FROM students;                          -- 数据来自源表 students
FAILED: SemanticException [Error 10096]: Dynamic partition strict mode requires at least one
static partition column. To turn this off set hive.exec.dynamic.partition.mode=nonstrict
```

但是这里会报错,这是因为动态分区功能没有开启。下面开启动态分区功能,并设置允许所有分区都是动态的,之后,再做一遍上面的实验:

```
hive> set hive.exec.dynamic.partition=true;                          -- 开启动态分区功能
hive> set hive.exec.dynamic.partition.mode=nonstrict;                -- 允许所有分区都是动态的
hive> set hive.exec.max.dynamic.partitions.pernode=1000;  -- 这步可不做,主要是设置最大分
                                                                        区个数
hive> INSERT OVERWRITE TABLE students_04
    > PARTITION (gender)
    > SELECT id,name,age,gender
    > FROM students;
...
OK
Time taken: 4.735 seconds
```

接下来,查看表 students_04 中的内容:

```
hive> SHOW PARTITIONS students_04;                     -- 查看表 students_04 的分区情况
OK
gender=F
gender=M
Time taken: 0.177 seconds, Fetched: 2 row(s)
hive> SELECT * FROM students_04                        -- 查看表中 gender = 'M'分区中的内容
    > WHERE gender = 'M';
OK
12005    Frank    21    M
12006    Tony     23    M
12007    Peter    21    M
Time taken: 0.555 seconds, Fetched: 3 row(s)
hive> SELECT * FROM students_04                        -- 查看表中 gender = 'F'分区中的内容
    > WHERE gender = 'F';
OK
12008    Angela   21    F
12009    Alice    20    F
12010    Amy      22    F
Time taken: 0.277 seconds, Fetched: 3 row(s)
```

Hive 根据 SELECT 语句中最后一列来确定分区字 gender 的值。这就是为什么在表 students 中我们使用了不同的命名,就是为了强调源表字段值和输出分区值之间的关系是根据位置而不是根据命名来匹配的。假设源表中共有 100 个分区,那么执行完上面这个查询后,目标表就将会有 100 个分区!

其实，用户也可以混合使用动态和静态分区。之前创建过表 employees，其中的分区字段有 country 和 state 两个字段。所以在动态插入数据的时候，可以这样使用：

```
INSERT OVERWRITE TABLE employees
PAPTITION ( country = 'US',state)
SELECT . . . ,se.cnty,se.st
FROM staged_employees
WHERE country = 'US';
```

这里 employees 是目标表，其分区字段为 country 和 state，要从表 staged_employees 中获取数据，动态插入到表 employees 中。在 PARTITION（country＝'US',state)命令中，可以看出，分区字段 country 是静态插入方式，而分区字段 state 是动态插入方式。需要注意的是，静态分区键必须出现在动态分区键之前。

动态分区功能默认情况下没有开启。开启后，默认是以"严格"模式执行的，在这种模式下要求至少有一列分区字段是静态的。这有助于阻止因设计错误导致查询产生大量的分区。例如，用户可能错误地使用时间戳作为分区字段，然后导致每秒都对应一个分区！而用户也许是期望按照天或者按照小时进行划分的。还有一些其他相关属性值用于限制资源利用。表 5.1 描述了这些属性。

表 5.1　动态分区属性

属 性 名 称	默认值	描　　　　述
hive.exec.dynamic.partition	false	设置成 true，表示开启动态分区功能
hive.exec.dynamic.partition.mode	Strict	设置成 nonstrict，表示允许所有分区都是动态的
Hive.exec.max.dynamic.partitions.pernode	100	每个 mapper 或 reducer 可以创建的最大动态分区个数。如果某个 mapper 或者 reducer 尝试创建大于这个值的分区，则会抛出一个致命错误信息
hive.exec.max.dynamic.partitions	+1000	一个动态分区创建语句可以创建的最大动态分区个数。如果超过这个值，则会抛出一个致命错误信息
hive.exec.max.created.files	100 000	全局可以创建的最大文件个数。有一个 Hadoop 计数器会跟踪记录创建了多少个文件，如果超过这个值，则会抛出一个致命错误信息

5.1.3　单个查询语句中创建表并加载数据

用户同样可以在一个语句中完成创建并将查询结果载入这个表的操作，下面的命令实现了创建表 students_05，与此同时，从表 students 中选取满足条件 gender＝'M'的数据，插入到表 students_05 中。

```
hive > CREATE TABLE students_05
    > AS SELECT id ,name ,age
    > FROM students
    > WHERE gender = 'M' ;
...
OK
Time taken: 2.59 seconds
```

Hadoop 核心技术与实战

```
hive> SELECT * FROM students_05 ;
OK
12005    Frank    21
12006    Tony     23
12007    Peter    21
Time taken: 0.169 seconds, Fetched: 3 row(s)
```

这张表只含有 students 表中来自 gender='M' 的三条信息。新表的模式是根据 SELECT 语句生成的。使用这个功能的常见情况是从一个大的宽表中选取部分需要的数据集。

这个功能不能用于外部表，可以回想一下使用 ALTER TABLE 语句可以为外部表"引用"到一个分区，这里本身就没有进行数据"装载"，而是在元数据中指定一个指向数据的路径。

5.1.4 导出数据

我们如何从表中导出数据呢？如果数据文件恰好是用户需要的格式，那么只需要简单地复制文件夹或者文件就可以了：使用 hadoop dfs-cp source_path target_path 可以实现。

下面使用该命令实现数据的导出：

在 HDFS 上建立一个新的文件夹 target_directory，并将路径/usr/hive/warehouse/school.db/stu_table/stu_table.txt 文件导入到 target_directory，实验结果如图 5.2 所示。

在 shell 命令行输入下面的命令：

```
# hadoop dfs -mkdir /target_directory/
# hadoop dfs -cp /usr/hive/warehouse/school.db/stu_table/stu_table.txt /target_directory/
# hadoop dfs -ls /target_directory/
```

```
[root@VM-75bb3788-78e8-401f-8006-305dce57df42 hive]# hadoop dfs -ls /target_directory/
DEPRECATED: Use of this script to execute hdfs command is deprecated.
Instead use the hdfs command for it.

16/08/02 17:57:35 WARN util.NativeCodeLoader: Unable to load native-hadoop library for your platform... using builtin-java classes where applicable
Found 1 items
-rw-r--r--   1 root supergroup         57 2016-08-02 17:57 /target_directory/stu_table.txt
```

图 5.2 命令运行

另外，还可以使用 INSERT…DIRECTORY…命令来实现数据的导出。如下面的例子所示：首先在本地/usr/tmp/下新建一个文件夹 target_directory，然后在 Hive 的 CLI 命令行下输入下列命令导出数据到本地文件夹/usr/tmp/target_directory 下，实验结果如图 5.3 所示。

在 shell 命令行输入下面的命令新建文件夹 target_directory：

```
# cd /usr/tmp
# mkdir target_directory
```

在 CLI 命令行中将数据导出到本地文件夹 target_directory：

```
#hive
hive> USE school;
OK
Time taken:1.198 seconds
hive> INSERT OVERWRITE LOCAL DIRECTORY '/usr/tmp/target_directory '
    > SELECT id,name,age
    > FROM students WHERE gender = 'M';
...
OK
Time taken:6.317 seconds
```

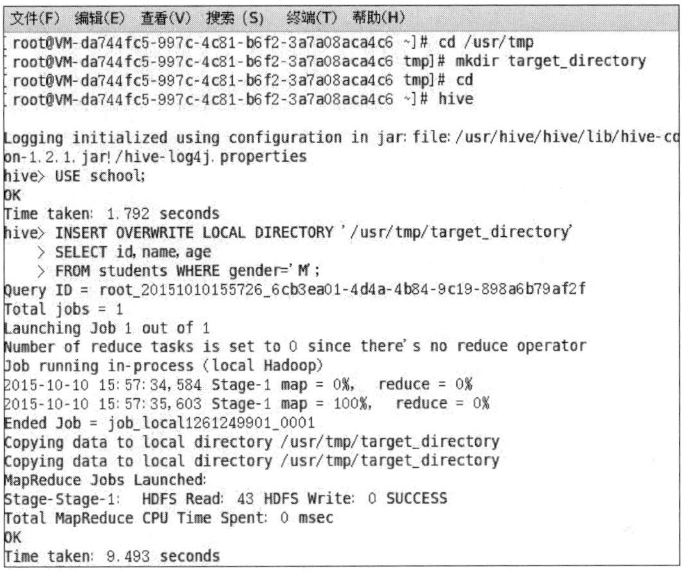

```
文件(F)  编辑(E)  查看(V)  搜索(S)  终端(T)  帮助(H)
[root@VM-da744fc5-997c-4c81-b6f2-3a7a08aca4c6 ~]# cd /usr/tmp
[root@VM-da744fc5-997c-4c81-b6f2-3a7a08aca4c6 tmp]# mkdir target_directory
[root@VM-da744fc5-997c-4c81-b6f2-3a7a08aca4c6 tmp]# cd
[root@VM-da744fc5-997c-4c81-b6f2-3a7a08aca4c6 ~]# hive

Logging initialized using configuration in jar:file:/usr/hive/hive/lib/hive-co
on-1.2.1.jar!/hive-log4j.properties
hive> USE school;
OK
Time taken: 1.792 seconds
hive> INSERT OVERWRITE LOCAL DIRECTORY '/usr/tmp/target_directory'
    > SELECT id,name,age
    > FROM students WHERE gender='M';
Query ID = root_20151010155726_6cb3ea01-4d4a-4b84-9c19-898a6b79af2f
Total jobs = 1
Launching Job 1 out of 1
Number of reduce tasks is set to 0 since there's no reduce operator
Job running in-process (local Hadoop)
2015-10-10 15:57:34,584 Stage-1 map = 0%,  reduce = 0%
2015-10-10 15:57:35,603 Stage-1 map = 100%,  reduce = 0%
Ended Job = job_local1261249901_0001
Copying data to local directory /usr/tmp/target_directory
Copying data to local directory /usr/tmp/target_directory
MapReduce Jobs Launched:
Stage-Stage-1:  HDFS Read: 43 HDFS Write: 0 SUCCESS
Total MapReduce CPU Time Spent: 0 msec
OK
Time taken: 9.493 seconds
```

图 5.3　数据导出

关键字 OVERWRITE 和 LOCAL 与前面的说明是一致的，路径格式也遵循通常的规则。一个或者多个文件将会被写入到/usr/tmp/target_directory，具体个数取决于调用的 reducer 个数。

不管在源表中数据实际是怎么存储的，Hive 都会将所有的字段序列化成字符串写入到文件中。Hive 使用和 Hive 内部存储的表相同的编码方式来生成输出文件。打开/usr/tmp/target_directory 目录下的 000000_0 文件，使用命令：vi /usr/tmp/target_directory/000000_0 查看其中的内容，结果如图 5.4 所示。

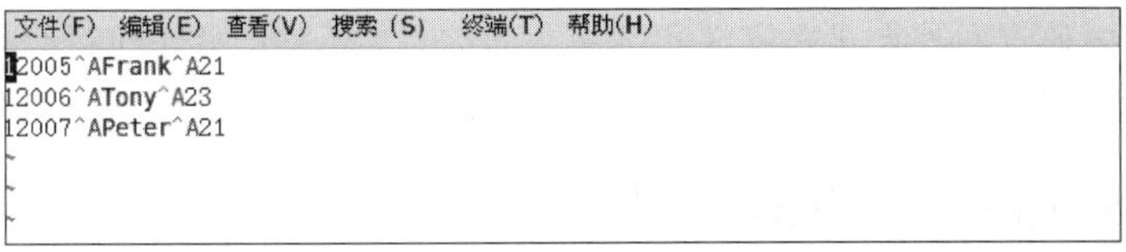

```
文件(F)  编辑(E)  查看(V)  搜索(S)  终端(T)  帮助(H)
12005^AFrank^A21
12006^ATony^A23
12007^APeter^A21
~
~
~
```

图 5.4　查看文件内容

可以看到，其分隔符同之前设置的"♯"已经不同了。所以，表的字段分隔符可能是需要考量的。例如，如果其使用的是默认的^A分隔符，而用户又经常导出数据的话，那么可能使用逗号或者制表键作为分隔符会更合适。另一种变通的方式是定义一个"临时"表，这个表的存储方式配置成期望的输出格式（例如，使用制表键作为字段分隔符）。然后再从这个临时表中查询数据，并使用INSERT OVERWTITE DIRECTORY将查询结果写入到这个表中。与很多关系型数据库不同的是，Hive中没有临时表的概念。需要手动删除任何已创建但又不想长期保留的表。

与向表中插入数据一样，也是可以通过如下方式指定多个输出文件夹目录：

```
♯ cd
♯ hive
hive> USE school;
OK
Time taken: 1.174 seconds
hive> FROM students
> INSERT OVERWRITE DIRECTORY '/target_directory/M'
> SELECT * WHERE gender = 'M'
> INSERT OVERWRITE DIRECTORY '/target_directory/F'
> SELECT * WHERE gender = 'F';
...
OK
Time taken: 11.094 seconds
```

也可以在Hive CLI中查看结果文件内容：

```
hive> dfs -ls /target_directory;
Found s items
drwxr-xr-x   - root supergroup          0 2015-08-24 13:36 /target_directory/F
drwxr-xr-x   - root supergroup          0 2015-08-24 13:36 /target_directory/M
-rw-r--r--   1 root supergroup         56 2015-08-24 10:41 /target_directory/stu_table.txt
hive> dfs -ls /target_directory/F;
Found 1 items
-rw-r--r--   1 root supergroup         50 2015-08-24 13:36 /target_directory/F/000000_0
hive> dfs -cat /target_directory/F/000000_0;
12008     Angela     21     F
12009     Alice      20     F
12010     Amy        22     F
```

5.2 数据查询

在了解了可以通过多种方式来定义和格式化表之后，再来介绍如何运行查询。其实在之前的学习中，为了说明一些概念，已经使用了一些查询语句。现在进一步学习一些细节部分。

5.2.1 SELECT…FROM 语句

SELECT是SQL中的射影算子。FROM子句标识了从哪个表、视图或嵌套查询中选择记录。

对于一个给定的记录，SELECT 指定了要保存的列以及输出函数需要调用的一个或多个列（例如 count(*)这样的聚合函数）。这里回想一下之前说明过的分区表 employees。

```
hive> CREATE TABLE employees (              -- 创建表 employees
    > name STRING,
    > salary INT,
    > subordinates ARRAY<STRING>,
    > deductions MAP<STRING, FLOAT>,
    > address STRUCT<street:STRING,city:STRING,state:STRING,zip:INT>
    > )
    > PARTITIONED BY (country STRING,state STRING)    -- 分区字段为 country 和 state
    > ROW FORMAT DELIMITED FIELDS TERMINATED BY '♯'
    > COLLECTION ITEMS TERMINATED BY '*'
    > MAP KEYS TERMINATED BY ':'
    > LINES TERMINATED BY '\n';
```

表 employees 中分别有六名员工信息，分别是：

```
(country = 'US', state = 'HI')的 John Doe、Boss Man、Fred Finance
(country = 'US', state = 'FL')的 Mary Smith
(country = 'CA', state = 'YT')的 Todd Jones
(country = 'CA', state = 'MB')的 Bill King
```

请运行 Hadoop 和 Hive 后开始实验，下面是对这个表进行查询的语句及其输出内容：

```
hive> USE mydatabase;                   -- 查询表 employees 的时候，必须到存放该表的目录下
                                           进行查询
OK
Time taken: 0.079 seconds
hive> SELECT name,salary FROM employees;    -- 查询表中的 name、salary 两列内容
OK
Bill    King     60000
Todd    Jones    70000
Mary    Smith    80000
John    Doe      100000
Boss    Man      200000
Fred    Finance  150000
Time taken:4.783 seconds, Fetched:6 row(s)
```

当用户选择的列是集合数据类型时，Hive 会使用 JSON（Java 脚本对象表示法）语法应用于输出。首先，选择 subordinates 列，该列为一个数组，其值使用一个被括在［…］内的以逗号分隔的列表进行表示。注意，集合的字符串元素是加上引号的，而基本数据类型 STRING 的列值是不加引号的。

```
hive> USE mydatabase;                   -- 若当前就是在数据库 mydatabase 下,则不用
                                           执行该命令,接下来的查询便不再叙述
OK
Time taken: 0.079 seconds
hive> SELECT name,subordinates FROM employees;  -- 查询表中的 name、subordinates 两列内容
OK
```

```
Bill King          []
Todd Jones         []
Mary Smith         ["Bill King"]
John Doe           ["Mary Smith","Todd Jones"]
Boss Man           ["Fred Finance"]
Fred Finance       ["Stacy Accountant"]
Time taken:0.282 seconds, Fetched:6 row(s)
```

deductions 列是一个 Map，其使用 JSON 格式来表达 Map，即使用一个被括在{…}内的以逗号分隔的键值对列表进行表示：

```
hive> SELECT name,deductions FROM employees;      -- 查询表中的 name、deductions 两列内容
OK
Bill King       {"Federal Taxes":0.3,"State Taxes":0.03,"Insurance":0.1}
Todd Jones      {"Federal Taxes":0.15,"State Taxes":0.03,"Insurance":0.1}
Mary Smith      {"Federal Taxes":0.2,"State Taxes":0.05,"Insurance":0.1}
John Doe        {"Federal Taxes":0.2,"State Taxes":0.05,"Insurance":0.1}
Boss Man        {"Federal Taxes":0.2,"State Taxes":0.05,"Insurance":0.1}
Fred Finance    {"Federal Taxes":0.2,"State Taxes":0.05,"Insurance":0.1}
Time taken:0.208 seconds,Fetched: 6 row(s)
```

最后，address 列是一个 STRUCT，其也是使用 JSON Map 格式进行表示的：

```
hive> SELECT name, address FROM employees;       -- 查询表中的 name、address 两列内容
OK
Bill King       {"street":"300 Obscure Dr.","city":"Winninpeg","state":"MB","zip":60100}
Todd Jones      {"street":"200 Chicago AVE.","city":"Whitehouse","state":"YT","zip":60700}
Mary Smith      {"street":"100 Ontario St.","city":"Miami","state":"FI","zip":60601}
John Doe        {"street":"1 Michigan Ave.","city":"Hilo","state":"HI","zip":60500}
Boss Man        {"street":"1 Pretentious Drive.","city":"Maui","state":"HI","zip":60500}
Fred Finance    {"street":"2 Pretentious Drive","city":"Kona","state":"HI","zip":60500}
Time taken:0.178 seconds, Fetched:6 row(s)
```

接下来看看如何引用集合数据类型中的元素。

首先，数组索引是基于 0 的，这个与在 Java 中是一样的。这里是一个选择 subordinates 数组中的第一个元素的查询：

```
hive> SELECT name, subordinates[0] FROM employees;
-- 查询表中的 name、subordinates 中的第一项内容两列内容
OK
Bill    King       NULL
Todd    Jones      NULL
Mary    Smith      Bill King
John    Doe        Mary Smith
Boss    Man        Fred Finance
Fred    Finance    Stacy Accountant
Time taken:0.228 seconds, Fetched:6 row(s)
```

注意，引用一个不存在的元素将会返回 NULL。同时，提取出的 STRING 数据类型的值将不再加引号！

为了引用一个 Map 元素,用户还可以使用 ARRAY [...] 语法,但是使用的是键值而不是整数索引:

```
hive> SELECT name,deductions["State Taxes"] FROM employees;  -- 查询 name、State Taxes 一项内容
OK
Bill    King       0.03
Todd    Jones      0.03
Mary    Smith      0.05
John    Doe        0.05
Boss    Man        0.05
Fred    Finance    0.05
Time    taken:     0.144 seconds, Fetched: 6 row(s)
```

最后,为了引用 STRUCT 中的一个元素,用户可以使用"点"符号,类似于前面提到的"表的别名,列名"这样的用法:

```
hive> SELECT name,address.city FROM employees;     -- 查询 name、和 address 里面的 city 的内容
OK
Bill    King       Winninpeg
Todd    Jones      Whitehouse
Mary    Smith      Miami
John    Doe        Hilo
Boss    Man        Maui
Fred    Finance    Kona
Time    taken:     0.262 seconds,Fetched:6 row(s)
```

1. 使用列值进行计算

用户不但可以选择表中的列,还可以使用函数调用和算术表达式来操作列值。例如,可以查询得到转换为大写的雇员姓名、雇员对应的薪水、需要缴纳的税收比例以及扣除税收后再进行取整所得的税后薪资。甚至可以通过调用内置函数 map_values 提取出 deductions 字段 Map 类型值的所有元素,然后使用内置的 sum 函数对 Map 中所有元素进行求和运算。

下面这个例子查询使用了算术运算。

因为该查询语句太长,所以将它分成两行显示。注意第二行 Hive 所使用的提示符,那是一个缩进了的大于符号(>)。

```
hive> SELECT upper(name),salary,deductions["Federal Taxes"],
    > round (salary * (1 - deductions["Federal Taxes"])) FROM employees;
OK
BILL    KING       60000     0.3    42000.0
TODD    JONES      70000     0.15   59500.0
MARY    SMITH      80000     0.2    64000.0
JOHN    DOE        100000    0.2    80000.0
BOSS    MAN        200000    0.2    160000.0
FRED    FINANCE    150000    0.2    120000.0
Time taken:0.579 seconds,Fetched: 6 row(s)
```

2. 算术运算符

Hive 中支持所有的典型的算术运算符,表 5.2 描述了具体细节。

表 5.2 算术运算符

运 算 符	类 型	描 述
A+B	数值	A 和 B 相加
A-B	数值	A 减去 B
A*B	数值	A 和 B 相乘
A/B	数值	A 除以 B。如果能够整除,那么返回商数
A%B	数值	A 除以 B 的余数
A&B	数值	A 和 B 按位取与
A|B	数值	A 和 B 按位取或
A^B	数值	A 和 B 按位取异或
~A	数值	A 按位取反

算术运算符可以是任意的数据类型。不过,如果数据类型不同,那么两种类型中值范围较小的那个数据类型将转换为其他范围更广的数据类型(范围更广在某种意义上就是指一个类型具有更多的字节从而可以容纳更大范围的值)。例如,对于 INT 和 BIGINT 运算,INT 会将类型转换提升为 BIGINT。对于 INT 和 FLOAT 运算,INT 将提升为 FLOAT。可以注意到我们的查询语句中包含(1-deductions[…])这个运算。因为字段 deductions 是 FLOAT 类型的,因此数字 1 会提升为 FLOAT 类型。

当进行算术运算时,需要注意数据溢出或数据下溢问题。Hive 遵循的是底层 Java 中数据类型的规则,因此当溢出或下溢发生时计算结果不会自动转换为更广泛的数据类型。乘法和除法最有可能会引发这个问题。如果担心产生溢出和下溢,那么可以考虑在表模式中定义使用范围更广的数据类型。不过这样做的缺点是每个数据值会占用更多额外的内存。而且,也可以使用特定的表达式将值转换为范围更广的数据类型。

需要注意所使用的数值数据的范围,并确认实际数据是否接近表模式中定义的数据类型所规定的数值范围上限或者下限,还需要确认人们可能对这些数据进行什么类型的计算。

有时使用函数将数据值按比例从一个范围缩放到另一个范围也是很有用的,例如按照 10 次方幂进行除法运算或取 log 值(指数值),等等。这种数据缩放也适用于某些机器学习计算中,用以提高算法的准确性和数值稳定性。

3. 使用函数

在前面的示例中还使用到了一个内置数学函数 round(),这个函数会返回一个 DOUBLE 类型的最近整数。

1)数学函数

表 5.3 中描述了 Hive 内置数学函数。

表 5.3 数学函数

返回值类型	样 式	描 述
BIGINT	round(DOUBLE d)	返回 BOUBLE 类型的 BIGINT 类型的近似值
DOUBLE	round(DOUBLE d,INT n)	返回 BOUBLE 类型 d 的保留 n 位小数的 DOUBLE 类型近似值
BIGINT	floor(DOUBLE d)	d 是 DOUBLE 类型的,返回<=d 的最大 BIGINT 类型值

续表

返回值类型	样　式	描　述
BIGINT	ceil(DOUBLE d) ceiling(DOUBLE d)	d是DOUBLE类型的,返回>=d的最小BIGINT类型值
DOUBLE	rand() rand(INT seed)	每行返回一个DOUBLE类型随机数,整数seed是随机因子
DOUBLE	exp(DOUBLE d)	返回e的d幂次方,返回的是个DOUBLE类型值
DOUBLE	ln(DOUBLE d)	以自然数为底d的对数,返回DOUBLE类型值
DOUBLE	log10(DOUBLE d)	以10为底d的对数,返回DOUBLE类型值
DOUBLE	log2(DOUBLE d)	以2为底d的对数,返回DOUBLE类型值
DOUBLE	log(DOUBLE base,DOUBLE d)	以base为底d的对数,返回DOUBLE类型值,其中base和d都是DOUBLE类型的
DOUBLE	pow(DOUBLE d,DOUBLE p) power(DOUBLE d,DOUBLE p)	计算d的p次幂,返回DOUBLE类型值,其中d和p都是DOUBLE类型的
DOUBLE	sqrt(DOUBLE d)	计算d的平方根,其中d是DOUBLE类型的
STRING	bin(DOUBLE i)	计算二进制i的STRING类型值,其中i是BIGINT类型的
STRING	hex(BIGINT i)	计算十六进制值i的STRING类型值,其中i是BIGINT类型的
STRING	hex(STRING str)	计算十六进制表达的值b的STRING类型值
STRING	hex(BINARY b)	计算二进制表达的值b的STRING类型值
STRING	unhex(STRING i)	hex(DTRING str)的逆方法
STRING	conv(BIGINT num,INT from_base,INT to_base)	将BIGINT类型的num从from_base进制转换成to_base进制,并返回STRING类型的结果
DOUBLE	conv(STRING num,INT from_base,INT to_base)	将STRING类型的num从from_basse进制转换成to_base进制,并返回STRING类型的结果
DOUBLE	abs(DOUBLE d)	计算DOUBLE类型值d的绝对值,返回结果也是DOUBLE类型的
INT	pmod(INT i1,INT i2)	INT值i1对INT值i2取模,结果也是INT类型的
DOUBLE	pmod(DOUBLE i1, DOUBLE i2)	DOUBLE值i1对DOUBLE值i2取模,结果也是DOUBLE类型的
DOUBLE	sin(DOUBLE d)	在弧度度量中,返回DOUBLE类型值d的正弦值,结果也是DOUBLE类型的
DOUBLE	asin(DOUBLE d)	在弧度度量中,返回DOUBLE类型值d的反正弦值,结果也是DOUBLE类型的
DOUBLE	cos(DOUBLE d)	在弧度度量中,返回DOUBLE类型值d的余弦值,结果也是DOUBLE类型的
DOUBLE	acos(DOUBLE d)	在弧度度量中,返回DOUBLE类型值d的反余弦值,结果也是DOUBLE类型的
DOUBLE	tan(DOUBLE d)	在弧度度量中,返回DOUBLE类型值d的正切值,结果是DOUBLE类型的
DOUBLE	atan(DOUBLE d)	在弧度度量中,返回DOUBLE类型值d的反正切值,结果是DOUBLE类型的
DOUBLE	degrees(DOUBLE d)	将DOUBLE类型弧度值d转换成角度值,结果是DOUBLE类型的

续表

返回值类型	样 式	描 述
DOUBLE	radians(DOUBLE d)	将 DOUBLE 类型角度值 d 转换成弧度值,结果是 DOUBLE 类型的
INT	positive(INT i)	返回 INT 类型值 i(其等效表达式是\+i)
DOUBLE	positive(DOUBLE d)	返回 DOUBLE 类型值 d(其等效的有效表达式是\+d)
INT	negative(INT i)	返回 INT 类型值 i 的负数(其等效表达式是\-i)
DOUBLE	negative(DOUBLE d)	返回 DOUBLE 类型值 d 的负数(等效的有效表达式是\-d)
FLOAT	sign(DOUBLE d)	如果 DOUBLE 类型值 d 是正数,则返回 FLOAT 类型值 1.0;如果 d 是负数,则返回-1.0;否则返回 0.0
DOUBLE	e()	数学常数 e,也就是超越数的 DOUBLE 类型值
DOUBLE	pi()	数学常数 pi,也就是圆周率的 DOUBLE 类型值

需要注意的是,函数 floor、round 和 ceil 输入的是 DOUBLE 类型的值,而返回值是 BIGINT 类型的,也就是将浮点型数转换成整型了。在进行数据类型转换时,这些函数是首选的处理方式,而不是使用前面提到过的 cast 类型转换操作符。

同样,也存在基于不同的底(例如十六进制)将整数转换为字符串的函数。

2) 聚合函数

聚合函数是一类比较特殊的函数,其可以对多行进行一些计算,然后得到一个结果值。更确切地说,这是用户自定义的聚合函数。这类函数中最有名的两个例子就是 count 和 avg,函数 count 用于计算有多少行数据(或者某列有多少值),而函数 avg 可以返回指定列的平均值。

这里是一个查询示例表 employees 中有多少雇员,以及计算这些雇员平均薪水的 HiveQL 语句:

```
hive> SELECT count(*),avg(salary)FROM employees;
…
OK
6    110000.0
Time taken: 29.991 seconds, Fetched: 1 row(s)
hive>
```

下面接着介绍 Hive 的内置函数。

表 5.4 列举了 Hive 的内置聚合函数。

表 5.4 聚合函数

返回值类型	样 式	描 述
BIGINT	count(*)	计算总行数,包括含有 NULL 值的行
BIGINT	count(expr)	计算提供的 expr 表达式的值非 NULL 的行数
BIGINT	count(DISTINCT expr[,expr_.])	计算指定行的值的和

续表

返回值类型	样式	描述
DOUBLE	sum(col)	计算指定行的值的平均值
DOUBLE	sum(DISTINCT col)	计算排重后的值的和
DOUBLE	avg(col)	计算指定行的平均值
DOUBLE	avg(DISTINCT col)	计算排重后的值的平均数
DOUBLE	min(col)	计算指定行的最小值
DOUBLE	max(col)	计算指定行的最大值
DOUBLE	variance(col),var_pop(col)	返回集合 col 中的一组数值的方差
DOUBLE	var_samp(col)	返回集合 col 中的一组数值的样本方差
DOUBLE	stddev_pop(col)	返回一组数值的标准偏差
DOUBLE	stddev_samp(col)	返回一组数值的标准样本偏差
DOUBLE	covar_pop(col1,col2)	返回一组数值的协方差
DOUBLE	covar_samp(col1,col2)	返回一组数值的样本协方差
DOUBLE	corr(col1,co12)	返回两组数值的相关系数
DOUBLE	percentile(BIGINT int_expr,p)	int_expr 在 p(范围是:[0,1])处的对应的百分比,其中 p 是一个 DOUBLE 类型的数组
DOUBLE	percentile(BIGINT int_expr,ARRAY(p1[,p2]…))	int_expr 在 p(范围是:[0,1])处的对应的百分比,其中 p 是一个 DOUBLE 类型的数组
DOUBLE	percentile_approx(DOUBLE col,p[,NB])	col 在 p(范围是:[0,1])处的对应的百分比,其中 p 是一个 DOUBLE 类型的数值,NB 是用于估计的直方图的仓库数量(默认是 10 000)
ARRAY < DOUBLE >	percentile_approx(DOUBLE col,ARRAY(p1[,p2]…)[,NB])	col 在 p(范围是:[0,1])处的对应的百分比,其中 p 是一个 DOUBLE 类型的数值,NB 是用于估计的直方图的仓库数量(默认是 10 000)
ARRAY < STRUCT {'x','y'}>	histogram_numeric(col,NB)	返回 NB 数量的直方图仓库数组,返回结果中的值 x 是中心,值 y 是仓库的高
ARRAY	collect_set(col)	返回集合 col 元素排重后的数组

通常,可以通过设置属性 hive.map.aggr 值为 true 来提高聚合的性能,如下所示:

```
hive> SET hive.map.aggr = true;
hive> SELECT count( * ),avg(salary)FROM employees;
…
OK
6    110000.0
Time taken:12.124 seconds,Fetched:1 row(s)
hive>
```

正如这个例子所展示的,这个设置会触发在 Map 阶段进行的"顶级"聚合过程(非顶级的聚合过程将会在执行一个 GROUP BY 后进行）。不过，这个设置将需要更多的内存。

如表 5.4 所示，多个函数都可以接受 DISTINCT…表达式。例如，可以通过这种方式计算排重后的孤僻交易码个数：

```
hive> SELECT count(country)FROM employees;        -- 不使用 DISTINCT,查询结果是 6
...
OK
6
Time taken: 5.241 seconds, Fetched: 1 row(s)
hive> SELECT count(DISTINCT country) FROM employees;   -- 使用 DISTINCT,查询结果是 2
...
OK
2
Time taken: 8.704 seconds, Fetched: 1 row(s)
hive>
```

注意，目前不允许在一个查询语句中使用多于一个的函数(DISTINCT...)表达式。例如下面这个查询语句按理说是不允许的，但是实际上是可以执行的：

```
hive> SELECT count(DISTINCT country),count(DISTINCT state)FROM employees;
-- 在一个函数中使用两个 DISTINCT
...
OK
2   4
Time taken: 4.77 seconds, Fetched: 1 row(s)
hive>
```

因此，从查询结果中可以看到有 2 个国家名、4 个州名。

3）表生成函数

与聚合函数"相反的"一类函数就是所谓的表生成函数，其可以将单列扩展成多列或者多行。这里简要地讨论一下，然后列举出 Hive 目前所提供的一些内置表生成函数。

下面通过一个例子来进行讲解。如下的这个查询语句将 employees 表中每行记录中的 subordinates 字段内容转换成 0 个或者多个新的记录行。如果某行雇员记录 suborinates 字段内容为空，那么将不会产生新的记录；如果不为空，那么这个数组的每个元素都将产生一行新记录：

```
hive> SELECT explode(subordinates) AS sub FROM employees;
OK
Bill King
Mary Smith
Todd Jones
Fred Finance
Stacy Accountant
Time taken: 0.268 seconds, Fetched: 5 row(s)
```

在上面的查询语句中，我们使用了 AS sub 子句定义了列别名 sub。当使用表生成函数时，Hive 要求使用列别名。

表 5.5 列举了 Hive 内置的表生成函数。

表 5.5 表生成函数

返回值类型	样 式	描 述
N 行结果	explode(ARRAY array)	返回 0 到多行结果,每行都对应输入的 array 数组中的一个元素
N 行结果	explode(MAP map)	返回 0 到多行结果,每行都对应每个 map 键值对,其中一个字段是 map 的键,另一个字段对应 map 的值
数组的类型	explode(ARRAY<TYPE> a)	对于 a 中的每个元素,explode()会生成一行记录包含这个元素
将结果插入表中	inline(ARRAY<STRUCT,STRUCT]>)	将结构体数组提取出来并插入到表中
TUPLE	json_tuple(STRING jsonStr,p1,p2,…,pn)	本函数可以接收多个标签名称,对输入的 JSON 字符串进行处理,与 get_json_object 与 UDF 类似,不过更高效,其通过一次调用就可以获得多个键值
TUPLE	parse_url_tuple(url,partname1,partname2,…,partnameN)其中 N>=1	从 URL 中解析出 N 个部分信息,其输入参数是:URL,以及多个要抽取的部分的名称,所有输入的参数的类型都是 STRING。部分名称是大小写敏感的,而且不应该包含空格:HOST、PATH、QUERY、RFF、PROTOCOL、AUTHORITY、FILE、USERINFO、QUERY:<KEY_NAME>
N 行结果	stack(INT n,col1,…,colM)	把 M 列转换成 N 行,每行有 M/N 个字段。其中 n 必须是个常数

4) 其他内置函数

表 5.6 描述了 Hive 中其余的内置函数,这些函数用于处理字符串、Map、数组、JSON 和时间戳,包含最近引入的 TIMESTAMP 数据类型。

表 5.6 其他内置函数

返回值类型	样 式	描 述
STRING	ascii(STRING s)	返回字符串 s 中首个 ASCII 字符
STRING	base64(BINARY bin)	将二进制值 bin 转换成基于 64 位的字符串
BINARY	binary(STRING s)binary(BINARY b)	将输入的值转换成二进制值
返回类型就是 type 定义的类型	cast(<expr> as <type>)	将 expr 转换成 type 类型的,例如 cast('1' as BIGINT)将会将字符串'1'转换成 BIGINT 数值类型,如果转换过程失败,则返回 NULL
STRING	concat(BINARY s1,BINARY s2,…)	将二进制字节码按次序拼接成一个字符串
STRING	concat(STRING s1,STRING s2,…)	将字符串 s1、s2 等拼接成一个字符串,例如 concat('ab','cd')的结果就是'abcd'

续表

返回值类型	样　式	描　述
STRING	concat_ws(STRING separator, STRING s1,STRING s2,…)	和concat类似,不过是使用指定的分隔符进行拼接的
STRING	concat_ws(BINARY separator, BINARY s1,BINARY s2,…)	和concat类似,不过是使用指定的分隔符进行拼接的
ARRAY<STRUCT<STRING,DOUBLE>>	context_ngrams(array<array<string>>,array<string>,int K,int pf)	和ngrams类似,但是从每个外层数组的第二个单词数组来查找前K个字符
STRING	decode(BINARY bin,STRING charset)	使用指定的字符集charset将二进制值bin解码成字符串(支持的字符集包括'US_ACII'、'ISO-8859-1'、'UTF-8'、'UTF-16BE'、'UTF-16LE'、'UTF-816')。如果任一输入参数为NULL,则结果为NULL
BINARY	encode(STRING src,STRING charset)	使用指定的字符集charset将字符串src编码成二进制值(支持的字符集包括'US_ASCII'、'ISO-8859-1'、'UTF-8'、'UTF-16BE'、'UTF-16LE'、'UTF-16')。如果任一输入参数为NULL,则结果为NULL
INT	find_in_set(STRING s,STRING commaSeparatedString)	返回在以逗号分隔的字符串中s出现的位置,如果没有找到,则返回NULL
STRING	format_number(NUMBER x, INT d)	将数值x转换成"#,###,###.##"格式的字符串,并保留d位小数。如果d为0,那么输出值就没有小数点后面的值
STRING	get_json_object(STRING json_string,STRING path)	从给定路径上的JSON字符串中抽取JSON对象,并返回这个对象的JSON字符串形式。如果输入的JSON字符串是非法的,则返回NULL
BOOLEAN	in	例如,test in(vall,vall2,…),表示如果test值等于后面列表示的任一值,则返回true
BOOLEAN	in_file(STRING s,STRING filename)	如果文件名为filename的文件中有完整的一行数据和字符串s完全匹配,则返回true
INT	instr(STRINGstr,STRING ss)	查找字符串str中字符串ss第一次出现的位置
INT	length(STRING s)	计算字符串s的长度
INT	locate(STRING substr,STRING str [,INT pos])	查找字符串str中的pos位置后字符串substr第一次出现的位置
STRING	lower(STRING s)	将字符串中所有的字母转换成小写字母。例如,lower('hIvE')的结果是'hive'
STRING	lcase(STRING s)	和lower()一样
STRING	lpad(STRING s,INT len,STRING pad)	从左边开始对字符串s使用字符串pad进行填充,最终达到len长度为止,如果字符串s本身长度比len大,那么多余的部分会被去除

续表

返回值类型	样 式	描 述
STRING	ltrim(STRING s)	将字符串 s 前面出现的空格全部去除
ARRAY < STRUCT < STRING,DOUBLE >>	ngrams（ARRAY < string >>, INT N,INT K ,INT pf)	估算文件中前 K 个字符,pf 是精度系数
STRING	parse_url(STRING url,STRING partname［,STRING key])	从 URL 中抽取指定部分的内容,参数 url 表示一个 URL 字符串,参数 part 那么表示要抽取的部分名称,它是大小写敏感的,可选的值有 HOST、PATH、QUERY、REF、PROTOCOL、AUTHORITY、FILE、USERINFO、QUERY：< key >。如果 partname 是 QUERY,那么还需要指定第三个参数 key
STRING	printf（STRING format,Obj…args)	按照 printf 风格格式化输出输入的字符串
STRING	regexp_extract（STRINGsubject,STRING regex_pattern,STRING index)	抽取字符串 subject 中符合正则表达式 regex_pattern 的第 index 个部分的子字符串
STRING	regexp_replace(STRING s,STRING regex,STRING replacement)	按照 Java 正则表达式 regex 将字符串 s 中的符合条件的部分替换成 replacement 所指定的字符串 a。如果 replacement 部分为空,那么符合正则表达式的部分就会被除掉。regexp_replace('hive',' [ie]','z')的结果是'hzvz'
STRING	repeat(STRING s,INT n)	重复输出 n 次字符串
STRING	reverse(STRING s)	反转字符串
STRING	rpad（STRING s, INT len, STRING pad)	从右边开始对字符串 s 使用字符串 pad 进行填充,最终达到 len 长度为止。如果字符串 s 本身长度比 len 大,那么多余的部分会被删除
STRING	rtrim(STRING s)	将字符串 s 后面出现的空格全部去掉。例如,rtrim('hive')的结果是'hive'
ARRAY < ARRAY < STRING >>	sentences（STRING s,STRING lang,STRING locale)	将输入字符串 s 转换成句子数组,每个句子又由一个单词数构成。参数 lang 和 local 是可选的,如果没有,则使用默认本地化信息
INT	size(MAP< K, V>)	返回 MAP 中元素的个数
INT	size(ARRAY< T>)	返回数组 ARRAY 的元素个数
STRING	space(INT n)	返回 n 个空格
ARRAY< STRING >	split(STRING s,STRING pattern)	按照正则表达式 pattern 分隔字符串 s,并将分隔后的部分以字符串数组的方式返回

续表

返回值类型	样　式	描　述
MAP < STRING, STRING >	str_to_map(STRING s, STRING deliml, STRING delim2)	将字符串 s 按照指定分隔符转换成 Map,第一个参数是输入的字符串,第二个参数是键值对之间的分隔符,第三个分隔符是键和值之间的分隔符
STRING	substr(STRING s, STRING start_index) substring(STRING s, STRING start_index)	对于字符串 s,从 start 位置开始截取 length 长度的字符串,作为子字符串。例如 substr('abcdefg',3,2)的结果是'cd'
STRING	substr(BINARY s, STRING start_index) substring(BINARY s, STRING start_index)	对于二进制字节值 s,从 start 位置开始截取 length 长度的字符串,作为子字符串
STRING	translate(STRING input, STRING from,STRING to	将 input 中出现在 from 中的字符串替换为 to 中的字符串
STRING	trim(STRING A)	将字符串 A 前后出现的空格全部去掉,例如 trim('hive')的结果是 hive
BINAARY	unbase64(STRING str)	将基于 64 位的字符串 str 转换成二进制值
STRING	upper(STRING A) ucase(STRING A)	将字符串中所有的字母转换成大写字母。例如,upper('hIvE')的结果是'HIVE'
STRING	from_unixtime(BIGINT unixtime [,STRING format])	将时间戳秒数转换成 UTC 时间,并用字符串表示,可以通过 format 规定的时间格式,指定输出的时间格式
BIGINT	unix_timestamp()	获取当前本地时区下的当前时间戳
BIGINT	unix_timestamp(STRING data)	输入的时间字符串格式必须是 yyyy-MM-dd HH:mm:ss,如果符合,则将此时间字符串转换成 UNIX 时间戳。例如:unix_timestamp('2009-03-20 11:30:01')＝1237573801
BIGINT	unix_timestamp(STRING date, STRING pattern)	将指定的时间字符串格式字符转换成 UNIX 时间戳,如果格式不对,则返回 0。 例如:unix_timestamp('2009-03-20','yyyyMM--dd')=1237532400
STRING	to_data(STRING timestamp)	返回时间字符串的日期部分,例如:to_data("1970-01-01 00:00:00")＝"1970-01-01"
INT	year(STRING date)	返回时间字符串中的年份并使用 INT 类型表示。例如:year("1970-01-01 00:00:00")＝1970,year("1970-01-01")＝1970
INT	month(STRING date)	返回时间字符串中的月份并使用 INT 类型表示。例如 month("1970-11-01 00:00:00")＝11,month("1970-11-01")＝11

续表

返回值类型	样　式	描　　述
INT	day(STRING data) dayofmonth (STRING date)	返回时间字符串中的天并使用 INT 类型表示,例如:day("1970-11-01" 00:00:00)=1,day("1970-11-01")=1
INT	hour(STRING date)	返回时间戳字符串中的小时并使用 INT 类型表示。例如:hour('2009-07-30 12:58:59')=12
INT	minute(STRING date)	返回时间字符串中的分钟数
INT	second(STRING date)	返回时间字符串中的秒数
INT	weekofyear(STRING date)	返回时间字符串位于一年中的第几周内。例如:weekofyear("1970-11-01 00:00:00")=44,weekofyear("1970-11-01")=44
INT	datediff(STRING enddate, STRING atartdate)	计算开始时间 startdate 到结束时间 enddate 相差的天数,例如:datediff('2009-03-01','2009-02-27')=2
STRING	date_add(STRING startdate, INT days)	为开始时间 startdata 到结束时间 enddata 相差的天数。例如:datediff('2009-03-01','2009-02-27')=2
STRING	date_sub(STRING startdata, INT days)	为开始时间 startdataa 减去 days 天。例如:date_sub('2008-12-31',1)='2008-12-30'
TIMESTAMP	from_utc_timestamp(TIMESTMP timestamp, STRING timezone)	如果给定的时间戳并非 UTC,则将其转化成指定时区下的时间戳
TIMESTAMP	to_utc_timestamp(TIMESTAMP timestamp, STRING timezone)	如果给定的时间戳是指定的时区下的时间戳,则将其转化成 UTC 下的时间戳

需要注意的是,与时间相关的函数输入的是整型或者字符串类型参数。对于 Hive 0.8.0 版本。这些函数同样接受 TIMESTAMP 类型的参数,同时为了向后兼容,它们还将继续支持之前的整型和字符串类型参数。

4. LIMIT 语句

典型的查询会返回多行数据。LIMIT 子句用于限制返回的行数:

```
hive> SELECT upper(name),salary,deductions["Federal Taxes"],
    > round (salary * (1 - deductions["Federal Taxes"])) FROM employees;
                                                -- 不对返回行数进行限制
OK
BILL    KING    60000    0.3     42000.0
TODD    JONES   70000    0.15    59500.0
MARY    SMITH   80000    0.2     64000.0
JOHN    DOE     100000   0.2     80000.0
BOSS    MAN     200000   0.2     160000.0
```

```
FRED         FINANCE     150000      0.2       120000.0
Time taken: 0.236 seconds, Fetched: 6 row(s)
hive> SELECT upper(name),salary,deductions["Federal Taxes"],       -- 限制返回行数为2行
    > round (salary * (1 - deductions["Federal Taxes"])) FROM employees
    > LIMIT 2;
OK
BILL         KING        60000       0.3       42000.0
TODD         JONES       70000       0.15      59500.0
Time taken: 0.346 seconds, Fetched: 2 row(s)
```

5. 列别名

前面的示例查询语句可以认为是返回一个由新列组成的新的关系,其中有些新产生的结果列对于表 employees 来说是不存在的。通常有必要给这些新产生的列起一个名字,也就是别名。下面这个例子对之前的那个查询进行了修改,为第 3 个和第 4 个字段起了别名,分别为 fed_taxes 和 salary_minus_fed_taxes。

```
hive> SELECT upper(name),salary,deductions["Federal Taxes"] as fed_taxes ,
    > round(salary * (1 - deductions["Federal Taxes"])) as salary_minus_fed_taxes
    > FROM employees LIMIT 2;
OK
BILL         KING        60000       0.3       42000.0
TODD         JONES       70000       0.15      59500.0
Time taken: 0.119 seconds, Fetched: 2 row(s)
```

6. 嵌套 SELECT 语句

对于嵌套查询语句来说,使用别名是非常有用的。下面使用前面的示例作为一个嵌套查询:

```
hive> FROM (
    > SELECT name,salary,deductions["Federal Taxes"] as fed_taxes,
    > round(salary * (1 - deductions["Federal Taxes"])) as salary_minus_fed_taxes
    > FROM employees
    > ) e
    > SELECT e.name,e.salary_minus_fed_taxes
    > WHERE e.salary_minus_fed_taxes > 60000;
OK
Mary      Smith      64000.0
John      Doe        80000.0
Boss      Man        160000.0
Fred      Finance    120000.0
Time taken:0.237 seconds,Fetched:4 row(s)
```

从这个嵌套查询语句中可以看到,我们将前面的结果集起了个别名,称之为 e,在这个语句外面嵌套查询了 name 和 salary_minus_fed_taxes 两个字段,同时约束后者的值要大于 60 000。

7. CASE…WHEN…THEN 句式

CASE…WHEN…THEN 语句和 if 条件语句类似,用于处理单个列的查询结果。

下面的这个例子就展示了 CASE…WHERE…THEN 句式的用法。

```
hive> SELECT name,salary,
    > CASE
    > WHEN salary < 70000 THEN 'low'
    > WHEN salary >= 70000 AND salary < 100000 THEN 'middle'
    > WHEN salary >= 100000 AND salary < 160000 THEN 'high '
    > ELSE 'very high'
    > END AS bracket FROM employees;
OK
Bill    King    60000    low
Todd    Jones   70000    middle
Mary    Smith   80000    middle
John    Doe     100000   high
Boss    Man     200000   very high
Fred    Finance 150000   high
Time taken: 0.158 seconds, Fetched: 6 row(s)
```

8. 什么情况下 Hive 可以避免进行 MapReduce

对于本书中的查询,如果用户执行过,那么可能会注意到大多数情况下查询都会触发一个 MapReduce 任务。如下面执行 SELECT count(DISTINCT country) FROM employees 语句时便触发了 MapReduce 任务,不过为了书写简便,本书中将其提示语句用…代替了。

```
hive> SELECT count(DISTINCT country) FROM employees;
Query ID = root_20150820153914_038f0eeb-eb43-43b1-9116-093accc7a9dc
Total jobs = 1
Launching Job 1 out of 1
Number of reduce tasks determined at compile time: 1
In order to change the average load for a reducer (in bytes):
   set hive.exec.reducers.bytes.per.reducer=<number>
In order to limit the maximum number of reducers:
   set hive.exec.reducers.max=<number>
In order to set a constant number of reducers:
   set mapreduce.job.reduces=<number>
Job running in-process (local Hadoop)
2015-08-20 15:39:19,317 Stage-1 map = 100%, reduce = 100%
Ended Job = job_local1371436096_0005
MapReduce Jobs Launched:
Stage-Stage-1: HDFS Read: 16110 HDFS Write: 0 SUCCESS
Total MapReduce CPU Time Spent: 0 msec
OK
2
Time taken: 4.931 seconds, Fetched: 1 row(s)
```

但是,Hive 中对某些情况的查询可以不必使用 MapReduce,也就是所谓的本地模式,例如:

```
hive> SELECT name,salary FROM employees;
OK
Bill    King    60000
```

```
Todd      Jones     70000
Mary      Smith     80000
John      Doe       100000
Boss      Man       200000
Fred      Finance   150000
Time taken: 0.127 seconds, Fetched: 6 row(s)
```

在这种情况下，Hive 可以简单地读取 employees 对应的存储目录下的文件，然后输出格式化后的内容到控制台。

对于 WHERE 语句中过滤条件只是分区字段这种情况（无论是否使用 LIMIT 语句限制输出记录条数），也是无须 MapReduce 过程的。

```
hive > SELECT name, salary FROM employees
    > WHERE country = 'US' AND state = 'HI';
OK
John      Doe       100000
Boss      Man       200000
Fred      Finance   150000
Time taken: 1.222 seconds, Fetched: 3 row(s)
```

此外，如果属性 hive.exec.mode.local.auto 的值设置为 true，那么 Hive 还会尝试使用本地模式执行其他的操作：

```
set hive.exec.mode.local.auto = true;
```

否则，Hive 使用 MapReduce 来执行其他所有的查询。

5.2.2　WHERE 语句

SELECT 语句用于选取字段，WHERE 语句用于过滤条件，两者结合使用可以查找到符合过滤条件的记录。与 SELECT 语句一样，在介绍 WHERE 语句之前已经在很多简单的例子中使用过它，之前都是假定读者是见过这样的语句的，现在将探讨一些细节。

WHERE 语句使用谓词表达式。有几种谓词表达式可以使用 AND 和 OR 相连接。当谓词表达式计算结果为 true 时，相应的行将被保留并输出。

下面这个例子限制了查询结果必须是美国夏威夷州的：

```
hive > USE mydatabase;
OK
Time taken: 0.04 seconds
hive > SELECT name, country, state FROM employees
    > WHERE country = 'US' AND state = 'HI';
OK
John      Doe       US    HI
Boss      Man       US    HI
Fred      Finance   US    HI
Time taken: 0.195 seconds, Fetched: 3 row(s)
```

谓词可以引用与 SELECT 语句中相同的各种对于列值的计算,这里修改之前的对于税收的查询。过滤保留那些工资减去税后总额大于 100 000 的查询结果:

```
hive> SELECT name,salary,deductions["Federal Taxes"],
    > salary * (1 - deductions["Federal Taxes"])
    > FROM employees
    > WHERE round (salary * (1 - deductions["Federal Taxes"]))>100000;
OK
Boss      Man       200000     0.2     160000.0
Fred      Finance150000        0.2     120000.0
Time taken: 0.196 seconds, Fetched: 2 row(s)
```

这个查询语句有点难理解,因为第 2 行的那个复杂的表达式和 WHERE 后面的表达式是一样的。下面的查询语句通过使用一个列别名消除了这里表达式重复的问题,但不幸的是它不是有效的:

```
hive> SELECT name,salary,deductions["Federal Taxes"],
    > salary * (1 - deductions["Federal Taxes"]) as salary_minus_fed_taxes
    > FROM employees
    > WHERE round (salary_minus_fed_taxes)>100000;
FAILED: SemanticException [Error 10004]: Line 4:12 Invalid table alias or column reference '
salary_minus_fed_taxes': (possible column names are: name, salary, subordinates, deductions,
address, country, state)
```

正如错误信息所提示的,不能在 WHERE 语句中使用列别名。不过,可以使用一个嵌套的 SELECT 语句:

```
hive> SELECT e. * FROM
    > (SELECT name,salary,deductions["Federal Taxes"] as ded,
    > salary * (1 - deductions["Federal Taxes"]) as salary_minus_fed_taxes
    > FROM employees
    > ) e
    > WHERE round (e.salary_minus_fed_taxes)>100000;
OK
Boss      Man       200000     0.2     160000.0
Fred      Finance150000        0.2     120000.0
Time taken: 0.123 seconds, Fetched: 2 row(s)
```

1. 谓词操作符

表 5.7 描述了谓词操作符,这些操作符同样可以用于 JOIN…ON 和 HAVING 语句中。

表 5.7　谓词操作符

操　作　符	支持的数据类型	描　　述
A＝B	基本数据类型	如果 A 等于 B 则返回 TRUE,反之返回 FALSE
A<＝>B	基本数据类型	如果 A 和 B 都为 NULL,则返回 TRUE,其他的与等号(＝)操作符的结果一致,如果任一为 NULL,则结果为 NULL

续表

操 作 符	支持的数据类型	描 述
A==B	没有	这是个错误的语法！SQL使用=,而不是==
A<>B,A!=B	基本数据类型	A或者B为NULL则返回NULL;如果A不等于B,则返回TRUE;反之则返回FALSE
A<B	基本数据类型	A或者B为NULL,则返回NULL;如果A小于B,则返回TRUE;反之则返回FALSE
A<=B	基本数据类型	A或者B为NULL,则返回NULL;如果A小于或等于B,则返回TRUE,反之则返回FALSE
A>B	基本数据类型	A或者B为NULL,则返回NULL;如果A大于B,则返回TRUE;反之则返回FALSE
A>=B	基本数据类型	A或者B为NULL,则返回NULL;如果A大于或等于B,则返回TRUE;反之则返回FALSE
A [NOT] BETWEEN B AND C	基本数据类型	如果A、B或者C任一为NULL,则结果为NULL。如果A的值大于或等于B而且小于或等于C,则结果为TRUE,反之为FALSE。如果使用NOT关键字,则可达到相反的效果
A IS NULL	所有数据类型	如果A等于NULL,则返回TRUE;反之则返回FALSE
A IS NOT NULL	所有数据类型	如果A不等于NULL,则返回TRUE;反之返回FALSE
A [NOT] LIKE B	STRING类型	B是一个SQL下的简单正则表达式,如果A与其匹配,则返回TRUE;反之则返回FAISE。B的表达式说明如下:'x%'表示A必须以字母'x'开头,'%x'表示A必须以字母'x'结尾;而'%x%'表示A包含有字母'x',可以位于开头、结尾或者字符串中间。类似地,下画线'_'匹配单个字符。B必须要和整个字符串A相匹配才行,如果使用NOT关健字则可达到相反的效果
A RLIKE B,A REGEXP B	STRING类型	B是一个正则表达式,如果A与其相匹配,则返回TRUE;反之则返回FALSE。匹配使用的是JDK中的正则表达式接口实现的,因为正则规则也依据其中的规则。例如,正刻表达式必须和整个字符串A相匹配,而不是只需与其子字符串匹配

接下来详细讨论 LIKE 和 RLIKE。

2. LIKE 和 RLIKE

表 5.7 描述了 LIKE 和 RLIKE 谓词操作符。大家可能在之前已经见过 LIKE 的使用了,因为它是一个标准的 SQL 操作符,可以让我们通过字符串的开头或结尾,以及指定特定的子字符串,或当子字符串出现在字符串内的任何位置时进行匹配。

例如,下面三个查询依次分别选择出了住址中街道是以字符串"Ave."结尾的雇员名称和住址,城市是以 H 开头的雇员名称和住址和街道名称中包含有 Chi 的雇员名称和住址:

```
hive> SELECT name,address.street FROM employees
    > WHERE address.street LIKE '% AVE.';
hive> SELECT name,address.street FROM employees
```

```
    > WHERE address.street LIKE '%AVE.';
OK
Todd Jones   200 Chicago AVE.
Time taken: 0.136 seconds, Fetched: 1 row(s)

hive> SELECT name,address.city FROM employees
    > WHERE address.city LIKE 'H%';
OK
John DoeHilo
Time taken: 0.109 seconds, Fetched: 1 row(s)

hive> SELECT name,address.street FROM employees
    > WHERE address.street LIKE '%Chi%';
OK
Todd Jones   200 Chicago AVE.
Time taken: 0.109 seconds, Fetched: 1 row(s)
```

RLIKE 子句是 Hive 中 LIKE 功能的一个扩展,其可以通过 Java 的正则表达式这种更强大的机制来指定匹配条件。这里通过一个例子来展示它的用法,这个例子会从 employees 表中查找所有住址的街道名称中含有单词 Chicago 或 Ontario 的雇员名称和街道信息:

```
hive> SELECT name,address.street FROM employees
    > WHERE address.street RLIKE '.*(Chicago|Ontario).*';
OK
Todd     Jones      200 Chicago AVE.
Mary     Smith      100 Ontario St.
Time taken: 0.147 seconds, Fetched: 2 row(s)
```

关键字 RLIKE 后面的字符串表达如下含义:字符串中的点号(.)表示和任意的字符匹配,星号(*)表示重复"左边的字符串"(在以上所示的两个例子中为点号)零次到无数次。表达式(x|y)表示和 x 或者 y 匹配。

不过,"Chicago"或者"Ontario"字符串前可能没有其他任何字符,而且它们后面也可能不含有其他任何字符。当然,也可以通过两个 LIKE 子句来改写这个例子为如下形式:

```
hive> SELECT name,address FROM employees
    > WHERE address.street LIKE '%Chicago' OR address.street LIKE '%Ontario%';
OK
Mary Smith{"street":"100 Ontario St.","city":"Miami","state":"FI","zip":60601}
Time taken: 0.128 seconds, Fetched: 1 row(s)
```

通过正则表达式可以比上面这种通过多个 LIKE 子句进行过滤表达更丰富的匹配条件。

关于 Hive 中通过 Java 实现的正则表达式的更详细信息,请查看如下链接中关于 Java 的正则表达式语法部分的介绍:http://docs.oracle.com/javase/6/docs/api/java/util/regex/Pattern.html,或者参考 Tony Stubblebine(O'Reilly)所著的《正则表达式参考手册》以及 Jan Goyvaerts 和 steven Levithan (O'Reilly)所著的《正则表达式 Cookbook》,还可以

参考 Jeffrey E. F. Friedl(O'Reilly)所著的《精通正则表达式(第三版)》。

5.2.3 GROUP BY 语句

在下面的学习中，会使用到表 stocks。这里再次回想一下之前说明过的表 stocks。

```
hive> CREATE TABLE stocks (
    > bourse STRING,              -- 交易所的名称
    > symbol STRING,              -- 公司的标准、名称
    > ymd STRING,                 -- 交易时间 year、month、day
    > price_open FLOAT,
    > price_high FLOAT,
    > price_low FLOAT,
    > price_close FLOAT,
    > volume INT,
    > price_adj_close FLOAT)
    > ROW FORMAT DELIMITED FIELDS TERMINATED BY '#'
    > LINES TERMINATED BY '\n';
OK
Time taken: 0.148 seconds
```

该表中的数据包括 NASDAQ 交易所提供的 Apple 公司 2008—2010 年提供的每个月 1 日的交易数据、NYSE 交易所提供的 IBM 公司的 2007—2010 年提供的每个月 1 日的交易数据、NYSE 交易所提供的 GE 公司的 2008—2010 年提供的每个月 1 日的交易数据。

具体的数据这里不详细说明，存放数据的 stocks.txt 存放在本地目录/usr/tmp/下。

GROUP BY 语句通常会和聚合函数一起使用，按照一个或者多个列对结果进行分组，然后对每个组执行聚合操作。

下面是对 stocks 表进行查询的语句及其输出内容，如下这个查询语句按照 Apple 公司股票(股票代码 APPL)的年份对股票记录进行分组，然后计算每年的平均收盘价：

```
hive> USE mydatabase;
OK
Time taken: 0.04 seconds
hive> SELECT year(ymd),avg(price_close) FROM stocks
    > WHERE bourse = 'NASDAQ' AND symbol = 'AAPL'
    > GROUP BY year(ymd);
...
OK
2008    204.09499867757162
2009    210.54916508992514
2010    198.12000020345053
Time taken: 11.325 seconds, Fetched: 3 row(s)
```

5.2.4 HAVING 语句

HAVING 子句允许用户通过一个简单的语法完成原本需要通过子查询才能对 GROUP BY 语句产生的分组进行条件过滤的任务。如下是对前面的查询语句增加一个

HAVING 语句来限制输出结果中年平均收盘价要大于 $50.0：

```
hive> SELECT year(ymd),avg(price_close) FROM stocks
    > WHERE bourse = 'NASDAQ' AND symbol = 'AAPL'
    > GROUP BY year(ymd)
    > HAVING avg(price_close)>50.0;
...
OK
2008    204.09499867757162
2009    210.54916508992514
Time taken: 25.836 seconds, Fetched: 2 row(s)
```

如果没使用 HAVING 子句，那么这个查询将需要使用一个嵌套 SELECT 子查询：

```
hive> SELECT s.year,s.avg FROM
    > (SELECT year(ymd) AS year,avg(price_close)AS avg FROM stocks
    > WHERE bourse = 'NASDAQ' AND symbol = 'AAPL'
    > GROUP BY year(ymd)
    >) s
    > WHERE s.avg>50.0;
...
OK
2008    204.09499867757162
2009    210.54916508992514
Time taken: 6.593 seconds, Fetched: 2 row(s)
```

5.2.5 JOIN 语句

Hive 支持通常的 SQL JOIN 语句，但是只支持等值连接。

1. INNER JOIN

本节需要使用到表 dividends。下面是建表语句：

```
hive> CREATE TABLE dividends (
    > bourse STRING,
    > symbol STRING,
    > ymd STRING,
    > dividend FLOAT)
    > ROW FORMAT DELIMITED FIELDS TERMINATED BY '#'
    > LINES TERMINATED BY '\n';
OK
Time taken: 1.436 seconds
```

该表中的数据包括 NASDAQ 交易所提供的 Apple 公司 2008—2010 年提供的每个季度 1 日的股息、NYSE 交易所提供的 IBM 公司的 2007—2010 年提供的每个季度 1 日的股息、NYSE 交易所提供的 GE 公司的 2008—2010 年提供的每个季度 1 日的股息。

具体的数据这里不详细说明，存放数据的 dividends.txt 存放在本地目录/usr/tmp/下。下面开始讲解本节的内容。

在内连接(INNER JOIN)中，只有进行连接的两个表中都存在与连接标准相匹配的数据才会被保留下来。例如，如下这个查询对 Apple 公司的股价（股票代码 AAPL）和 IBM 公司的股价（股票代码 IBM）进行比较。股票表 stocks 进行自连接，连接条件是 ymd 字段（也就是 year-month-day）内容必须相等。我们也称 ymd 字段是这个查询语句中的连接关键字。

```
hive > SELECT a.ymd, a.price_close, b.price_close
     > FROM stocks a JOIN stocks b ON a.ymd = b.ymd
     > WHERE a.symbol = 'AAPL' AND b.symbol = 'IBM';
. OK
2008 - 01 - 01      214.01      129.0
2008 - 02 - 01      210.73      130.25
2008 - 03 - 01      211.64      134.14
2008 - 04 - 01      209.1       131.78
2008 - 05 - 01      211.61      132.31
2008 - 06 - 01      209.04      130.23
2008 - 07 - 01      202.1       130.51
2008 - 08 - 01      200.36      129.48
2008 - 09 - 01      198.23      130.85
2008 - 10 - 01      195.43      129.55
2008 - 11 - 01      191.86      130.0
2008 - 12 - 01      195.03      130.85
2009 - 01 - 01      208.07      132.45
2009 - 02 - 01      211.73      130.9
2009 - 03 - 01      215.04      132.57
2009 - 04 - 01      205.93      131.85
2009 - 05 - 01      209.43      132.31
2009 - 06 - 01      210.65      130.57
2009 - 07 - 01      207.72      130.0
2009 - 08 - 01      210.11      129.93
2009 - 09 - 01      211.98      128.65
2009 - 10 - 01      210.58      127.91
2009 - 11 - 01      210.97      127.4
2009 - 12 - 01      214.38      128.71
2010 - 01 - 01      194.12      128.49
2010 - 02 - 01      195.46      129.93
2010 - 03 - 01      192.05      129.68
2010 - 04 - 01      199.23      129.34
2010 - 05 - 01      195.86      128.39
2010 - 06 - 01      194.73      126.8
2010 - 07 - 01      192.06      127.04
2010 - 08 - 01      199.29      127.25
2010 - 09 - 01      207.88      127.55
2010 - 10 - 01      205.94      127.21
2010 - 11 - 01      203.07      127.94
2010 - 12 - 01      197.75      126.35
Time taken: 30.818 seconds, Fetched: 36 row(s)
```

ON 子句指定了两个表间数据进行连接的条件。WHERE 子句限制了左边表是 AAPL 的记录，右边表是 IBM 的记录，同时用户可以看到这个查询中需要为两个表分别指定表别名。

众所周知，IBM 要比 Apple 历史长。IBM 也比 Apple 具有更久的股票交易记录。不过，既然这是一个内连接（INNER JOIN），那么 IBM 的 2008 年 1 月 1 日前的记录就会被过滤掉，也就是 Apple 股票交易日的第一天算起（stocks 表 IBM 公司的数据是 2007—2010 年的）。

标准 SQL 是支持对连接关键词进行非等值连接的。例如下面这个显示 Apple 和 IBM 对比数据的例子,连接条件是 Apple 的股票交易日期要比 IBM 的股票交易日期早。这个将会返回很少的数据(如下例所示)。

例如,Hive 中不支持的查询语句:

```
hive> SELECT a.ymd,a.price_close,b.price_close
    > FROM stocks a JOIN stocks b
    > ON a.ymd <= b.ymd
    > WHERE a.symbol = 'AAPL' AND b.symbol = 'IBM';
FAILED: SemanticException [Error 10017]: Line 3:3 Both left and right aliases encountered in JOIN 'ymd'
```

这个语句在 Hive 中是非法的,主要原因是通过 MapReduce 很难实现这种类型的连接。不过因为 Pig 提供了一个交叉生产功能,所以在 Pig 中是可以实现这种连接的,尽管 Pig 的原生连接功能并不支持这种连接。

同时,Hive 目前还不支持在 ON 子句中的谓词间使用 OR。

下面这个例子就是 Apple 公司的 stocks 表和 dividods 表按照字段 ymd 和字段 symbol 作为等值连接键的内连接(INNER JOIN):

```
hive> SELECT s.ymd,s.symbol,s.price_close,d.dividend
    > FROM stocks s JOIN dividends d ON s.ymd = d.ymd AND s.symbol = d.symbol
    > WHERE s.symbol = 'AAPL';
...
OK
2008 - 01 - 01    AAPL    214.01    0.03
2008 - 04 - 01    AAPL    209.1     0.03
2008 - 07 - 01    AAPL    202.1     0.02
2008 - 10 - 01    AAPL    195.43    0.02
2009 - 01 - 01    AAPL    208.07    0.02
2009 - 04 - 01    AAPL    205.93    0.02
2009 - 07 - 01    AAPL    207.72    0.03
2009 - 10 - 01    AAPL    210.58    0.03
2010 - 01 - 01    AAPL    194.12    0.03
2010 - 04 - 01    AAPL    199.23    0.03
2010 - 07 - 01    AAPL    192.06    0.03
2010 - 10 - 01    AAPL    205.94    0.03
Time taken: 76.637 seconds, Fetched: 12 row(s)
```

应注意,因为使用了内连接,所以只看到每隔 3 个月的记录。通常支付股息的时间表会在发布季度业绩报告时进行公布。

用户可以对多于两张表的多张表进行连接操作。下面来对 Apple 公司、IBM 公司和 GE 公司并排进行比较:

```
hive> SELECT a.ymd,a.price_close,b.price_close,c.price_close
    > FROM stocks a JOIN stocks b ON a.ymd = b.ymd
    > JOIN stocks C ON a.ymd = c.ymd
    > WHERE a.symbol = 'AAPL' AND b.symbol = 'IBM' AND c.symbol = 'GE';
```

```
...
OK
2008 - 01 - 01    214.01    129.0     34.88
2008 - 02 - 01    210.73    130.25    34.78
2008 - 03 - 01    211.64    134.14    35.02
2008 - 04 - 01    209.1     131.78    35.11
2008 - 05 - 01    211.61    132.31    35.32
2008 - 06 - 01    209.04    130.23    35.29
2008 - 07 - 01    202.1     130.51    35.36
2008 - 08 - 01    200.36    129.48    35.55
2008 - 10 - 01    195.43    129.55    35.79
2008 - 11 - 01    191.86    130.0     36.0
2008 - 12 - 01    195.03    130.85    35.82
2009 - 01 - 01    208.07    132.45    36.86
2009 - 02 - 01    211.73    130.9     36.84
2009 - 03 - 01    215.04    132.57    35.84
2009 - 04 - 01    205.93    131.85    35.41
2009 - 05 - 01    209.43    132.31    34.76
2009 - 06 - 01    210.65    130.57    34.8
2009 - 07 - 01    207.72    130.0     35.13
2009 - 08 - 01    210.11    129.93    35.0
2009 - 09 - 01    211.98    128.65    35.13
2009 - 10 - 01    210.58    127.91    35.2
2009 - 11 - 01    210.97    127.4     35.36
2009 - 12 - 01    214.38    128.71    35.38
2010 - 01 - 01    194.12    128.49    36.6
2010 - 02 - 01    195.46    129.93    16.25
2010 - 03 - 01    192.05    129.68    36.64
2010 - 04 - 01    199.23    129.34    36.97
2010 - 05 - 01    195.86    128.39    36.78
2010 - 06 - 01    194.73    126.8     37.26
2010 - 07 - 01    192.06    127.04    37.08
2010 - 08 - 01    199.29    127.25    37.24
2010 - 09 - 01    207.88    127.55    37.15
2010 - 10 - 01    205.94    127.21    37.34
2010 - 11 - 01    203.07    127.94    37.31
2010 - 12 - 01    197.75    126.35    37.1
Time taken: 20.302 seconds, Fetched: 35 row(s)
```

大多数情况下,Hive 会对每对 JOIN 连接对象启动一个 MapReduce 任务。本例中,会首先启动一个 MapReduce 任务对表 a 和表 b 进行连接操作,然后会再启动一个 MapReduce 任务将第一个 MapReduce 任务的输出和表 c 进行连接操作。

2. LEFT OUTER JOIN

左外连接(LEFT OUTER JOIN)通过关键字 LEFT OUTER 进行标识:

```
hive > SELECT s.ymd, s.symbol, s.price_close, d.dividend
    > FROM stocks s LEFT OUTER JOIN dividends d ON s.ymd = d.ymd AND s.symbol = d.symbol
    > WHERE s.symbol = 'AAPL';
...
OK
```

```
2008 - 01 - 01    AAPL    214.01    0.03
2008 - 02 - 01    AAPL    210.73    NULL
2008 - 03 - 01    AAPL    211.64    NULL
2008 - 04 - 01    AAPL    209.1     0.03
2008 - 05 - 01    AAPL    211.61    NULL
2008 - 06 - 01    AAPL    209.04    NULL
2008 - 07 - 01    AAPL    202.1     0.02
2008 - 08 - 01    AAPL    200.36    NULL
2008 - 09 - 01    AAPL    198.23    NULL
2008 - 10 - 01    AAPL    195.43    0.02
2008 - 11 - 01    AAPL    191.86    NULL
2008 - 12 - 01    AAPL    195.03    NULL
2009 - 01 - 01    AAPL    208.07    0.02
2009 - 02 - 01    AAPL    211.73    NULL
2009 - 03 - 01    AAPL    215.04    NULL
2009 - 04 - 01    AAPL    205.93    0.02
2009 - 05 - 01    AAPL    209.43    NULL
2009 - 06 - 01    AAPL    210.65    NULL
2009 - 07 - 01    AAPL    207.72    0.03
2009 - 08 - 01    AAPL    210.11    NULL
2009 - 09 - 01    AAPL    211.98    NULL
2009 - 10 - 01    AAPL    210.58    0.03
2009 - 11 - 01    AAPL    210.97    NULL
2009 - 12 - 01    AAPL    214.38    NULL
2010 - 01 - 01    AAPL    194.12    0.03
2010 - 02 - 01    AAPL    195.46    NULL
2010 - 03 - 01    AAPL    192.05    NULL
2010 - 04 - 01    AAPL    199.23    0.03
2010 - 05 - 01    AAPL    195.86    NULL
2010 - 06 - 01    AAPL    194.73    NULL
2010 - 07 - 01    AAPL    192.06    0.03
2010 - 08 - 01    AAPL    199.29    NULL
2010 - 09 - 01    AAPL    207.88    NULL
2010 - 10 - 01    AAPL    205.94    0.03
2010 - 11 - 01    AAPL    203.07    NULL
2010 - 12 - 01    AAPL    197.75    NULL
Time taken: 26.327 seconds, Fetched: 36 row(s3)
```

在这种 JOIN 连接操作中,JOIN 操作符左边表中符合 WHERE 子句的所有记录将会被返回。JOIN 操作符右边表中如果没有符合 ON 后面连接条件的记录时,那么从右边表指定选择的列的值将会是 NULL。

因此,在这个结果集中,可以看到 Apple 公司的股票记录都被返回了,而 d.dividend 字段的值通常是 NULL。除了当天有支付股息的那条记录(也就是输出中的 2008 年 1 月 1 日那天的记录)。

在讨论其他外连接之前,先来讨论一个用户应该明白的问题。

回想一下,前面我们说过,通过在 WHERE 子句中增加分区过滤器可以加快查询速度。为了提高前面那个查询的执行速度,可以对两张表的 bourse 字段增加谓词限定:

```
hive> SELECT s.ymd,s.symbol,s.price_close,d.dividend
```

```
    > FROM stocks s LEFT OUTER JOIN dividends d ON s.ymd = d.ymd AND s.symbol = d.symbol
    > WHERE s.symbol = 'AAPL'
    > AND s.bourse = 'NASDAQ' AND d.bourse = 'NASDAQ';
...
OK
2008-04-01      AAPL    209.1    0.03
2008-07-01      AAPL    202.1    0.02
2008-10-01      AAPL    195.43   0.02
2009-01-01      AAPL    208.07   0.02
2009-04-01      AAPL    205.93   0.02
2009-07-01      AAPL    207.72   0.03
2009-10-01      AAPL    210.58   0.03
2010-01-01      AAPL    194.12   0.03
2010-04-01      AAPL    199.23   0.03
2010-07-01      AAPL    192.06   0.03
2010-10-01      AAPL    205.94   0.03
Time taken: 22.164 seconds, Fetched: 11 row(s)
```

不过,可以发现输出结果改变了!

重新获得每年 4 条左右的股票交易记录,可以发现每年对应的股息值都是非 NULL 的。换句话说,这个效果和之前的内连接(INNER JOIN)是一样的!

在大多数的 SQL 实现中,这种现象实际上比较常见。之所以发生这种情况,是因为会先执行 JOIN 语句,然后再将结果通过 WHERE 语句进行过滤。在到达 WHERE 语句时,d.exchange 字段中大多数值为 NULL,因此这个"优化"实际上过滤掉了那些非股息支付日的所有记录。

3. RIGHT OUTER JOIN

右外连接(RIGHT OUTER JOIN)会返回右边表所有符合 WHERE 语句的记录。左表中匹配不上的字段值用 NULL 代替。

这里调整 stocks 表和 divideneds 表的位置来执行右外连接,并保留 SELECT 语句不变:

```
hive> SELECT s.ymd, s.symbol, s.price_close, d.dividend
    > FROM dividends d RIGHT OUTER JOIN stocks s ON d.ymd = s.ymd AND d.symbol = s.symbol
    > WHERE s.symbol = 'AAPL';
...
OK
2008-01-01      AAPL    214.01   0.03
2008-02-01      AAPL    210.73   NULL
2008-03-01      AAPL    211.64   NULL
2008-04-01      AAPL    209.1    30.03
2008-05-01      AAPL    211.61   NULL
2008-06-01      AAPL    209.04   NULL
2008-07-01      AAPL    202.1    0.02
2008-08-01      AAPL    200.36   NULL
2008-09-01      AAPL    198.23   NULL
2008-10-01      AAPL    195.43   0.02
2008-11-01      AAPL    191.86   NULL
2008-12-01      AAPL    195.03   NULL
2009-01-01      AAPL    208.07   0.02
```

```
2009 - 02 - 01        AAPL       211.73      NULL
2009 - 03 - 01        AAPL       215.04      NULL
2009 - 04 - 01        AAPL       205.93      0.02
2009 - 05 - 01        AAPL       209.43      NULL
2009 - 06 - 01        AAPL       210.65      NULL
2009 - 07 - 01        AAPL       207.72      0.03
2009 - 08 - 01        AAPL       210.11      NULL
2009 - 09 - 01        AAPL       211.98      NULL
2009 - 10 - 01        AAPL       210.58      0.03
2009 - 11 - 01        AAPL       210.97      NULL
2009 - 12 - 01        AAPL       214.38      NULL
2010 - 01 - 01        AAPL       194.12      0.03
2010 - 02 - 01        AAPL       195.46      NULL
2010 - 03 - 01        AAPL       192.05      NULL
2010 - 04 - 01        AAPL       199.23      0.03
2010 - 05 - 01        AAPL       195.86      NULL
2010 - 06 - 01        AAPL       194.73      NULL
2010 - 07 - 01        AAPL       192.06      0.03
2010 - 08 - 01        AAPL       199.29      NULL
2010 - 09 - 01        AAPL       207.88      NULL
2010 - 10 - 01        AAPL       205.94      0.03
2010 - 11 - 01        AAPL       203.07      NULL
2010 - 12 - 01        AAPL       197.75      NULL
Time taken: 11.993 seconds, Fetched: 36 row(s)
```

4. FULL OUTER JOIN

完全外连接(FULL OUTER JOIN)将会返回所有表中符合 WHERE 语句条件的所有记录。如果任一表的指定字段没有符合条件的值的话,那么就使用 NULL 值替代。

如果将前面的查询改写成一个完全外连接查询,那么事实上获得的结果和之前的一样。这是因为不可能存在有股息支付记录而没有对应的股票交易记录的情况。

```
hive> SELECT s.ymd,s.symbol,s.price_close,d.dividend
    > FROM dividends d FULL OUTER JOIN stocks s ON d.ymd = s.ymd AND d.symbol = s.symbol
    > WHERE s.symbol = 'AAPL';
...
OK
2008 - 01 - 01        AAPL       214.01      0.03
2008 - 02 - 01        AAPL       210.73      NULL
2008 - 03 - 01        AAPL       211.64      NULL
2008 - 04 - 01        AAPL       209.1       0.03
2008 - 05 - 01        AAPL       211.61      NULL
2008 - 06 - 01        AAPL       209.04      NULL
2008 - 07 - 01        AAPL       202.1       0.02
2008 - 08 - 01        AAPL       200.36      NULL
2008 - 09 - 01        AAPL       198.23      NULL
2008 - 10 - 01        AAPL       195.43      0.02
2008 - 11 - 01        AAPL       191.86      NULL
2008 - 12 - 01        AAPL       195.03      NULL
2009 - 01 - 01        AAPL       208.07      0.02
2009 - 02 - 01        AAPL       211.73      NULL
```

```
2009 - 03 - 01      AAPL      215.04      NULL
2009 - 04 - 01      AAPL      205.93      0.02
2009 - 05 - 01      AAPL      209.43      NULL
2009 - 06 - 01      AAPL      210.65      NULL
2009 - 07 - 01      AAPL      207.72      0.03
2009 - 08 - 01      AAPL      210.11      NULL
2009 - 09 - 01      AAPL      211.98      NULL
2009 - 10 - 01      AAPL      210.58      0.03
2009 - 11 - 01      AAPL      210.97      NULL
2009 - 12 - 01      AAPL      214.38      NULL
2010 - 01 - 01      AAPL      194.12      0.03
2010 - 02 - 01      AAPL      195.46      NULL
2010 - 03 - 01      AAPL      192.05      NULL
2010 - 04 - 01      AAPL      199.23      0.03
2010 - 05 - 01      AAPL      195.86      NULL
2010 - 06 - 01      AAPL      194.73      NULL
2010 - 07 - 01      AAPL      192.06      0.03
2010 - 08 - 01      AAPL      199.29      NULL
2010 - 09 - 01      AAPL      207.88      NULL
2010 - 10 - 01      AAPL      205.94      0.03
2010 - 11 - 01      AAPL      203.07      NULL
2010 - 12 - 01      AAPL      197.75      NULL
Time taken: 2.6 seconds, Fetched: 36 row(s)
```

5. LEFT SEMI-JOIN

左半开连接(LEFT SEMI-JOIN)会返回左边表的记录,前提是其记录对于右边表满足 ON 语句中的判定条件。对于常见的内连接(INNER JOIN)来说,这是一个特殊的、优化了的情况。大多数的 SQL 语言会通过 IN…EXISTS 结构来处理这种情况。例如例 5-1 中的查询,其将试图返回限定的股息支付日内的股票交易记录,不过 Hive 是不支持这个查询的。

例 5-1　Hive 中不支持的查询。

```
hive> SELECT s.ymd, s.symbol, s.price_close FROM stocks s
    > WHERE s.ymd, s.symbol IN
    > (SELECT d.ymd, d.symbol FROM dividends d);
FAILED: ParseException line 2:11 missing EOF at ',' near 'ymd'
```

不过,用户可以使用如下的 LEFT SEMI JOIN 语法达到同样的目的:

```
hive> SELECT s.ymd,s.symbol,s.price_close
    > FROM stocks s LEFT SEMI JOIN dividends d ON s.ymd = d.ymd AND s.symbol = d.symbol;
...
OK
2008 - 01 - 01      AAPL      214.01
2008 - 04 - 01      AAPL      209.1
2008 - 07 - 01      AAPL      202.1
2008 - 10 - 01      AAPL      195.43
2009 - 01 - 01      AAPL      208.07
2009 - 04 - 01      AAPL      205.93
2009 - 07 - 01      AAPL      207.72
```

```
2009 - 10 - 01    AAPL    210.58
2010 - 01 - 01    AAPL    194.12
2010 - 04 - 01    AAPL    199.23
2010 - 07 - 01    AAPL    192.06
2010 - 10 - 01    AAPL    205.94
2007 - 01 - 01    IBM     121.88
2007 - 04 - 01    IBM     125.66
2007 - 07 - 01    IBM     122.39
2007 - 10 - 01    IBM     125.75
2008 - 01 - 01    IBM     129.0
2008 - 04 - 01    IBM     131.78
2008 - 07 - 01    IBM     130.51
2008 - 10 - 01    IBM     129.55
2009 - 01 - 01    IBM     132.45
2009 - 04 - 01    IBM     131.85
2009 - 07 - 01    IBM     130.0
2009 - 10 - 01    IBM     127.91
2010 - 01 - 01    IBM     128.49
2010 - 04 - 01    IBM     129.34
2010 - 07 - 01    IBM     127.04
2010 - 10 - 01    IBM     127.21
2008 - 01 - 01    GE      34.88
2008 - 04 - 01    GE      35.11
2008 - 07 - 01    GE      35.36
2008 - 10 - 01    GE      35.79
2009 - 01 - 01    GE      36.86
2009 - 04 - 01    GE      35.41
2009 - 10 - 01    GE      35.2
2010 - 01 - 01    GE      36.6
2010 - 04 - 01    GE      36.97
2010 - 07 - 01    GE      37.08
2010 - 10 - 01    GE      37.34
Time taken: 27.226 seconds, Fetched: 39 row(s)
```

请注意，SELECT 和 WHERE 语句中不能引用到右边表中的字段。

SEMI-JOIN 比通常的 INNER JOIN 要更高效，原因如下：对于左边表中一条指定的记录，在右边表中一旦找到匹配的记录，Hive 就会立即停止扫描。从这点来看，左边表中选择的列是可以预测的。

6. 笛卡儿积 JOIN

笛卡儿积是一种连接，表示左边表的行数乘以右边表的行数等于笛卡儿结果集的大小。也就是说，如果左边表有 5 行数据，而右边表有 6 行数据，那么产生的结果将是 30 行数据：

```
hive > SELECT stocks.ymd FROM stocks JOIN dividends;
...
2008 - 01 - 01
2008 - 01 - 01
2008 - 01 - 01
...
Time taken: 15.476 seconds, Fetched: 4800 row(s)
```

如上面的查询，以 stocks 表和 dividends 表为例，实际上很难找到合适的理由来执行这类连接，因为一只股票的股息通常并非和另一只股票配对。此外，笛卡儿积会产生大量的数据。与其他连接类型不同，笛卡儿积不是并行执行的，而且如果使用 MapReduce 计算架构，那么任何方式都无法进行优化。

这里非常有必要指出，如果使用了错误的连接（JOIN）语法可能会导致产生一个执行时间长、运行缓慢的笛卡儿积查询。例如，如下这个查询在很多数据库中会被优化成内连接（INNER JOIN），但是在 Hive 中没有此优化：

```
hive> SELECT stocks.ymd FROM stocks JOIN dividends
    > WHERE stocks.symbol = dividends.symbol and stocks.symbol = 'AAPL';
...
OK
2008 - 01 - 01
2008 - 01 - 01
2008 - 01 - 01
2008 - 01 - 01
2008 - 01 - 01
...
Time taken. 14.774 seconds, Fetched: 432 row(s)
```

在 Hive 中，这个查询在应用 WHERE 语句中的谓词条件前会先进行完全笛卡儿积计算，这个过程将会消耗很长的时间。如果设置属性 hive.mapred.mode 值为 strict，那么 Hive 会阻止用户执行笛卡儿积查询。

7. Map-side JOIN

如果所有表中只有一张表是小表，那么可以在最大的表通过 Mapper 的时候将小表完全放到内存中。Hive 可以在 Map 端执行连接过程（称为 Map-side JOIN），这是因为 Hive 可以和内存中的小表进行逐一匹配，从而省略掉常规连接操作所需要的 Reduce 过程。

即使对于很小的数据集，这个优化也明显快于常规的连接操作。其不仅减少了 Reduce 过程，而且有时还可以同时减少 Map 过程的执行步骤。

stocks 表和 dividends 表之间的连接操作也可以利用这个优化，因为 dividends 表中的数据集很小，已经可以全部放在内存中缓存起来了。

在 Hive v0.7 之前的版本中，如果想使用这个优化，需要在查询语句中增加一个标记来进行触发。如下面的这个内连接（INNER JOIN）的例子所示：

```
hive> SELECT /* + MAPJOIN(d) */ s.ymd, s.symbol, s.price_close, d.dividend
    > FROM stocks s JOIN dividends d ON s.ymd = d.ymd AND s.symbol = d.symbol
    > WHERE s.symbol = 'AAPL';
...
OK
2008 - 01 - 01    AAPL    214.01    0.03
2008 - 04 - 01    AAPL    209.1     0.03
2008 - 07 - 01    AAPL    202.1     0.02
2008 - 10 - 01    AAPL    195.43    0.02
2009 - 01 - 01    AAPL    208.07    0.02
009 - 04 - 01     AAPL    205.93    0.02
2009 - 07 - 01    AAPL    207.72    0.03
```

```
2009 - 10 - 01      AAPL      210.58     0.03
2010 - 01 - 01      AAPL      194.12     0.03
2010 - 04 - 01      AAPL      199.23     0.03
2010 - 07 - 01      AAPL      192.06     0.03
2010 - 10 - 01      AAPL      205.94     0.03
Time taken: 14.908 seconds, Fetched: 12 row(s)
```

从 Hive v0.7 版本开始,废弃了这种标记的方式,不过如果增加了这个标记,同样是有效的。如果不加上这个标记,那么这时用户需要设置属性 hive.auto.convert.JOIN 的值为 true,这样 Hive 才会在必要的时候启动这个优化,默认情况下这个属性的值是 false。

```
hive > set hive.auto.convert.join = true;
hive > SELECT s.ymd, s.symbol, s.price_close, d.dividend
    > FROM stocks s JOIN dividends d ON s.ymd = d.ymd AND s.symbol = d.symbol
    > WHERE s.symbol = 'AAPL';
...
OK
2008 - 01 - 01      AAPL      214.01     0.03
2008 - 04 - 01      AAPL      209.1      0.03
2008 - 07 - 01      AAPL      202.1      0.02
2008 - 10 - 01      AAPL      195.43     0.02
2009 - 01 - 01      AAPL      208.07     0.02
2009 - 04 - 01      AAPL      205.93     0.02
2009 - 07 - 01      AAPL      207.72     0.03
2009 - 10 - 01      AAPL      210.58     0.03
2010 - 01 - 01      AAPL      194.12     0.03
2010 - 04 - 01      AAPL      199.23     0.03
2010 - 07 - 01      AAPL      192.06     0.03
2010 - 10 - 01      AAPL      205.94     0.03
Time taken: 13.064 seconds, Fetched: 12 row(s)
```

需要注意的是,用户也可以配置能够使用这个优化的小表的大小。如下是这个属性的默认值(单位是字节):hive.mapjoin.smalltable.filesize = 25000000(说明:输入小表的文件大小的阈值,如果小于该值,就采用普通的连接操作)。如果用户期望 Hive 在必要的时候自动启动这个优化,那么可以将这一个(或两个)属性设置在 $HOME/.hiverc 文件中。

但是,Hive 对于右外连接(RIGHT OUTER JOIN)和全外连接(FULL OUTER JOIN)不支持这个优化。

5.2.6 ORDER BY 和 SORT BY

Hive 中 ORDER BY 语句和其他的 SQL 语言中的定义是一样的,都会对查询结果集执行一个全局排序。也就是说,会有一个所有的数据都通过一个 Reducer 进行处理的过程。对于大数据集,这个过程可能会消耗太过漫长的时间来执行。

Hive 增加了一个可供选择的方式,也就是 SORT BY,其只会在每个 Reducer 中对数据进行排序,也就是执行一个局部排序过程。这可以保证每个 Reducer 的输出数据都是有序的(但并非全局有序)。这样可以提高后面进行的全局排序的效率。

对于这两种情况,语法区别仅仅是:一个关键字是 ORDER,另一个关键字是 SORT。用户可以指定任意希望进行排序的字段,并可以在字段后面加上 ASC 关键字(默认的),表示按升序排序;或加 DESC 关键字,表示按降序排序。

下面是一个使用 ORDER BY 的例子:

```
hive> SELECT s.ymd, s.symbol, s.price_close
    > FROM stocks s
    > ORDER BY s.ymd ASC, s.symbol DESC;
...
2007-01-01      IBM     121.88
2007-02-01      IBM     123.52
2007-03-01      IBM     123.0
2007-04-01      IBM     125.66
2007-05-01      IBM     125.53
2007-06-01      IBM     124.67
2007-07-01      IBM     122.39
2007-08-01      IBM     123.75
2007-09-01      IBM     126.33
2007-10-01      IBM     125.75
2007-11-01      IBM     126.12
2007-12-01      IBM     125.5
2008-01-01      IBM     129.0
2008-01-01      GE      34.88
2008-01-01      AAPL    214.01
2008-02-01      IBM     130.25
2008-02-01      GE      34.78
2008-02-01      AAPL    210.73
2008-03-01      IBM     134.14
2008-03-01      GE      35.02
...
Time taken: 3.102 seconds, Fetched: 120 row(s)
```

下面是一个类似的例子,不过使用的是 SORT BY:

```
hive> SELECT s.ymd, s.symbol, s.price_close
    > FROM stocks s
    > SORT BY s.ymd ASC, s.symbol DESC;
...
OK
2007-01-01      IBM     121.88
2007-02-01      IBM     123.52
2007-03-01      IBM     123.0
2007-04-01      IBM     125.66
2007-05-01      IBM     125.53
2007-06-01      IBM     124.67
2007-07-01      IBM     122.39
2007-08-01      IBM     123.75
2007-09-01      IBM     126.33
2007-10-01      IBM     125.75
...
Time taken: 1.537 seconds, Fetched: 120 row(s)
```

上面介绍的两个查询看上去几乎一样,不过如果使用的 Reducer 的个数大于 1,那么输出结果的排序就大不一样了。既然只保证每个 Reducer 的输出是局部有序的,那么不同 Reducer 的输出就可能会有重叠。

因为 ORDER BY 操作可能会导致运行时间过长,如果属性 hive.mapred.mode 的值是 strict,那么 Hive 要求这样的语句必须加上 LIMIT 语句进行限制。默认情况下,这个属性的值是 nonstrict,也就是不会有这样的限制。

5.2.7　含有 SORT BY 的 DISTRIBUTE BY

DISTRIBUTE BY 控制 Map 的输出在 Reducer 中是如何划分的。MapReduce 任务中传输的所有数据都是按照键值对的方式进行组织的,因此 Hive 在将用户的查询语句转换成 MapReduce 任务时,其必须在内部使用这个功能。

通常,用户不需要担心这个特性。不过对于使用了流特性以及一些状态为 UDAF(用户自定义聚合函数)的查询是个例外。还有,在另外一个场景下,使用这些语句是有用的。

默认情况下,MapReduce 计算框架会依据 Map 输入的键计算相应的哈希值,然后按照得到的哈希值将键值对均匀分发到多个 Reducer 中去。这也就意味着当我们使用 SORT BY 时,不同 Reducer 的输出内容会有明显的重叠,至少对于排列顺序而言是这样(即使每个 Reducer 的输出的数据都是有序的)。

假设我们希望具有相同股票交易码的数据在一起处理。那么可以使用 DISTRIBUTE BY 来保证具有相同股票交易码的记录会分发到同一个 Reducer 中进行处理,然后使用 SORT BY 来按照我们的期望对数据进行排序。下面的例子就演示了这种用法:

```
hive> SELECT s.ymd,s.symbol,s.price_close
    > FROM stocks s
    > DISTRIBUTE BY s.symbol
    > SORT BY s.symbol ASC,s.ymd ASC;
...
2008-01-01      AAPL    214.01
2008-02-01      AAPL    210.73
2008-03-01      AAPL    211.64
2008-04-01      AAPL    209.1
2008-05-01      AAPL    211.61
2008-06-01      AAPL    209.04
2008-07-01      AAPL    202.1
2008-08-01      AAPL    200.36
2008-09-01      AAPL    198.23
2008-10-01      AAPL    195.43
...
Time taken: 4.529 seconds, Fetched: 120 row(s)
```

当然,上面例子中的 ASC 关键字是可以省略掉的,因为它本身就是默认值。

DISTRIBUTE BY 和 GROUP BY 在其控制 Reducer 是如何接收一行行数据进行处理这方面是类似的,而 SORT BY 则控制 Reducer 内的数据是如何进行排序的。

需要注意的是,Hive 要求 DISTRIBUTE BY 语句要写在 SORT BY 语句之前。

5.2.8 CLUSTER BY

在前面的例子中,s.symbol 列被用在了 DISTRIBUTE BY 语句中,而 s.symbol 列和 s.ymd 位于 SORT BY 语句中,如果这两个语句中涉及的列完全相同,而且采用的是升序排序方式(也就是默认的排序方式),那么在这种情况下,CLUSTER BY 就等价于前面的两个语句,相当于是前面两个句子的一个简写方式。

如下面的例子所示,我们将前面的查询语句中 SORT BY 后面的 s.ymd 字段去掉而只对 s.symbol 字段使用 CLUSTER BY 语句:

```
hive> SELECT s.ymd, s.symbol, s.price_close
    > FROM stocks s
    > CLUSTER BY s.symbol;
...
OK
2008-02-01      AAPL    210.73
2008-03-01      AAPL    211.64
2008-04-01      AAPL    209.1
2008-05-01      AAPL    211.61
2008-06-01      AAPL    209.04
2008-07-01      AAPL    202.1
2008-08-01      AAPL    200.36
2008-09-01      AAPL    198.23
2008-10-01      AAPL    195.43
2008-11-01      AAPL    191.86
...
Time taken: 3.556 seconds, Fetched: 120 row(s)
```

因此排序限制中去除了 s.ymd 字段,所以输出中展示的是股票数据的原始排序方式,也就是降序排列。

使用 DISTRIBUTE BY…SORT BY 语句或其简化版的 CLUSTER BY 语句会剥夺 SORT BY 的并行性,但这样可以使输出文件的数据是全局排序的。

5.2.9 类型转换

这里讨论 cast() 函数,用户可以使用这个函数对指定的值进行显式的类型转换。

回想一下,前面介绍过的 employees 表中 salary 列是使用 FLOAT 数据类型的。现在假设这个字段使用的数据类型是 STRING,那么如何才能将其作为 FLOAT 值进行计算呢?

下面的例子会先将值转换为 FLOAT 类型,然后才会执行数值大小比较过程:

```
hive> SELECT name,salary FROM employees
    > WHERE cast(salary AS FLOAT)< 80000.0;
OK
Bill    King    60000
Todd    Jones   70000
Time taken: 0.158 seconds, Fetched: 2 row(s)
```

类型转换函数的语法是 cast(value AS TYPE)。如果例子中的 salary 字段的值不是合法的浮点数字符串,那么 Hive 会返回 NULL。

需要注意的是,将浮点数转换成整数的推荐方式是使用表 5.3 中列举的 round() 或者 floor() 函数,而不是使用类型转换操作符 cast。

类型转换 BINARY 值

Hive 0.8.0 版本中新引入的 BINARY 类型只支持将 BINARY 类型转换为 STRING 类型。不过,如果用户知道其值是数值,那么可以通过嵌套 cast() 的方式对其进行类型转换:

```
cast(cast (b as  string) as  double)
```

其中 b 字段类型原本是 BINARY。同样,也可以将 STRING 类型转换为 BINARY 类型。

5.2.10 抽样查询

对于非常大的数据集,有时用户需要使用的是一个具有代表性的查询结果而不是全部结果。Hive 可以通过对表进行分桶抽样来满足这个需求。

下面创建一个 number 表。该表非常简单,表中值包含 number 字段。

(1) 在 /usr/tmp 目录下新建一个 numbers.txt 文件,在该文件中写入以下数据:
vi numbers.txt

```
1
2
3
4
5
6
7
8
9
10
```

(2) 退出到根目录下,输入 hive 命令,进入 CLI 命令行,在数据库 mydatabase 中创建表 numbers。

```
hive> USE mydatabase;
OK
Time taken: 1.002 seconds
hive> CREATE TABLE numbers(number INT);
OK
Time taken: 0.565 seconds
```

(3) 将刚刚创建的 numbers.txt 文件中的数据装载到表 numbers 中。

```
hive> LOAD DATA LOCAL INPATH '/usr/tmp/numbers.txt' INTO TABLE numbers;
Loading data to table mydatabase.numbers
Table mydatabase.numbers stats: [numFiles = 1, totalSize = 21]
```

```
OK
Time taken: 3.06 seconds
```

（4）可以使用"SELECT * FROM numbers;"语句来查看 numbers 表是否正确。

```
hive> SELECT * FROM numbers;
OK
1
2
3
4
5
6
7
8
9
10
Time taken: 0.432 seconds, Fetched: 10 row(s)
```

我们知道，numbers 表只有 number 字段，其值是 1～10。

可以使用 rand() 函数进行抽样，这个函数会返回一个随机值。前两个查询都返回了两个不相等的值，而第三个查询语句无返回结果：

```
hive> SELECT * FROM numbers TABLESAMPLE(BUCKET 3 OUT OF 10 ON rand()) s;      --第一次执行
OK
1
Time taken: 0.348 seconds, Fetched: 2 row(s)
hive> SELECT * FROM numbers TABLESAMPLE(BUCKET 3 OUT OF 10 ON rand()) s;      --第二次执行
OK
5
6
hive> SELECT * FROM numbers TABLESAMPLE(BUCKET 3 OUT OF 10 ON rand()) s;      --第三次执行
OK
Time taken: 0.158 seconds
```

如果按照指定的列而非 rand() 函数进行分桶，那么同一语句多次执行的返回值是相同的：

```
hive> SELECT * FROM numbers TABLESAMPLE(BUCKET 3 OUT OF 10 ON number) s;
OK
2
Time taken: 0.121 seconds, Fetched: 1 row(s)
hive> SELECT * FROM numbers TABLESAMPLE(BUCKET 3 OUT OF 10 ON number) s;
OK
2
Time taken: 0.135 seconds, Fetched: 1 row(s)
hive> SELECT * FROM numbers TABLESAMPLE(BUCKET 3 OUT OF 10 ON number) s;
OK
2
Time taken: 0.135 seconds, Fetched: 1 row(s)
```

分桶语句中分母表示的是数据将会被散列的桶的个数,而分子表示将会选择的桶的个数:

```
hive> SELECT * FROM numbers TABLESAMPLE(BUCKET 1 OUT OF 2 ON number) s;
OK
2
4
6
8
10
Time taken: 0.123 seconds, Fetched: 5 row(s)
hive> SELECT * FROM numbers TABLESAMPLE(BUCKET 2 OUT OF 2 ON number) s;
OK
1
3
5
7
9
Time taken: 0.129 seconds, Fetched: 5 row(s)
```

1. 数据块抽样

Hive 提供了另外一种按照抽样百分比进行抽样的方式,这种是基于行数的,按照输入路径下的数据块百分比进行的抽样:

```
hive> SELECT * FROM numbers TABLESAMPLE(10 PERCENT) s;
OK
1
2
Time taken: 0.079 seconds, Fetched: 2 row(s)
hive> SELECT * FROM numbers TABLESAMPLE(50 PERCENT) s;
OK
1
2
3
4
5
6
Time taken: 0.104 seconds, Fetched: 6 row(s)
```

2. 分桶表的输入裁剪

第一次看 TABLESAMPLE 语句,大家应该会得出"如下的查询和 TABLESAMPLE 操作相同"的结论。

```
hive> SELECT * FROM numbers WHERE number % 2 = 0;
OK
2
4
6
8
10
```

```
Time taken: 0.157 seconds, Fetched: 5 row(s)
```

对于大多数类型的表确实是这样的。抽样会扫描表中所有的数据,然后在每 N 行中抽取一行数据。不过,如果 TABLESAMPLE 语句中指定的列和 CLUSTERED BY 语句中指定的列相同,那么 TABLESAMPLE 查询就只会扫描所涉及的表的分区下的数据。

5.2.11　UNION ALL

UNION ALL 可以将两个或多个表进行合并。每一个 union 子查询都必须具有相同的列,而且对应的每个字段的字段类型必须是一致的。例如,如果第二个字段是 FLOAT 类型的,那么所有其他子查询的第二个字段也必须都是 FLOAT 类型的。

下面的例子将表 stocks 和表 dividends 进行了合并。

```
hive> SELECT stocks.ymd,stocks.symbol
    > FROM (
    > SELECT s.ymd,s.symbol,'stocks' AS source FROM stocks s
    > UNION ALL
    > SELECT d.ymd,d.symbol,'dividends' AS source FROM dividends d
    > ) stocks
    > SORT by stocks.ymd ASC;
...
OK
2007-01-01    IBM
2007-01-01    IBM
2007-02-01    IBM
2007-03-01    IBM
...
Time taken: 5.837 seconds, Fetched: 160 row(s)
```

5.3　Hive 实战

5.3.1　背景

本节主要讲述使用 Hive 的实践,本次的任务是对学校的成绩以及教师信息等不同的表进行连接、计算、查询。其中表 5.8 是 18 位同学的各科测试成绩信息表,表 5.9 各个班级的授课老师信息。

使用 Hive,从导入到分析、排序、结果输出,这些操作都可以运用 HQL 语句来解决,一条语句经过处理被解析成几个任务来运行,即使是不同表上的数据也能通过表关联这样的语句自动完成,节省了大量工作量。

5.3.2　实战数据及要求

下面根据表 5.8 和表 5.9 中的数据,实现下列要求:

表 5.8 成绩表

学号 sid	姓名 sname	性别 sgender	班级 sclass	期中 m_grade			期末 f_grade		
				Chinese	Math	English	Chinese	Math	English
12001	David	M	301	91	89	79	89	75	99
12002	Denny	M	301	92	99	87	97	89	91
12003	Mary	F	301	87	76	79	91	92	89
12004	Evan	M	301	85	89	91	89	87	85
12005	Ella	F	301	93	87	99	92	75	93
12006	Frank	M	301	76	93	86	78	94	92
12007	Tom	M	302	89	79	92	89	96	97
12008	Angela	F	302	74	99	91	97	88	89
12009	Carry	F	302	95	91	83	75	96	90
12010	Alice	F	302	87	85	86	97	79	87
12011	Tony	M	302	88	91	94	89	86	94
12012	Robert	M	302	83	85	89	83	88	92
12013	Amy	F	303	94	91	97	95	91	89
12014	Peter	M	303	79	87	96	91	88	95
12015	Emma	F	303	88	89	86	94	85	85
12016	Scott	M	303	92	91	97	86	92	79
12017	Eva	F	303	85	79	77	79	85	90
12018	Sam	M	303	87	83	91	93	98	87

表 5.9 教师信息表

班级 tclass	学科 tsubject	姓名 tname
301	Chinese	Mr Wang
301	Math	Miss Cheng
301	English	Mr Li
302	Chinese	Mr Hu
302	Math	Miss Ma
302	English	Miss Wu
303	Chinese	Mr Tang
303	Math	Miss Cheng
303	English	Miss Wu

(1) 在 Hive 中创建数据库 mydatabase1。

(2) 根据表 5.8 中的信息,在数据库 mydatabase1 中新建一个分区表 mystudents,根据班级分成三个区,列与列之间用"♯"分隔。

(3) 修改表名,将表 mystudents 的表名修改为 students。

(4) 将表 5.8 中的信息装载到表 students 中。

(5) 根据表 5.9 中的信息,在数据库 mydatabase2 中新建表 teachers,列与列之间用"♯"分隔。

(6) 将表 5.9 中的信息装载到表 teachers 中。

(7) 按照期中成绩 0.4、期末成绩 0.6 的权重来计算 301 班的平均英语成绩，要求显示出学号、姓名、期中英语成绩、期末英语成绩、平均英语成绩，并要求按照平均英语成绩从高到低排序。

(8) 计算出每次成绩都大于或等于 85 分的同学的人数。

(9) 算出每位同学平均成绩，并按照下面的要求分成三个等级：大于或等于 90 为 excellent、小于 90 且大于或等于 80 的为 fine、小于 80 的为 qualified。要求显示出学号、姓名、班级、平均成绩、等级。

(10) 计算各班语文期末成绩的平均分，选择出各班语文期末成绩大于或等于 80 分的同学，对这些同学的成绩求平均分。要求显示出班级、语文平均分。

(11) 查询出 Mr Wu 所带班级的学习成绩情况，要求显示出学号、姓名、各次成绩、班级、学科以及老师姓名。

(12) 通过创建视图的方法来完成题(11)的要求。

5.3.3 实验步骤

(1) 输入用户名、密码进入到实验平台。

(2) 在 shell 命令行中输入下列命令，新建 students_301.txt、students_302.txt 和 students_303.txt，分别向其中输入 301 班、302 班和 303 班的信息，输入完毕后保存并退出。

```
# mkdir /usr/yc-in
# cd /usr/yc-in
# vi students_301.txt
# vi students_302.txt
# vi students_303.txt
```

输入到 studnets_301.txt 中的数据：

```
12001#David#M#91#89#79#89#75#99
12002#Denny#M#92#99#87#97#89#91
12003#Mary#F#87#76#79#91#92#89
12004#Evan#M#85#89#91#89#87#85
12005#Ella#F#93#87#99#92#75#93
12006#Frank#M#76#93#86#78#94#92
```

输入到 students_302.txt 中的数据：

```
12007#Tom#M#89#79#92#89#96#97
12008#Angela#F#74#99#91#97#88#89
12009#Carry#F#95#91#83#75#96#90
12010#Alice#F#87#85#86#97#79#87
12011#Tony#M#88#91#94#89#86#94
12012#Robert#M#83#85#89#83#88#92
```

输入到 students_303.txt 中的数据：

```
12013#Amy#F#94#91#97#95#91#89
12014#Peter#M#79#87#96#91#88#95
12015#Emma#F#88#89#86#94#85#85
12016#Scott#M#92#91#97#86#92#79
12017#Eva#F#85#79#77#79#85#90
12018#Sam#M#87#83#91#93#98#87
```

（3）在 shell 命令行输入下面的命令，启动 Hadoop，并从根目录下进入 Hive 的 CLI 命令行模式。

```
#cd /usr/hadoop/hadoop/sbin
#./start-all.sh
#cd
#hive
hive>
```

（4）创建数据库 mydatabase1。

```
hive> CREATE DATABASE mydatabase1;
OK
Time taken: 0.811 seconds
hive>
```

（5）在数据库 mydatabase1 中创建分区表 mystudents。

```
hive> USE mydatabase1;
OK
Time taken: 0.022 seconds
hive> CREATE TABLE mystudents (
    > sid INT,
    > sname STRING,
    > sgender STRING,
    > m_Chinese INT,
    > m_Math INT,
    > m_English INT,
    > f_Chinese INT,
    > f_Math INT,
    > f_English INT)
    > PARTITIONED BY (sclass INT)
    > ROW FORMAT DELIMITED FIELDS TERMINATED BY '#';
OK
Time taken: 1.663 seconds
hive> SHOW TABLES;
OK
mystudents
Time taken: 0.079 seconds, Fetched: 1 row(s)
hive>
```

（6）修改表名，将表 mystudents 修改为 students。

```
hive> ALTER TABLE mystudents RENAME TO students;
```

```
OK
Time taken: 2.282 seconds
hive> SHOW TABLES;
OK
students
Time taken: 0.049 seconds, Fetched: 1 row(s)
hive>
```

(7) 向分区表 students 中装载数据并查询。

```
hive> LOAD DATA LOCAL INPATH '/usr/yc-in/students_301.txt' INTO TABLE students
    > PARTITION(sclass='301');
Loading data to table mydatabase1.students partition (sclass=301)
Partition mydatabase1.students{sclass=301} stats: [numFiles=1, numRows=0, totalSize=189, rawDataSize=0]
OK
Time taken: 15.854 seconds
hive> LOAD DATA LOCAL INPATH '/usr/yc-in/students_302.txt' INTO TABLE students
    > PARTITION(sclass='302');
Loading data to table mydatabase1.students partition (sclass=302)
Partition mydatabase1.students{sclass=302} stats: [numFiles=1, numRows=0, totalSize=191, rawDataSize=0]
OK
Time taken: 1.107 seconds
hive> LOAD DATA LOCAL INPATH '/usr/yc-in/students_303.txt' INTO TABLE students
    > PARTITION(sclass='303');
Loading data to table mydatabase1.students partition (sclass=303)
Partition mydatabase1.students{sclass=303} stats: [numFiles=1, numRows=0, totalSize=185, rawDataSize=0]
OK
Time taken: 0.801 seconds
hive> SELECT * FROM students;
OK
12001   David    M   91   89   79   89   75   99   301
12002   Denny    M   92   99   87   97   89   91   301
12003   Mary     F   87   76   79   91   92   89   301
12004   Evan     M   85   89   91   89   87   85   301
12005   Ella     F   93   87   99   92   75   93   301
12006   Frank    M   76   93   86   78   94   92   301
12007   Tom      M   89   79   92   89   96   97   302
12008   Angela   F   74   99   91   97   88   89   302
12009   Carry    F   95   91   83   75   96   90   302
12010   Alice    F   87   85   86   97   79   87   302
12011   Tony     M   88   91   94   89   86   94   302
12012   Robert   M   83   85   89   83   88   92   302
12013   Amy      F   94   91   97   95   91   89   303
12014   Peter    M   79   87   96   91   88   95   303
```

```
12015    Emma    F    88    89    86    94    85    85    303
12016    Scott   M    92    91    97    86    92    79    303
12017    Eva     F    85    79    77    79    85    90    303
12018    Sam     M    87    83    91    93    98    87    303
Time taken: 3.514 seconds, Fetched: 18 row(s)
hive>
```

(8) 退出到 shell 命令行,输入下面的命令,在目录/usr/yc-in 下新建 teachers.txt 文档,分别向其中输入表 5.9 中的教师信息,输入完毕后保存并退出。

```
# cd /usr/yc-in
# vi teachers.txt
```

输入到 teachers.txt 中的数据:

```
301#Chinese#Mr Wang
301#Math#Miss Cheng
301#English#Mr Li
302#Chinese#Mr Hu
302#Math#Miss Ma
302#English#Miss Wu
303#Chinese#Mr Tang
303#Math#Miss Cheng
303#English#Miss Wu
```

(9) 从根目录下进入 Hive 的 CLI 命令行下,在数据库 mydatabase1 中新建表 teachers。

```
# cd
# hive
hive> USE mydatabase1;
OK
Time taken: 1.18 seconds
hive> CREATE TABLE teachers(
    > tclass INT,
    > tsubject STRING,
    > tname STRING)
    > ROW FORMAT DELIMITED FIELDS TERMINATED BY '#';
OK
Time taken: 1.43 seconds
```

(10) 向表 teachers 中装载数据并查询。

```
hive> LOAD DATA LOCAL INPATH '/usr/yc-in/teachers.txt' INTO TABLE teachers;
Loading data to table mydatabase1.teachers
Table mydatabase1.teachers stats: [numFiles=1, totalSize=173]
OK
Time taken: 2.59 seconds
hive> SELECT * FROM teachers;
OK
301     Chinese     Mr Wang
301     Math        Miss Cheng
```

```
301     English     Mr Li
302     Chinese     Mr Hu
302     Math        Miss Ma
302     English     Miss Wu
303     Chinese     Mr Tang
303     Math        Miss Cheng
303     English     Miss Wu
Time taken: 0.631 seconds, Fetched: 9 row(s)
hive >
```

(11) 按照期中成绩 0.4、期末成绩 0.6 的权重来计算 301 班每位同学的平均英语成绩。

```
hive > SELECT t.* FROM
    > (SELECT sid, sname, m_English, f_English, m_English * 0.4 + f_English * 0.6 as avg_English
    > FROM students
    > WHERE sclass = '301' ) t
    > SORT BY t.avg_English DESC;
...
OK
12005   Ella    99    93    95.4
12001   David   79    99    91.0
12006   Frank   86    92    89.6
12002   Denny   87    91    89.4
12004   Evan    91    85    87.4
12003   Mary    79    89    85.0
Time taken: 12.053 seconds, Fetched: 6 row(s)
```

在该查询语句中，使用了嵌套 SELECT 语句。先将表 students 中的列 sid、sname、m_English、f_English 中的内容查询出来并通过 m_English * 0.4 + f_English * 0.6 as avg_English 计算出 avg_English。

借助表 t 是为了能够使用 t.avg_English 列名，对这 18 条数据按照 avg_English 从高到低的顺序进行排序。

该命令的第一行 SELECT t.* FROM…便是查询表 t 中的所有列，一共有 5 列数据项。

(12) 计算出每次成绩都大于 85 分的人数。

```
hive > SELECT count(sid) FROM students
    > WHERE m_Chinese > 85
    > AND m_Math > 85
    > AND m_English > 85
    > AND f_Chinese > 85
    > AND f_Math > 85
    > AND f_English > 85;
...
OK
3
Time taken: 26.7 seconds, Fetched: 1 row(s)
```

(13) 算出每位同学的平均成绩，并按照下面的要求分成三个等级：大于或等于 90 分的为 excellent，小于 90 分且大于或等于 85 分的为 fine，小于 85 分的为 qualified。

```
hive > FROM (
    > SELECT sid,sname,sclass,
    > (m_Chinese + m_Math + m_English + f_Chinese + f_Math + f_English)/6 as avg_grade
    > FROM students ) s
    > SELECT sid,sname,sclass,avg_grade,
    > CASE
    > WHEN avg_grade > = 90 THEN 'excellent'
    > WHEN avg_grade < 90 AND avg_grade > = 85 THEN 'fine'
    > ELSE 'qualified'
    > END AS bracke;
OK
12001    David    301    87.0                  fine
12002    Denny    301    92.5                  excellent
12003    Mary     301    85.66666666666667     fine
12004    Evan     301    87.66666666666667     fine
12005    Ella     301    89.83333333333333     fine
12006    Frank    301    86.5                  fine
12007    Tom      302    90.33333333333333     excellent
12008    Angela   302    89.66666666666667     fine
12009    Carry    302    88.33333333333333     fine
12010    Alice    302    86.83333333333333     fine
12011    Tony     302    90.33333333333333     excellent
12012    Robert   302    86.66666666666667     fine
12013    Amy      303    92.83333333333333     excellent
12014    Peter    303    89.33333333333333     fine
12015    Emma     303    87.83333333333333     fine
12016    Scott    303    89.5                  fine
12017    Eva      303    82.5                  qualified
12018    Sam      303    89.83333333333333     fine
Time taken: 0.218 seconds, Fetched: 18 row(s)
```

(14) 选择出各班语文期末成绩大于或等于 80 分的同学，对这些同学的成绩求平均分。

```
hive > SELECT sclass,avg(f_Chinese) FROM students
    > WHERE f_Chinese > = 80
    > GROUP BY sclass;
...
OK
301    91.6
302    91.0
303    91.8
Time taken: 34.9 seconds, Fetched: 3 row(s)
hive >
```

(15) 查询出 Miss Wu 所带班级的科目，以及该班级的学习成绩情况。

```
hive > SELECT sid,sname,m_Chinese,m_Math,m_English,f_Chinese,f_Math,f_English,tclass,
tsubject,tname
```

```
        > FROM students JOIN teachers ON students.sclass = teachers.tclass
        > WHERE tname = 'Miss Wu';
...
OK
12007   Tom     89   79   92   89   96   97   302   English     Miss Wu
12008   Angela  74   99   91   97   88   89   302   English     Miss Wu
12009   Carry   95   91   83   75   96   90   302   English     Miss Wu
12010   Alice   87   85   86   97   79   87   302   English     Miss Wu
12011   Tony    88   91   94   89   86   94   302   English     Miss Wu
12012   Robert  83   85   89   83   88   92   302   English     Miss Wu
12013   Amy     94   91   97   95   91   89   303   English     Miss Wu
12014   Peter   79   87   96   91   88   95   303   English     Miss Wu
12015   Emma    88   89   86   94   85   85   303   English     Miss Wu
12016   Scott   92   91   97   86   92   79   303   English     Miss Wu
12017   Eva     85   79   77   79   85   90   303   English     Miss Wu
12018   Sam 87  83   91   93   98   87   303   English Miss Wu
Time taken: 17.832 seconds, Fetched: 12 row(s)
hive >
```

（16）通过视图的方法完成题（11）的要求。

```
hive > CREATE VIEW view AS
     > SELECT sid, sname, m_Chinese, m_Math, m_English, f_Chinese, f_Math, f_English, tclass,
tsubject, tname
     > FROM students JOIN teachers ON students.sclass = teachers.tclass
     > WHERE tname = 'Miss Wu';
OK
Time taken: 2.756 seconds
hive > SELECT * FROM view;
...
OK
12007   Tom     89   79   92   89   96   97   302   English     Miss Wu
12008   Angela  74   99   91   97   88   89   302   English     Miss Wu
12009   Carry   95   91   83   75   96   90   302   English     Miss Wu
12010   Alice   87   85   86   97   79   87   302   English     Miss Wu
12011   Tony    88   91   94   89   86   94   302   English     Miss Wu
12012   Robert  83   85   89   83   88   92   302   English     Miss Wu
12013   Amy     94   91   97   95   91   89   303   English     Miss Wu
12014   Peter   79   87   96   91   88   95   303   English     Miss Wu
12015   Emma    88   89   86   94   85   85   303   English     Miss Wu
12016   Scott   92   91   97   86   92   79   303   English     Miss Wu
12017   Eva     85   79   77   79   85   90   303   English     Miss Wu
12018   Sam 87  83   91   93   98   87   303   English Miss Wu
Time taken: 15.278 seconds, Fetched: 12 row(s)
hive >
```

第 6 章

Hadoop数据库：HBase

6.1 HBase 概述

HBase(Hadoop Database)是一个高可靠、高性能、面向列、可伸缩的分布式数据库,利用 HBase 技术可在廉价 PC 上搭建起大规模结构化存储集群。HBase 参考 Google 的 BigTable 建模,使用类似 GFS 的 HDFS 作为底层文件存储系统,在其上可以运行 MapReduce 批量处理数据,使用 ZooKeeper 作为协同服务组件。

HBase 的整个项目使用 Java 语言实现,它是 Apache 基金会的 Hadoop 项目的一部分,既是模仿 Google BigTable 的开源产品,同时又是 Hadoop 的衍生产品。Hadoop 作为批量离线计算系统已经得到了业界的普遍认可,并经过了工业上的验证,所以 HBase 具备"站在巨人肩膀之上"的优势,其发展势头非常迅猛。HBase 还是一种非关系型数据库,即 NoSQL 数据库。在 Eric Brewer 的 CAP 理论中,HBase 属于 CP 类型的系统,其 NoSQL 的特性非常明显,这些特性也决定了其独特的应用场景。接下来的内容将详细介绍 HBase 的发展历史、发行版本和特性。

6.1.1 HBase 的发展历史

Apache HBase 最初是 Powerset 公司为了处理自然语言搜索产生的海量数据而开展的项目,由 Chad Walters 和 Jim Kellerman 两人发起,经过两年的发展之后被 Apache 基金会收录为顶级项目,同时成为非常活跃、影响重大的项目。

2006 年 11 月,Google 开放了论文 BigTable: A Distributed Storage System for structured Data,该论文介绍的就是 HBase 的原型。2007 年 2 月,倡导者提出作为 Hadoop

的模块的 HBase 原型,该原型包含 HBase 的基本介绍、表设计、行键(Rowkey)设计和底层数据存储结构设计等内容。

经过一段时间的酝酿和开发工作,在 2007 年 10 月第一个可用的、简单的 HBase 版本发布,该版本只实现了最基本的模块和功能,因为只是初始开发阶段,此时的 HBase 版本发展很不完善。2008 年 1 月,Hadoop 升级为顶级项目,HBase 作为 Hadoop 的一个子项目存在,HBase 的活跃度非常高,在短短不到 2 年的时间经历了多个版本的发布,并且其中包含了版本号的大"跳跃"。下面是一些版本发布的信息:

- HBase-0.18.0 于 2008 年 9 月 21 日发布。
- HBase-0.20.6 于 2010 年 7 月 10 日发布。
- HBase-0.89.20100621 于 2010 年 6 月 25 日发布。

其中,从 HBase-0.20.6 到 HBase-0.89.20100621 版本,经历了版本的大"跳跃"。在 2009 年秋季发布 0.20 系列版本后,HBase 经历了发展历史上的一次版本大变动,在此之前的版本都追随 Hadoop 的主版本,例如,HBase 0.X.* 版本都会伴随着 Hadoop 0.X.* 版本,之所以出现版本跳跃,官方给出的解释有两点。

- Hadoop 的版本更新速度已经放缓,而 HBase 相比 Hadoop 开发来讲更加活跃,发布版本更加频繁,并且 Hadoop 已经有多个分支,HBase 也需要兼容多个分支,所以不再需要与 Hadoop 的版本更新步伐保持一致。
- 从 HBase 的功能实现上来讲,已经基本实现 BigTable 论文中实现的功能,也就是 HBase 的实现已经接近 1.0,应该赋予一个更接近 1.0 的版本。

"跳跃"之后的版本发布比较规律,先后经历了 0.90.*、0.92.*、0.94.*、0.96.*、0.98.*、1.0.*、1.1* 七个大的版本,现在的稳定版本是 1.0.1.1。

6.1.2 HBase 的发行版本

本节主要介绍现有 HBase 的版本知识,从 0.90.0 之后,HBase 的版本更新是非常有规律的,可以从 0.90.0、0.91.0、0.92.0、0.93.0、0.94.0、0.95.0、0.96.0、0.97.0、0.98.0、0.99.0、1.0.0、1.1.0 这样的版本变化中发现一些规律。

这些版本都是大版本,其中偶数版本是稳定发布版,奇数版本都是开发版,基本不对外发布,但是可以在官方 JIRA 的项目管理系统中找到这些奇数版本对应的开发信息,并且可以在 SVN 上找到相关的最新开发代码。所以,偶数发布版本属于稳定版本,奇数开发版本属于不稳定版本,一般不建议用户在生产环境中使用开发版本,这些也是大版本的发布规律。

小版本一般基于当前大版本的问题进行修正,一般表示小版本的数字在 1~100 之间,例如,0.98.1、0.98.2、0.98.3、0.98.4。这些小版本都是基于 0.98 大版本的,截止到编写本书时,最新版本是 1.1.1。小版本都是从小到大依次递增,不存在版本跳跃的情况。对于小版本而言,原则上数值越大越稳定,因为小版本都是基于某一个大版本的,在小版本中并不会增加新特征,而是修正一些代码的漏洞和问题。

6.1.3　HBase 的特性

HBase 作为一个典型的 NoSQL 数据库，可以通过行键检索数据，仅支持行事务，主要用于存储非结构化和半结构化的松散数据。与 Hadoop 相同，HBase 主要依靠横向扩展，通过不断增加廉价的商用服务器来增加计算和存储能力。

HBase 有不少特性，这些特性都标志着 HBase 的特立独行、与众不同，同时其良好的出身和特性也奠定了其在大数据处理领域的地位。下面介绍 HBase 具备的一些非常显著的特点。

1．容量巨大

HBase 的单表可以有百亿行、百万列，数据矩阵横向和纵向两个维度所支持的数据量级都非常具有弹性。传统的关系型数据库，如 Oracle 和 MySQL 等，如果数据记录在亿级别，查询和写入的性能都会呈指数级下降，所以更大的数据量级对传统数据库来讲是一种灾难。而 HBase 对于存储百亿、千亿甚至更多的数据处理都不存在任何问题。对于高维数据，百万量级的列没有任何问题。有的读者可能关心更加多的列——千万和亿级别，这种非常特殊的应用场景，并不是说 HBase 不支持，而是这种情况下访问单个行键可能造成访问超时，如果限定某个列则不会出现这种问题。

2．面向列

HBase 是面向列的存储和权限控制，并支持列独立检索。有些读者可能不清楚什么是列式存储，下面进行简单介绍。列式存储不同于传统的关系型数据库，其数据在表中是按列存储的，这样在查询只需要少数几个字段的时候，能大大减少读取的数据量，比如一个字段的数据聚集存储，那就更容易为这种聚集存储设计更好的压缩和解压算法。下面是传统行式数据库与列式数据库的不同特性。

传统行式数据库的特性如下：
- 数据是按行存储的。
- 没有索引的查询使用大量 I/O。
- 建立索引和物化视图需要花费大量的时间和资源。
- 面对查询需求，数据库必须被大量膨胀才能满足需求。

列式数据库的特性如下：
- 数据按列存储，即每一列单独存放。
- 数据即索引。
- 只访问查询涉及的列，可以大量降低系统 I/O。
- 每一列由一个线索来处理，即查询的并发处理性能高。
- 数据类型一致，数据特征相似，可以高效压缩。

列式存储不但解决了数据稀疏性问题，最大程度上节省存储开销，而且在查询发生时，仅检索查询涉及的列，能够大量降低磁盘 I/O。这些特性也支撑 HBase 能够保证一定的读写性能。

3．稀疏性

在大多数情况下，采用传统行式存储的数据往往是稀疏的，即存在大量为空（NULL）的列,而这些列都是占用存储空间的,这就造成了存储空间的浪费。对于 HBase 来说,为空的列不占用存储空间,因此,表可以设计得非常稀疏。

4. 扩展性

HBase 底层文件存储依赖 HDFS,从"基因"上决定了其具备可扩展性。这种遗传的可扩展性就如同 OOP 中的继承,"父类"HDFS 的扩展性遗传到 HBase 框架中。这是最底层的关键点。同时,HBase 的 Region 和 RegionServer 的概念对应的数据可以分区,分区后数据可以位于不同的机器上,所以在 HBase 核心架构层面也具备可扩展性。HBase 的扩展性是热扩展,在不停止现有服务的前提下,可以随时添加或者减少节点。

5. 高可靠性

HBase 提供 WAL 和 Replication 机制。前者保证了数据写入时不会因集群异常而导致写入数据丢失;后者保证了在集群出现严重问题时,数据不会发生丢失或者损坏。而且 HBase 底层使用 HDFS,HDFS 本身的副本机制在很大程度上保证了 HBase 的高可靠性。同时,协调服务的 ZooKeeper 组件是经过工业验证的,具备高可用性和高可靠性。

6. 高性能

底层的 LSM 数据结构和行键有序排列等架构上的独特设计,使得 HBase 具备非常高的写入性能。Region 切分、主键索引和缓存机制使得 HBase 在海量数据下具备一定的随机读取性能,该性能针对行键的查询能够达到毫秒级别。同时,HBase 对于高并发的场景也具备很好的适应能力。该特性也是业界众多公司选取 HBase 作为存储数据库非常重要的一点。

6.1.4 HBase 与 Hadoop 的关系

HBase 参考了 Google 的 BigTable 建模,且将下面三篇博文作为 HBase 实现的理论基础:

- BigTable by Google(2006)
- HBase and HDFS Locality by Lars George(2010)
- No Relation:The Mixed Blessings of Non-Relational Databases by Ian Varley(2009)

HBase 和 HDFS 有着非常紧密的关系,更准确的说法是:HBase 严重依赖 Hadoop 的 HDFS 组件,HBase 使用 HDFS 作为底层存储系统。因此,如果要使用 HBase,前提是首先必须有 Hadoop 系统。Hadoop 的组件之一 MapReduce 可以直接访问 HBase,但是这不是必需的,因为 HBase 中最重要的访问方式是原生 Java API,而不是 MapReduce 这样的批量操作方式。图 6.1 展示了 HBase 在 Hadoop 生态系统中的位置。

6.1.5 HBase 的核心功能模块

Hadoop 框架包含两个核心组件:HDFS 和 MapReduce,其中 HDFS 是文件存储系统,负责数据存储;MapReduce 是计算框架,负责数据计算。它们之间分工明确、低度耦合、相关关联。对于 HBase 数据库的核心组件,即核心功能模块共有四个,分别是客户端 Client、

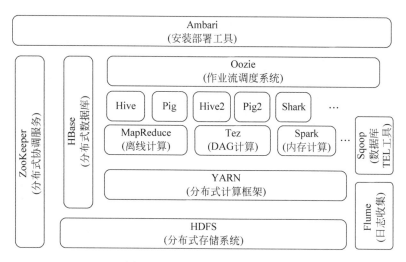

图 6.1　Hadoop 生态系统总图

协调服务模块 ZooKeeper、主节点 HMaster 和 Region 节点 RegionServer，这些组件相互之间的关联关系如图 6.2 所示。

6.1.5.1　客户端 Client

客户端 Client 是整个 HBase 系统的入口。使用者直接通过客户端操作 HBase。客户端使用 HBase 的 RPC 机制与 HMaster 和 RegionServer 进行通信。对于管理类操作，Client 与 HMaster 进行 RPC 通信；对于数据读写类操作，Client 与 RegionServer 进行 RPC 交互。这里客户端可以是多个，并不限定是原生 Java 接口，还有 Thrift、Avro、Rest 等客户端模式，甚至 MapReduce 也可以算作一种客户端。

6.1.5.2　协调服务组件 ZooKeeper

ZooKeeper Quorum（队列）负责管理 HBase 中多 HMaster 的选举、服务器之间状态同步等。再具体一些就是，HBase 中 ZooKeeper 实例负责的协调工作有：存储 HBase 元数据信息、实时监控 Regionserver、存储所有 Region 的寻址入口，当然还有最常见的功能就是保证 HBase 集群中只有一个 HMaster 节点。

6.1.5.3　主节点 HMaster

HMaster 没有单点故障问题，在 HBase 中可以启动多个 HMaster，通过 ZooKeeper 的 Master 选举机制保证总有一个 HMaster 正常运行并提供服务，其他 HMaster 作为备选时刻准备（当目前 HMaster 出现问题时）提供服务。HMaster 主要负责 Table 和 Region 的管理工作：

- 管理用户对 Table 的增、删、改、查操作。
- 管理 RegionServer 的负载均衡，调整 Region 分布。
- 在 Region 分裂后，负责新 Region 的分配。
- 在 RegionServer 死机后，负责失效 RegionServer 上的 Region 迁移。

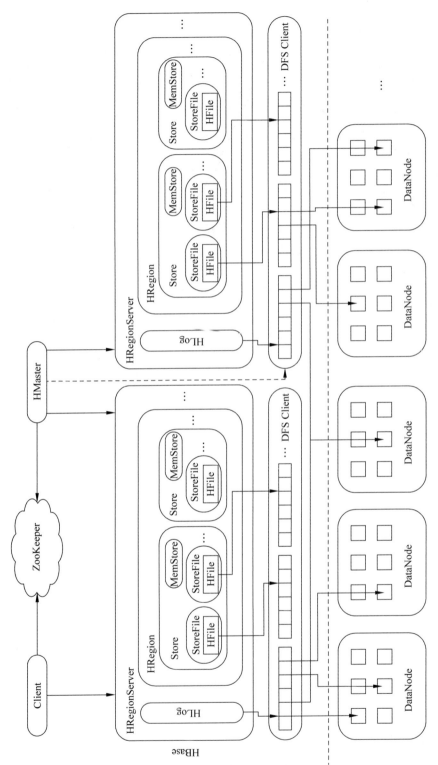

图 6.2 HBase 架构图

6.1.5.4 Region 节点 HRegionServer

HRegionServer 主要负责响应用户 I/O 请求，向 HDFS 文件系统中读写数据，是 HBase 中最核心的模块。HRegionServer 的组成结构如图 6.3 所示。HRegion 内部管理了一系列 HRegion 对象，每个 HRegion 对应了 Table 中的一个 Region。HRegion 由多个 HStore 组成，每个 HStore 对应了 Table 中的一个 Column Family 的存储。可以看出每个 Column Family 其实就是一个集中的存储单元，因此最好将具备共同 I/O 特性的列放在一个 Column Family 中，这样能保证读写的高效性。

HStore 存储是 HBase 存储的核心，由两部分组成：MemStore 和 StoreFile。Memstore 是 Sorted Memory Buffer，用户写入的数据首先会放入 MemStore 中，当 MemStore 满了以后会缓冲(flush)成一个 StoreFile(底层实现是 HFile)，当 StoreFile 文件数量增长到一定阈值，会触发 Compact 操作，将多个 StoreFiles 合并成一个 StoreFile，在合并过程中会进行版本合并和数据删除，因此可以看出 HBase 其实只有增加数据，所有的更新和删除操作都是在后续的 Compact 过程中进行的，这使得用户的写操作只要进入内存中就可以立即返回，保证了 HBase I/O 的高性能。

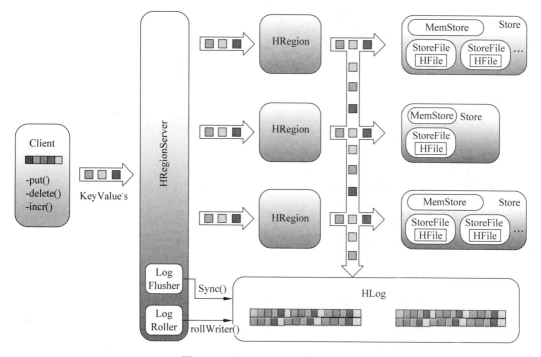

图 6.3　HRegionServer 的组成结构

StoreFiles 在触发 Compact 操作后，会逐步形成越来越大的 StoreFile，当单个 StoreFile 大小超过一定阈值后，会触发 Split 操作，同时把当前 Region 分裂成两个 Region，父 Region 会下线，新分裂的两个子 Region 会被 HMaster 分配到相应的 HRegionServer 上，使得原先一个 Region 的压力得以分流到两个 Region 上。

每个 HRegionServer 中都有一个 HLog 对象，HLog 是一个实现 Write Ahead Log 的类，在每次用户操作写入 MemStore 的同时，也会写一份数据到 HLog 文件中，HLog 文件定期会滚动出新文件,并删除旧的文件(已持久化到 StoreFile 中的数据)。在 HRegionServer 意外终止后,HMaster 会通过 ZooKeeper 感知到,首先处理遗留的 HLog 文件,将其中不同 Region 的 Log 数据进行拆分,分别放到相应 Region 的目录下,然后再将失效的 Region 重新分配,领取到这些 Region 的 HRegionServer 在加载 Region 的过程中,会发现有历史 HLog 需要处理,因此会将 HLog 中的数据回放到 MemStore 中,然后缓冲(flush)到 StoreFiles,完成数据恢复。

6.2 HBase 的安装和配置

本节将讲述如何安装、部署、启动和停止 HBase 集群,以及如何通过命令行的方式对 HBase 进行基本操作,如插入、查询、删除数据。

配置 HBase 的方式与 Hadoop 类似,都是通过修改安装包的 conf 目录下的配置文件完成的。在对一台机器修改配置文件后要记得同步到集群中的所有节点上,此时可以使用 scp 或 rsync 命令。在大多数情况下,需要重新启动 HBase 使配置生效。

在开始安装 HBase 之前需要做一些准备工作,这涉及操作系统设置、分布式模式 Hadoop 的部署及 HBase 自身的配置,因此要确保在运行 HBase 之前已经具备这些条件。下面将介绍 HBase 依赖的一些重要的中间件、系统服务或配置。

1. 运行时环境 JDK

与 Hadoop 一样,HBase 需要 JDK I.6 或者更高版本,推荐采用 Oracle 公司的版本,对于 JDK I.6 不要使用 u18 及以前的版本,因为这些版本 Java 的垃圾收集器会遇到 jvm crash 的问题(可以通过 Goolge 搜索 jvm crash,找到该问题的详细描述)。为了能够管理超过 4GB 内存空间,需要安装 64 位的 JDK。

2. SSH 服务

集群模式的 HBase 的启动和关闭依赖于 SSH 服务,所以操作系统必须安装该服务,sshd 进程必须处于运行状态,可执行以下命令查看服务状态:

```
service sshd status
```

HBase 通过 SSH 管理所有节点的守护进程,与 Hadoop 的 NameNode 一样,HMaster 必须能够免密登录到集群的所有节点,可以通过 Google 搜索"SSH 免密登录"找到配置方法。注意由于 HBase 通常有多个 HMaster 节点,所以需要每个 HMaster 到所有节点都可以免密登录。

3. 域名系统 DNS

HBase 通过本地主机名(Host Name)或域名(Domain Name)来获取 IP 地址,因此要确保正向和反向 DNS 解析是正常的。在进行 DNS 解析时会首先查询本地/etc/hosts 文件,因此建议通过配置该文件指定主机名或域名到 IP 地址的映射关系而不使用域名解析服务,这样做更容易维护,当出现主机无法识别的异常时也更加容易定位问题出现的位置,并且通过本地/etc/hosts 文件解析 IP 地址速度也会更快一些。

当决定使用 DNS 服务的时候,还可以通过如下设置更加精确地控制 HBase 的行为。

如果有多个网卡,则可以通过参数 hbase.regionservers.dns.interface 指定主网卡,该配置参数的默认值是 default，可以通过这个参数指定主网络接口，不过这要求集群所有节点的配置是相同的且每台主机都使用相同的网卡配置，可以修改这个配置参数为 eth0 或 eth1，这要视具体的硬件配置而定。

另外一个配置是指定 hbase.regionserver.dns.nameserver 可以选择一个不同的 DNS 的 name server。

4. 本地环回地址 Loopback IP

HBase 要求将本地回环接口配置成 127.0.0.1，可以在 /etc/hosts 文件配置，通常系统安装后就已经包含了该配置。

```
127.0.0.1 localhost
```

5. 网络时间协议 NTP

HBase 要求集群中节点间的系统时间要基本一致，可以容忍一些偏差，默认相差 30s 以内。可以通过设置参数 hbase.master.maxclockskew 属性值修改最大容忍偏差时间。偏差时间较多时集群会产生一些奇怪的行为。用户需要在集群中运行 NTP 服务来同步集群的时间，如果在运行正常的集群中读取数据时发生了一些莫名其妙的问题，例如读到的不是刚写进集群的数据而是旧的数据，这时需要检查集群各节点的时间是否同步。

6. 资源限制命令：ulimit 和 nproc

HBase 和其他的数据库软件一样会同时打开很多文件。Linux 中默认的 ulimit 值是 1024，这对 HBase 来说太小了。当使用诸如 bulkload 这种工具批量导入数据的时候会得到这样的异常信息：java.io.IOException：Too many open files。我们需要改变这个值，注意，这是对操作系统的参数调整，而不是通过 HBase 配置文件完成的。可以大致估算 ulimit 值需要配置为多大，例如，每个列族至少有一个存储文件（HFile），每个被加载的 Region 可能管理多达 5 或 6 个列族所对应的存储文件，用存储文件的个数乘以列族数再乘以每个 RegionServer 中的 Region 数量得到 RegionServer 主机管理的存储文件数量。假如每个 Region 有 3 个列族，每个列族平均有 3 个存储文件，每个 RegionServer 有 100 个 region，将至少需要 3×3×100＝900 个文件。这些存储文件会频繁被客户端调用，涉及大量的磁盘操作。应根据实际情况调整 ulimit 参数值的大小。

关于 ulimit 有两个地方需要调整，通过在 etc/security/limits.conf 追加参数进行设置，一个参数是 nofile，设置如下：

```
*    soft mofile 10240
*    hard mofile 10240
```

由于上面已经讲解了如何安装 Hadoop，故此处不再赘述。

6.2.1 HBase 的运行模式

6.2.1.1 单机模式

这是 HBase 默认的运行模式，在单机模式中，HBase 使用本地文件系统，而不是

HDFS,所有的服务和ZooKeeper都运行在一个JVM中。ZooKeeper监听一个端口,这样客户端就可以连接HBase了。安装步骤如下:

1. 安装HBase

(1) 在Apache镜像网站上下载一个稳定版本的安装包,可以在http://mirror.bit.edu.cn/apache/下载并安装,本书使用的版本是HBase-1.0.1.1。

(2) 在工作目录下解压已下载好的HBase-1.0.1.1-bin.tar.gz安装包。

```
# tar -zxvf HBase-1.0.1.1-bin.tar.gz
```

(3) 将解压后的目录HBase-1.0.1.1修改为HBase。

```
# mv HBase-1.0.1.1 HBase
```

(4) 配置系统环境变量,增加HBASE_HOME路径,并将HBase的bin文件夹添加到PATH中,修改后如图6.4所示。

图6.4 编辑配置文件

(5) 修改/usr/HBase/conf目录下的配置文件HBase-env.sh,设置JAVA_HOME的路径,并将HBASE_MANAGES_ZK的属性设置为true,修改后如图6.5所示。

注意:#true表示zookeeper交给HBase管理,启动HBase时,会自动启动HBase-site.xml的HBase.zookeeper.quorum属性中的所有ZooKeeper实例。

#false表示启动HBase时,要手动启动所有ZooKeeper实例。

(6) 修改/usr/HBase/conf目录下的配置文件HBase-site.xml,在文件的最后两行<configuration></configuration>之间添加配置信息,如下所示:

```
<configuration>
<property>
<name>HBase.rootdir</name>
<value>file:///usr/HBase/HBase-tmp</value>
```

```
</property>
<property>
<name>HBase.zookeeper.property.dataDir</name>
<value>/usr/HBase/zookeeper</value>
</property>
<configuration>
```

图 6.5 编辑配置文件

注意：

HBase.tootdir 代表 HBase 数据存放的位置，单机模式下存储到本地目录，即存储到 /usr/HBase/HBase-tmp 下面。

HBase.zookeeper.property.dataDir 代表 ZooKeeper 数据存放的位置，即存储到 /usr/HBase/zookeeper 下面。

2. 启动 HBase

进入到 HBase 目录下，运行如下脚本启动 HBase：

```
# bin/start-HBase.sh
```

如图 6.6 所示，成功启动 HBase。

图 6.6 启动 HBase

启动成功后通过 jps 命令可以看到如下信息：

```
# jps
```

如图 6.7 所示，成功启动 HMaster 端口。

```
[root@VM-6f567646-27a1-46a7-94a6-8e3a9c05e926 hbase]# jps
4099 Jps
3839 HMaster
```

图 6.7　查看进程信息

通过如下命令可以连接到 HBase 表示安装成功：

```
# bin/HBase shell
```

连接成功后显示如图 6.8 所示。

```
[root@VM-6f567646-27a1-46a7-94a6-8e3a9c05e926 hbase]# bin/hbase shell
2015-08-12 14:46:02,838 WARN  [main] util.NativeCodeLoader: Unable to load nativ
e-hadoop library for your platform... using builtin-java classes where applicabl
e
HBase Shell; enter 'help<RETURN>' for list of supported commands.
Type "exit<RETURN>" to leave the HBase Shell
Version 1.0.1.1, re1dbf4df30d214fca14908df71d038081577ea46, Sun May 17 12:34:26
PDT 2015

hbase(main):001:0>
```

图 6.8　进入命令行

6.2.1.2　伪分布式模式

HBase 分布式模式有两种。伪分布式模式是把所有进程运行在一台机器上，但不是一个 JVM 上；而完全分布式模式就是把整个服务分布在各个节点上。无论采用哪种分布模式，都需要使用 HDFS。在操作 HBase 之前，要确认 HDFS 可以正常运行。在安装 HBase 之后，需要确认伪分布式模式或完全分布式模式的配置是否正确，这两种模式可以使用相同的验证脚本。

此处只对 HBase 伪分布式的安装与配置进行介绍。伪分布式模式是一个相对简单的分布式模式，是用于测试的。不能把这个模式用于生产环节，也不能用于测试。

1. 在 HBase 单机模式的基础上修改 HBase 的配置文件

（1）修改 /usr/HBase/conf 目录下的配置文件 HBase-env.sh，在图 6.9 标出的位置添加 HBASE_CLASSPATH，HBASE_CLASSPATH 就是 Hadoop 配置文件的路径。

```
文件(F) 编辑(E) 查看(V) 搜索(S) 终端(T) 帮助(H)
# Set environment variables here.

# This script sets variables multiple times over the course of starting an hbase
 process.
# so try to keep things idempotent unless you want to take an even deeper look
# into the startup scripts (bin/hbase, etc.)

# The java implementation to use.  Java 1.7+ required.
export JAVA_HOME=/usr/java/jdk
export HBASE_MANAGES_ZK=true
export HBASE_CLASSPATH=/usr/hadoop/etc/hadoop

# Extra Java CLASSPATH elements.  Optional.
# export HBASE_CLASSPATH=

# The maximum amount of heap to use. Default is left to JVM default.
# export HBASE_HEAPSIZE=1G

# Uncomment below if you intend to use off heap cache. For example, to allocate
8G of
# offheap, set the value to "8G".
# export HBASE_OFFHEAPSIZE=1G
```

图 6.9　编辑配置文件

（2）修改/usr/HBase/conf 目录下的配置文件 HBase-site.xml，将之前单机模式下的配置信息删除，重新在<configuration></configution>之间添加如下所示的配置信息：

```
<configuration>
    <property>
        <name>HBase.rootdir</name>
        <value>hdfs://localhost:9000/HBase</value>
    </property>
    <property>
        <name>dfs.replication</name>
        <value>1</value>
    </property>
    <property>
        <name>HBase.cluster.distributed</name>
        <value>true</value>
    </property>
    <property>
        <name>HBase.zookeeper.property.dataDir</name>
        <value>/usr/HBase/zookeeper</value>
    </property>
</configuration>
```

注意：

HBase.rootdir——这个目录是 RegionServer 的共享目录，用来持久化 HBase。URL 需要是"完全正确的"，还要包含文件系统的 scheme。例如，"/HBase"表示 HBase 在 HDFS 中占用的实际存储目录，HDFS 的 NameNode 运行在主机名为 localhost 的 9000 端口，则 HBase 的设置应为 hdfs://localhost:9000/HBase。在默认情况下 HBase 是写到/tmp 中的。不修改这个配置，数据会在重启时丢失。

HBase.cluster.distributed——用来指定 HBase 的运行模式。为 false 表示单机模式，为 true 表示分布式模式。若为 false，则 HBase 和 ZooKeeper 会运行在同一个 JVM 中。默认值是 false。

HBase.zookeeper.property.dataDir——这个参数用于设置 ZooKeeper 快照的存储位置。默认值是/tmp，在操作重启的时候该目录会被清空，应该修改默认值到其他目录，这里修改值为/usr/HBase/zookeeper（这个路径需要运行 HBase 的用户拥有读写操作权限）。

（3）进入到 HBase 目录下，运行如下脚本启动 HBase。这里需要注意的是，如果启动之前 HBase 单机模式仍在运行，请先关闭单机模式。

```
# bin/start-HBase.sh
```

（4）启动成功后通过 jps 命令可以看到如图 6.10 所示的信息。

```
# jps
```

（5）若通过如下命令可以连接到 HBase，则表示安装成功：

```
[root@VM-3798830c-f00b-40d5-bcbd-2d91bac5c960 HBase]# bin/HBase shell
```

```
SLF4J: Class path contains multiple SLF4J bindings.
SLF4J: Found binding in [jar:file:/usr/HBase/lib/slf4j-log4j12-1.7.7.jar!/org/slf4j/
impl/StaticLoggerBinder.class]
SLF4J: Found binding in [jar:file:/usr/hadoop/share/hadoop/common/lib/slf4j-log4j12-1.7.
10.jar!/org/slf4j/impl/StaticLoggerBinder.class]
SLF4J: See http://www.slf4j.org/codes.html#multiple_bindings for an explanation.
SLF4J: Actual binding is of type [org.slf4j.impl.Log4jLoggerFactory]
2015-09-23 10:14:43,700 WARN [main] util.NativeCodeLoader: Unable to load native-hadoop
library for your platform... using builtin-java classes where applicable
HBase Shell; enter 'help<RETURN>' for list of supported commands.
Type "exit<RETURN>" to leave the HBase Shell
Version 1.0.1.1, re1dbf4df30d214fca14908df71d038081577ea46, Sun May 17 12:34:26 PDT 2015

Hbase(main):001:0>
```

图 6.10　查看进程

(6) 启动备份的 HMaster。

可以通过如下命令启动 HBase：

```
bin/start-HBase.sh
```

也可以在同一服务器启动额外备份 HMaster：

```
bin/local-master-backup.sh start 1
```

1 表示使用端口 60001 和 60011，该备份的 HMaster 及其日志文件放在 logs/HBase-${USER}-1-regionserver$-{HOSTNAME}.log 中。

在刚运行的 RegionServer 上增加 4 个额外的 RegionServer，最多可以支持 100 个。

```
bin/local-regionservers.sh start 2 3 4 5
```

(7) 停止 HBase。

假设想停止备份 HMaster 1，则运行如下命令：

```
cat /${PID_DIR}/HBase-${USER}-1-master.pid | xargs kill -9
```

在本书中，PID_DIR 存放在默认的 /tmp 目录下，具体命令如下：

```
cat /tmp/HBase-root-1-master.pid | xargs kill -9
```

停止 RegionServer，可以运行如下命令：

```
bin/local-regionservers.sh stop 1
```

6.2.2 HBase 的 Web UI

通过 HMater 的 16010 端口可以查看 HBase 的 Web UI 界面，图 6.11 列出了集群的一些关键信息，这些信息包括 HBase 的版本、ZooKeeper 集群的主机列表、HBase 根目录等。

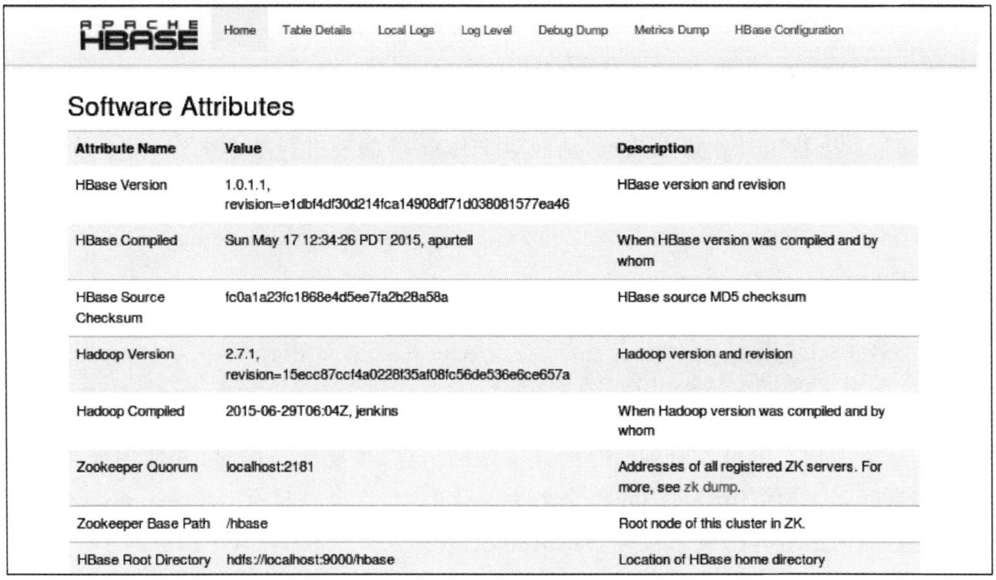

图 6.11　HBase Web UI 中的属性部分

图 6.12 列举出了目前 HBase 上有哪些表及表的一些基本信息，其中 hbase：meta 和 hbase：namespace 表是永远存在且成功加载的，如果这两张表加载不成功，那么集群虽然已经启动，却无法正常读写数据。

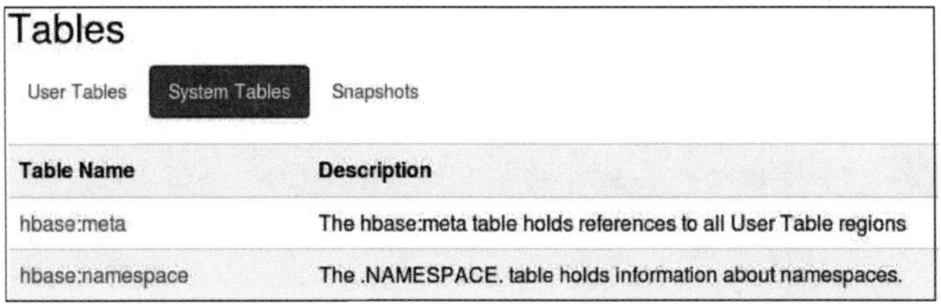

图 6.12　HBase Web UI 中表的描述部分

图 6.13 列举出了集群中的 RegionServer 主机，后面包含了每台 RegionServer 的重要信息，包括每秒请求次数、Region 的数量、当前 JVM 堆大小等信息。

图 6.13　HBase Web UI 中 RegionServer 主机的信息

6.2.3　Hbase Shell 工具使用

可以通过命令行工具连接 HBase，从而对 HBase 中的表进行基本操作，命令如下：

```
# bin/hbase shell
```

连接成功后将进入 HBase 的执行环境：

```
HBase Shell; enter 'help<RETURN>' for list of supported commands.
Type "exit<RETURN>" to leave the HBase Shell
```

输入 help，然后按回车键可以看到命令的详细帮助信息，需要注意的是，在使用命令引用到表名、行和列时需要加单引号。创建一个名为 test 的表，这个表只有一个 column family（列族）为 cf。然后列出所有的表来检查创建情况，之后加入一些数据，命令如下：

```
hbase(main):001:0> create 'test','cf'
```

如图 6.14 所示，成功创建了一张名称为 test 的表。

```
hbase(main):001:0> create 'test','cf'
0 row(s) in 3.8650 seconds

=> Hbase::Table - test
```

图 6.14　创建表

```
hbase(main):002:0> put 'test', 'row1', 'cf:a', 'value1'
hbase(main):003:0> put 'test', 'row2', 'cf:b', 'value'
hbase(main):004:0> put 'test', 'row3', 'cf:c', 'value'
```

如图 6.15 所示，以上命令分别插入了三行数据。第一行 rowkey 为 row1，列为"cf：a"，值为 value1。HBase 中的列是由 column family 前缀和列的名字组成的，以冒号分隔。

扫描整个表的数据使用 scan 命令，操作如下：

```
hbase(main):007:0> scan 'test'
```

查询结果如图 6.16 所示。

```
hbase(main):002:0> put 'test','row1','cf:a','value1'
0 row(s) in 0.4070 seconds

hbase(main):003:0> put 'test','row2','cf:b','value2'
0 row(s) in 0.0400 seconds

hbase(main):004:0> put 'test','row3','cf:c','value3'
0 row(s) in 0.4090 seconds
```

图 6.15 插入数据

```
hbase(main):007:0> scan 'test'
ROW                    COLUMN+CELL
 row1                  column=cf:a, timestamp=1439182587607, value=value1
 row2                  column=cf:b, timestamp=1439182616012, value=value2
 row3                  column=cf:c, timestamp=1439182637928, value=value3
3 row(s) in 0.2030 seconds
```

图 6.16 扫描表

获取单行数据使用 get 命令，操作如下：

```
hbase(main):0010:0> get 'test', 'cf'
```

如图 6.17 所示，成功获取一条数据。

```
hbase(main):010:0> get 'test','row1'
COLUMN                 CELL
 cf:a                  timestamp=1439182587607, value=value1
1 row(s) in 0.2900 seconds
```

图 6.17 获取数据

停用表使用 disable 命令，删除这个表可以通过 drop 命令实现，此时数据也随之删除，操作如下：

```
hbase(main):012:0> disable 'test'
hbase(main):013:0> drop 'test'
```

退出 shell 使用 exit 命令：

```
hbase(main):014:0> exit
```

6.2.4 停止 HBase 集群

通过 HBase 提供的脚本可以停止一个正在运行中的集群，命令如下：

```
# bin/stop-hbase.sh
```

在屏幕上可以看到类似下面的输出信息，请耐心等待，当 HMaster 和所有 RegionServer 进程正常退出后集群将停止服务。

```
stopping hbase…
```

第 7 章

HBase数据操作

7.1 Shell 工具的使用

HBase 的 Shell 工具是很常用的工具,运维过程的 DDL 和 DML 都会通过此进行,其具体实现是用 Ruby 语言编写的,并且使用了 JRuby 解释器。该工具有两种常用的模式:交互模式和命令批处理模式。交互模式用于实时随机访问,而命令批处理模式通过使用 Shell 编程来批量、流程化处理访问命令,常用于 HBase 集群运维和监控中的定时执行任务。本节主要介绍交互模式。

7.1.1 命令分类

选择一台 HBase 集群的节点(最好是客户端节点),进入 HBase 安装目录后执行下面的命令:

```
bin/hbase shell
```

然后会看到下面的输出信息:

```
HBase Shell; enter 'help<RETURN>' for list of supported commands.
Type "exit<RETURN>" to leave the HBase Shell
Version 1.0.1.1, re1dbf4df30d214fca14908df71d038081577ea46, Sun May 17 12:34:26 PDT 2015

hbase(main):001:0>
```

此时已经进入 HBase Shell 交互模式,在该模式中执行 help 命令,执行后部分输出信息如下:

```
COMMAND GROUPS:
  Group name: general
  Commands: status, table_help, version, whoami

  Group name: ddl
  Commands: alter, alter_async, alter_status, create, describe, disable, disable_all, drop,
  drop_all, enable, enable_all, exists, get_table, is_disabled, is_enabled, list, show_filters

  Group name: namespace
  Commands: alter_namespace, create_namespace, describe_namespace, drop_namespace, list_
  namespace, list_namespace_tables

  Group name: dml
  Commands: append, count, delete, deleteall, get, get_counter, incr, put, scan, truncate,
  truncate_preserve

  Group name: tools
  Commands: assign, balance_switch, balancer, catalogjanitor_enabled, catalogjanitor_run,
  catalogjanitor_switch, close_region, compact, compact_rs, flush, major_compact, merge_
  region, move, split, trace, unassign, wal_roll, zk_dump

  Group name: replication
  Commands: add_peer, append_peer_tableCFs, disable_peer, enable_peer, list_peers, list_
  replicated_tables, remove_peer, remove_peer_tableCFs, set_peer_tableCFs, show_peer_tableCFs

  Group name: snapshots
  Commands: clone_snapshot, delete_all_snapshot, delete_snapshot, list_snapshots, restore_
  snapshot, snapshot

  Group name: configuration
  Commands: update_all_config, update_config

  Group name: security
  Commands: grant, revoke, user_permission

  Group name: visibility labels
  Commands: add_labels, clear_auths, get_auths, list_labels, set_auths, set_visibility

SHELL USAGE:
Quote all names in HBase Shell such as table and column names. Commas delimit
command parameters. Type <RETURN> after entering a command to run it.
Dictionaries of configuration used in the creation and alteration of tables are
Ruby Hashes. They look like this:

  {'key1' => 'value1', 'key2' => 'value2', ...}
```

从上面的输出信息可以看到，Shell 的所有命令可以分为六组：常规（General）、DDL、DML、工具（Tools）、复制（Replication）和安全（security）。下面将详细讲解这些命令的含义和使用方法。

7.1.2　常规命令

常规命令只有两种：集群状态命令 status 和 HBase 版本命令 version，下面详细介绍这两种命令的使用方式。

1. 集群状态命令 status

该命令用于查看整个集群的状态信息，在交互模式下，执行 status 命令如下：

```
hbase(main):003:0> status
1 servers, 0 dead, 5.0000 average load
```

从上面代码的返回信息可以看出，该集群共有一台 RegionServer，在没有"死掉"的 RegionServer 中，每台 RegionServer 上有五个 Region（平均值，即 Region 总数除以 RegionServer 总数）。

2. HBase 版本命令 version

该命令用于查看集群的 HBase 版本信息，在交互模式下，执行 version 命令如下：

```
hbase(main):004:0> version
1.0.1.1, re1dbf4df30d214fca14908df71d038081577ea46, Sun May 17 12:34:26 PDT 2015
```

从上面代码返回的信息可以看出，一共包含由逗号分隔的三个部分：第一部分 1.0.1.1 是 HBase 的版本号，第二部分是版本修订号，第三部分是编译 HBase 的时间。

7.1.3　DDL 命令

DDL 命令，即数据定义语言命令，包含的命令非常丰富，用于管理表相关的操作，包括创建表、修改表、上线和下线表、删除表、罗列表等操作，这些命令的详细解释和使用实例如表 7.1 所示。

表 7.1　DDL 命令列表

命　　令	命令含义	命令使用含义
alter	修改表的列族的描述属性	alter 't1',NAME => 'f1',VERSIONS => 5
alter_async	异步修改表的列族的描述属性，并不需要等待所有 Region 都完成操作。用法与 alter 命令相同	alter_async 't1', NAME => 'f1', VERSIONS => 5
alter_status	获取 alter 命令的状态，会标注已经有多少 Region 更改了 Schema。命令的参数是表名	alter_status 't1'
create	创建表	create 't1', { NAME => 'f1', VERSIONS => 5} create 't1','f1','f2','f3'
describe	获取表的元数据信息和是否可用的状态	describe 't1'
disable	下线某个表	disable 't1'
disable_all	下线所有匹配正则表达式的表	disable_all 't.*'
drop	删除某个表	drop 't1'
drop_all	删除所有匹配正则表达式的表	drop_all 't.*'

命　　令	命 令 含 义	命令使用含义
enable	上线某个表	`enable 't1'`
enable_all	上线所有匹配正则表达式的表	`enable 't.*'`
exists	判断某个表是否存在	`exists 't1'`
is_disabled	判断某个表是否下线	`is_disabled 't1'`
is_enabled	判断某个表是否在线	`is_enabled 't1'`
show_filters	查看所支持的所有过滤器的名称	`show_filters`
list	罗列所有表名称	`list`

表7.1中罗列的命令的使用方法有些是比较简单的，如create命令的使用说明如下：

```
Here is some help for this command:
Creates a table. Pass a table name, and a set of column family
specifications (at least one), and, optionally, table configuration.
Column specification can be a simple string (name), or a dictionary
(dictionaries are described below in main help output), necessarily
including NAME attribute.
Examples:

Create a table with namespace = ns1 and table qualifier = t1
  hbase > create 'ns1:t1', {NAME => 'f1', VERSIONS => 5}

Create a table with namespace = default and table qualifier = t1
  hbase > create 't1', {NAME => 'f1'}, {NAME => 'f2'}, {NAME => 'f3'}
  hbase > # The above in shorthand would be the following:
  hbase > create 't1', 'f1', 'f2', 'f3'
  hbase > create 't1', {NAME => 'f1', VERSIONS => 1, TTL => 2592000, BLOCKCACHE => true}
  hbase > create 't1', {NAME => 'f1', CONFIGURATION => {'hbase.hstore.blockingStoreFiles' => '10'}}

Table configuration options can be put at the end.
Examples:

  hbase > create 'ns1:t1', 'f1', SPLITS => ['10', '20', '30', '40']
  hbase > create 't1', 'f1', SPLITS => ['10', '20', '30', '40']
  hbase > create 't1', 'f1', SPLITS_FILE => 'splits.txt', OWNER => 'johndoe'
  hbase > create 't1', {NAME => 'f1', VERSIONS => 5}, METADATA => { 'mykey' => 'myvalue' }
  hbase > # Optionally pre - split the table into NUMREGIONS, using
  hbase > # SPLITALGO ("HexStringSplit", "UniformSplit" or classname)
  hbase > create 't1', 'f1', {NUMREGIONS => 15, SPLITALGO => 'HexStringSplit'}
  hbase > create 't1', 'f1', {NUMREGIONS => 15, SPLITALGO => 'HexStringSplit', REGION_REPLICATION => 2, CONFIGURATION => { 'hbase.hregion.scan.loadColumnFamiliesOnDemand' => 'true'}}

You can also keep around a reference to the created table:

  hbase > t1 = create 't1', 'f1'
```

其中罗列了七种不用的使用实例，每一行都代表一种不同的使用方式，"create't1'，{NAME=>'f1',VERSIONS=>5}"样例将会创建一个名字为t1、列族名为f1，该列族版本数为 5 的表;"create 't1','f1','f2','f3'"样例将会创建一个名字为 t1,同时拥有三个列族 f1、f2、f3 的表。

这里需要重点说明的是,HBase Shell 命令中的表达式符号与其他语言不同,特别是列族中的符号使用,使用规则如下:
- =>表示赋值,如"NAME=>'f1'";
- 字符串必须使用单引号引起,如'f1';
- 如果指定列族的特定属性,需要使用花括号括起,如{NAME=>'f1',VERSIONS=>5}。

如果不确定某个命令的使用方法,可以直接查看该命令的使用说明,在交互模式下执行下面代码:

```
help "${COMMONED_NAME}"
```

7.1.4 DML 命令

DML 命令,即数据操纵语言命令,包含的命令非常丰富,用于数据的写入、删除、修改、查询、清空等操作,这些命令的详细解释和使用示例如表 7.2 所示。

表 7.2 DML 命令列表

命 令	命令含义	命令使用示例
count	统计表的总行数	counter 't1' counter 't1',INTERVAL => 100000 counter't1',CACHE => 1000 counter 't1',INTERVAL => 10,CACHE => 1000
delete	删除一个单元格	delete 't1','r1','c1',ts1
deleteall	删除一行或者一列	deleteall 't1','r1' deleteall 't1','r1','c1' deleteall 't1','r1','c1',ts1
get	单行读	get 't1','r1' get 't1','r1',{TIMERANGE =>[ts1,ts2]} get 't1','r1',{COLUMN =>'c1'} get't1','r1',{COLUMN =>['c1','c2','c3']} get 't1','r1', {COLUMN => 'c1',TIMESTAMP => ts1} get 't1','r1', {COLUMN => 'c1',TIMESTAMP =>[ts1,ts2],VERSIONS => 4} get 't1','r1', {COLUMN => 'c1',TIMESTAMP => ts1,VERSIONS => 4} get't1','r1',{FILTER =>"ValueFilter(= ,'binary:abc')"} get 't1','r1','c1' get 't1','r1','c1','c2' get 't1','r1',['c1','c2']

续表

命　　令	命 令 含 义	命令使用示例
get_counter	读取计数器	get_counter 't1','r1','c1'
incr	自增写入	get 't1','r1','c1' get 't1','r1','c1',1
put	数据写入	get 't1','r1','c1','value',ts1
scan	扫描读	scan 'hbase:meta' scan 'hbase:meta',{COLUMNS =>'info:regioninfo'} scan' t1 ', { COLUMNS = >' c1 ', TIMERANGE = >[1342956435, 1342957698]} scan' t1 ', { FILLTER = >"(PrefixFilter (' row1 ') AND) (Qualifier (> =, ' binary: xyz ')) AND (TimestanpsFilter (43657,47896))"} truncate
truncate	清空表	truncate 't1'

表 7.2 罗列了每种方法的使用实例，其中不少命令包含多种不同的使用方法，这些方法应用在不同的应用场景下，类似 get、scan 这样的命令，使用方法可能有十几种，这充分体现了 HBaseAPI 和 Shell 具有的多样性和灵活性。由于这两个命令的使用方法充分代表了 DML 命令组，并且这两个命令的使用方法非常类似，所以接下来重点讲解 scan 命令的使用方法，讲解的过程中使用数据表 table1(table1 在 HBase 模板 2 中已被创建好，可直接使用)。

1. 扫描全表

扫描全表最简单的使用方法是 scan 命令后直接跟表名，当然，表名需要使用单引号引起，该命令会罗列表的所有数据内容(对每个单元格只输出最新时间戳版本的数据)，命令的使用方法和部分结果输出如图 7.1 所示。

```
hbase(main):012:0> scan 'table1'
ROW                    COLUMN+CELL
 row1                  column=cf: a, timestamp=1443153491490, value=value1
 row1                  column=cf: b, timestamp=1443153504460, value=value2
 row1                  column=cf: c, timestamp=1443153516561, value=value3
 row2                  column=cf: a, timestamp=1443153534995, value=A
 row2                  column=cf: b, timestamp=1443153562538, value=B
 row2                  column=cf: c, timestamp=1443153572005, value=C
 row3                  column=cf: a, timestamp=1443153601323, value=2013
 row3                  column=cf: b, timestamp=1443153609499, value=2014
 row3                  column=cf: c, timestamp=1443153619570, value=2015
3 row(s) in 0.6170 seconds
```

图 7.1　扫描全表命令的部分输出结果

从图 7.1 中的输出结果中可以看到，表 table1 中该行数据包含一个列族 cf 以及三个字表段：a、b 和 c，下面的命令使用实例中也会使用这些信息。

2. 指定列名的全表扫描

使用扫描读的时候，往往会遇到这种情形：只需要查行表中某个字段。假设扫描 table1 表的 cf：a 字段，此时可以使用下面的命令：

```
scan 'table1',{COLUMNS => 'cf:a'}
```

结果如图 7.2 所示。

```
hbase(main):013:0> scan 'table1',{COLUMNS=>'cf:a'}
ROW                    COLUMN+CELL
 row1                  column=cf:a, timestamp=1443153491490, value=value1
 row2                  column=cf:a, timestamp=1443153534995, value=A
 row3                  column=cf:a, timestamp=1443153601323, value=2013
3 row(s) in 1.4330 seconds
```

图 7.2　指定列扫描全表命令的部分输出结果

3. 指定多列、限定返回行数、设置开始行的全表扫描

该全表扫描有三个约束条件：指定多列、限定返回行数、设置开始行。指定列名为 cf：a 和 cf：b，限定返回行数为 2，设置开始行为 row2，代码如下：

```
scan 'table1',{COLUMNS => ['cf:a','cf:b'],LIMIT => 1,STARTROW => 'row2'}
```

命令中所有的约束条件都放到了花括号中，每个条件之间使用逗号分隔。多列的指定使用数据结构：COLUMNS =>['cf：a','cf：b']。限定返回行数使用 LIMIT 关键字，后面的设置值不需要使用单引号引起。设置开始行使用 STARTROW 关键字。命令执行的部分结果如图 7.3 所示。

```
hbase(main):021:0> scan 'table1',{COLUMNS=>['cf:a','cf:b'],LIMIT=>1,STARTROW=>'row2'}
ROW                    COLUMN+CELL
 row2                  column=cf:a, timestamp=1443153534995, value=A
 row2                  column=cf:b, timestamp=1443153562538, value=B
1 row(s) in 0.1320 seconds
```

图 7.3　指定多列、限定返回行数、设置开始行的扫描全表命令的部分输出结果

4. 设定时间戳范围的全表扫描

该全表扫描包含两个约束条件：指定列和时间戳时间范围。指定列为 a，间戳范围是闭区间[1443153516565,1443153619571]，代码如下：

```
scan 'table1',{COLUMNS => 'cf:a',TIMERANGE => [1443153516565,1443153619571]}
```

命令执行结果如图 7.4 所示。

```
hbase(main):024:0> scan 'table1',{COLUMNS=>'cf:c',TIMERANGE=>[1443153516565,1443153619571]}
ROW                    COLUMN+CELL
 row2                  column=cf:c, timestamp=1443153572005, value=C
 row3                  column=cf:c, timestamp=1443153619570, value=2015
2 row(s) in 0.1310 seconds
```

图 7.4　设定时间戳范围的扫描全表命令的部分输出结果

5. 带有过滤条件的全表扫描

该全表扫描的约束条件是使用过滤器，下面的代码中使用了前缀过滤器、列名过滤器和时间戳过滤器，并且使用了组合过滤器，代码如下（至少要两个列族，每组至少两行）：

```
scan 'table1',{FILTER=>"(PrefixFilter('row2')
        AND ( QualifierFilter ( =, ' binary: b ' ))) AND ( TimestampsFilter ( 1443153562538,
1443153609499))"}
```

在约束条件中：过滤器使用关键字 FILTER；PrefixFilter 表示前缀过滤器，作用于行键上，行键以 row3 为前缀；QualifierFilter 表示列名过滤器，第一个参数"＝"表示比较器，即列名等于 c，其中的"binary："表示使用二进制比较，冒号是分隔符；TimestampsFilter 是时间戳过滤器，两个参数是时间戳，这两个时间戳并不是区间，而是数据组中的两个元素。执行上面的带有过滤器的全表扫描命令，输出结果如图 7.5 所示。

```
hbase(main):025:0> scan 'table1',{FILTER=>"(PrefixFilter(' row2') AND(QualifierFil
ter(=,' binary: b' ))) AND(TimestampsFilter(1443153562538,1443153609499))"}
ROW                        COLUMN+CELL
 row2                      column=cf: b, timestamp=1443153562538, value=B
1 row(s) in 5.7740 seconds
```

图 7.5　带有过滤条件的扫描全表命令的输出结果

由图 7.5 可以看出，整个 table1 表符合条件的只有一行，行键以 row2 为前缀，包含列 b，时间戳也符合过滤器中所定义的。

7.1.5　工具命令 Tools

HBase Shell 工具提供了一些工具命令，组名称为 Tools，这些命令多用于 HBase 集群管理和调优。这些命令涵盖了合并、分裂、负载均衡、日志回滚、Region 分配和移动以及 ZooKeeper 信息查看等方面。每种命令的使用方法有多种，适用于不同的场景。例如合并命令 compact，可以合并一张表、一个 Region 的某个列族或一张表的某个列族。命令的详细解释和使用实例如表 7.3 所示。

表 7.3　Tools 的命令列表

命　　令	命　令　含　义	命令使用实例
assign	分配 Region	assign 'region1'
balance_swicth	启用或关闭负载均衡器，返回结果是当前均衡器状态	balance_switch true balance_switch false
balancer	触发集群负载均衡器。如果成功运行则返回 true，很可能将所有 Region 重新分配。如果是 false，说明某些 Region 在 RIT 状态，不会执行该命令	balancer
close_region	关闭某个 Region	close_region 'REGIONNAME' close_region 'REGIONNAME','SERVER_NAME'
compact	合并表或 Region	compact 't1' compact 'r1','c1' compact 't1','c1'
flush	Flush 表或 Region	flush 'TABLENAME' flush 'REGIONNAME'
hlog_roll	HLog 日志回滚，参数是 RegionServer 的名字	hlog_roll 'REGIONSERVERNAME'

续表

命 令	命令含义	命令使用实例
major_compact	大合并表或 Region	compact 't1' compact 'r1','c1' compact 't1','c1'
move	移动 Region。如果没有目标 RegionServer，则随机选择一个	move 'ENCODED_REGIONNAME' move'ENCODED_REGIONNAME','SERVER_NAME'
split	分裂表或 Region	split 'tableName' split 'regionName' # format:'tableName, startKey,id' split 'tableName','splitKey' split 'regionName','splitKey'
unassign	解除指定某个 Region	unassign 'REGIONNAME' unassign 'REGIONNAME',true
zk_dump	打印输出 ZooKeeper 的信息，包括 HBase 主节点、RegionServer 状态，以及 ZooKeeper 节点的状态统计	zk_dump

7.1.6 复制命令

复制命令用于 HBase 高级特性——复制的管理，可以添加、删除、启动和停止复制功能相关操作，这里不展开讲解命令的细节，仅给出每个命令的介绍和使用案例以供读者参考，具体解释如表 7.4 所示。

表 7.4 复制命令列表

命 令	命令含义	命令使用实例
add_peer	添加对等集群，需要指定对等集群的 ID、主机名、端口号和 ZooKeeper 的根路径	add_peer '1',"server1.cie.com:2181:/hbase" add_peer'1',"zk1,zk2,zk3:2181:/hbase-prod"
disable_peer	停止到特定集群的复制流，但仍然保持对新改动的跟踪。参数是对等集群的 ID	disable_peer'1'
enable_peer	启动到对等集群的复制，从上次关闭的位置继续复制。参数是对等集群的 ID	enable_peer'1'
list_peer	罗列所有正在复制的对等集群	list_peers
remove_peer	停止某个复制流，并且删除其对应的元数据信息。参数是对等集群的 ID	list_peers
start_replication	重启所有复制流，只用在负载达到临界的情况下	remove_replication
stop_replication	关闭所有复制流，只用在负载达到临界的情况下	stop_replication

7.1.7 安全命令

安全命令属于 DCL(Data Control Languag,数据控制语言)的范畴,HBase shell 提供三种安全命令:grant、revoke 和 use_permission。这三种命令并不是直接执行如表 7.5 所示的使用实例就可以使用,还需要两个前提条件:使用附带 security 的 HBase 版本和配置完成 Kerberos 安全认证。

表 7.5 安全命令列表

命　令	命　令　含　义	命令使用实例
grant	赋给用户特定的权限,权限集合是 RWXCA:READ('R')、WRITE('W')、EXEC('X')、CREATE('C')和 ADMIN('A')	grant 'bobsmith','RWXCA' grant 'bobsmith','RW''t1',' f1',coll
revoke	撤销用户的特定权限	revoke 'bobsmith' revoke'bobsmith','t1', 'f1','coll'
usr_permission	显示某用户的所有权限,如果加上参数——表名,则表示该用户在该表上的所有权限	user_permission 'table1'

7.2 Java 客户端的使用

HBase 官方代码包中包含原生访问客户端,由 Java 语言实现,同时它也是最主要、最高效的客户端相关的类。在 org.apache.hadoop.hbase.client 包中,涵盖增、删、改、查等所有 API。主要的类包含 HTable、HBaseAdmin、Put、Get、Scan、Increment 和 Delate 等。

HBase 作为一个 NoSQL 数据库,最基本的操作就是 CRUD(即前面提到的增、删、改、查),其中的改(Update)操作没有实现,其他三类操作已经有丰富且成熟的实现。除了这些操作,原生 Java 客户端还实现了创建、删除和修改表等 DDL(数据定义语言)操作,同时还提供一些工具操作,类似合并、分裂 Region、分配 Region 等。接下来,将详细讲解原生 Java 客户端的基本知识和使用方法。

7.2.1 客户端配置

使用原生 Java 客户端之前首先要安装 Eclipse,安装完成之后再进行 Java 原生客户端的配置。

1. 简单的配置实例

(1) 在/usr 目录下解压系统已下载好的 eclipse-jee-mars-R-linux-gtk-x86_64.tar.gz 压缩包,Eclipse 安装包的下载地址为 http://www.eclipse.org/download/。

(2) 设置 Eclipse 桌面快捷方式。右击桌面,选中"创建启动器"命令,将"名称(N)"和"命令(A)"两栏补充完整,"命令(A)"一栏的内容填写 Eclipse 启动图标的路径。

(3) 上述步骤完成之后,如果 Eclipse 无法正常启动,则修改/usr/eclipse 目录下的 eclipse.ini 文件,在-vmargs 之前加入下面两行代码

```
- vm
/usr/java/jdk/bin
```

(4) 打开 Eclipse,在 Eclipse 中新建一个 Java 工程,工程名为 HBase,此工程的 workspace 是在/root 根目录下。然后选择项目属性,右击项目,选择 Libraries 命令,在如图 7.6 所示的对话框中单击 Add External JARs 按钮。

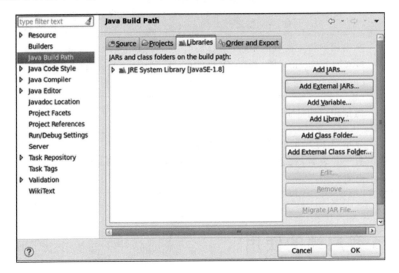

图 7.6 添加 jar 包

(5) 选择/usr/hbase/lib 下的相关 jar 包,如果只是测试用,可简单一点,将所有的 JAR 都选中,如图 7.7 所示。

图 7.7 添加 jar 包

(6) 在项目 HBase 下新建一个文件夹,命名为 conf,将 Hbase 集群的配置文件 hbase-site.xml 复制到该目录。

```
# cp /usr/hbase/conf/hbase-site.xml /root/workspace/HBase/conf
```

(7) 右击 conf 文件,选择 Bulid Path → Configure Bulid Path 命令。选择 Java Bulid Path→Libraries→Add Class Folder,将刚刚增加的 conf 目录选中,如图 7.8 所示。

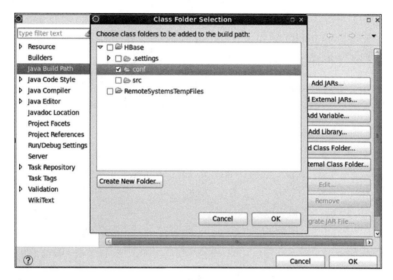

图 7.8　引入配置文件

(8) 单击 OK 按钮之后,可以看到如图 7.9 所示的 conf 图标。

(9) 为新建 HBase 工程配置好客户端之后,就可以在 HBase 工程中做相关的数据库操作。在后文中,我们会接着讲解具体的 HBase 表操作。

图 7.9　引入配置文件

注意:

(1) 每次新建工程时,都需要进行一次客户端配置。

(2) 这里选择了一种较为便捷的方式来配置原生 Java 客户端。除此之外,也可以直接在工程中用 Java 代码来配置客户端,代码需要指明 HMaster 地址、ZooKeeper 端口和 ZooKeeper 列队名称。有兴趣的读者可以尝试使用这种方式。

2. 表操作入口类 HTable

下面将详细介绍表操作入口类 HTable,主要从构造函数和主要实现方法两方面进行阐述。

了解 Java 语言的读者能够理解构造函数的含义,它是 Java 语言不可或缺的一部分。在多数情况下,构造函数都是创建实例的入口。类的构造函数有四类,分别具备如图 7.10 所示的参数类型或者参数个数。

在图 7.10 中,第一个使用 ExecutorService 类,即外部独立维护的线程池类,与其他构造函数区别很大。这个构造函数用于需要外部单独维护执行线程池的情况。

第二个构造函数最常用,上面的简单配置代码中使用的是第二个构造函数。一般情况

```
Constructor and Description
HTable(ClusterConnection conn, BufferedMutatorParams params)
For internal testing.
HTable(TableName tableName, ClusterConnection connection, TableConfiguration tableConfig,
RpcRetryingCallerFactory rpcCallerFactory, RpcControllerFactory rpcControllerFactory, ExecutorService pool)
Creates an object to access a HBase table.
```

图 7.10　HTable 类的构造函数

下,都会选用第二种方法作为创建实例的构造函数。

HTable 类中实现的方法有很多种,这里不会罗列所有的方法,只汇总分类介绍最常用的方法,具体包含四类:CRUD 操作方法、获取元数据信息方法、获取状态信息方法和设置属性信息方法,下面进行详细介绍。

1) CRUD 操作方法

CRUD 操作是 HTable 最基本的功能,其中的每类方法都包含多种实现,方法 API 如下:

```
// 删除
boolean     ckeckAndDelete(byte[ ] row, byte[ ] famlily, byte[ ] qualifier, byte[ ] value,Delete delete)
void        delete(Delete delete)
void        delete(List<Delete> deletes)
// 查询
Result get(Get get)
Result[ ]   get(List<Get> gets)
ResultScanner getScanner(byte[ ] famlily, byte[ ] qualifier)
ResultScanner getScanner(Scan scan)
// 更新
Result      increment(Increment increment)
long        incrementColumnValue(byte[ ] row, byte[ ] famlily, byte[ ] qualifier, long amount)
long        incrementColumnValue(byte[ ] row, byte[ ] famlily, byte[ ] qualifier, long amount, Boolean writeToWAL)
// 写入
boolean     checkAndPut(byte[ ] row, byte[ ] famlily, byte[ ] qualifier, byte[ ] value, Put put)
void        put(List<Put> puts)
// 验证是否存在
boolean     exists(Get get)
```

从上面的代码中可以看到,删除、查询、更新、写入的实现方法有多种重载:

(1) 删除、Get 查询和写入支持批量数据操作;

(2) 查询方法包含 Get 和 Scan 两种,分别有多种实现方法;

(3) 更新操作只针对长整型变量;

(4) 验证是否存在的方法 exists( ) 的参数是 Get 实例,本质上该方法是查询的一种,每类操作的不同实现方法应用在不同的场景下,例如 checkAndPut( ) 方法用在当处理行记录时,不希望其他客户端对该行进行其他处理,首先需要检测值是否发生改变,如果没有则写入,如果改变则写入失败。

检测后执行(Compare-And-Set,CAS)的方法作用强大,可以用于解耦多客户端操作,经常用在状态过渡和数据处理中。HTable 提供了两种 CAS 状态的方法:checkAndPut( )

和 checkAndelete( )，多客户端操作中会经常用到这些方法。

2）获取元数据信息方法

这里提到的元数据信息包含 Region 的位置信息、客户端配置信息、开始和结束行键、操作超时时间等。这些方法并不是最常用的，在某些特定的场景下才会使用。例如查看 Region 开始和结束位置，方法 API 如下：

```
HRegionLocation getRegionLocation(byte[ ], Boolean reload)
HRegRegionLocation getRegionLocation(String row)
NavigableMap<HRegionInfo, ServerName> getRegionLocations( )
org.apache.hadoop.conf.Configuration      getConfiguration()
byte[ ][ ]        getStartKeys( )
byte[ ][ ]        getEndKeys( )
pair<byte[ ][ ], byte[ ][ ]>       getStartEndKeys( )
int          getOperationTimeout( )
```

3）获取状态信息方法

状态信息包含是否自动 Flush、表是否可用等，方法 API 如下：

```
boolean        isAutoFlush
static         Boolean isTableEnabled ( org. apache. hadoop. conf. configuration conf, byte [ ],
tableName)
```

4）设置属性信息方法

该部分方法用于设置相关的状态和属性信息，例如是否自动 Flush、操作超时时间、客户端写缓存大小等，通过这些设置方法控制与连接和写入相关的一些属性，可以实现客户端性能调优。

```
void         setAutoFlush(boolean antoFlush)
void         setAutoFlush(boolean autoFlush, boolean clearBufferOnFair)
void         setOperationTimeout(int operationTimeout)
void         setWriteBufferSize(long writeBufferSize)
void         flushCommits( )
```

3. 管理入口类 HBaseAdmin

HBaseAdmin 类是 HBase 数据库的管理入口类，通过该类可以创建表、删除表、修改表、罗列表名、上线和下线表等表级别的操作，还可以管理 Region、负载均衡、分裂与合并等操作。

HBaseAdmin 类的构造函数有两类，分别具备不同的参数类型：如图 7.11 所示，图中的第一个构造函数的参数是 configuration，通过配置类初始化；第二个构造函数的参数是 HConnection，通过该对象初始化 HBaseAdmin。

HBaseAdmin 类中实现的方法比 HTable 多，这里不会罗列所有的方法，只汇总分类介绍比较常用的，共包含四类：表的创建、删除、修改方法，表的状态信息方法，Region 相关操作方法，快照相关操作方法和其他方法，详细介绍如下。

1）表的创建、删除、修改方法

该部分方法包含表的创建、删除和修改三类，每类方法又包含多种重载方法，从方法名

Constructor and Description
`HBaseAdmin(org.apache.hadoop.conf.Configuration c)`   **Deprecated.**   Constructing HBaseAdmin objects manually has been deprecated. Use `Connection.getAdmin()` to obtain an instance of `Admin` instead.
`HBaseAdmin(Connection connection)`   **Deprecated.**   Constructing HBaseAdmin objects manually has been deprecated. Use `Connection.getAdmin()` to obtain an instance of `Admin` instead.

图 7.11　HBaseAdmin 类的操作函数

称上能很好地区分这几种方法。例如，创建表的方法都以 create 开头，删除都以 delete 开头，修改操作相对特殊一些，因为涉及修改表描述信息和列族的描述信息，所以方法名字不是以某个单词开头，详细方法如下：

```
//创建
void        createTable(HTableDescriptor desc)
void        createTable(HTableDescriptor desc, byte[ ][ ] splitKeys)
void         createTable (HTableDescriptor desc, byte[ ] startkeys, byte[ ] endkey, int numRegions)
//删除
void        deleteColumn(byte[ ] tableName, byte[ ] columnName)
void        deleteColumn(String tableName, String columnName)
void        deleteSnapshot(byte[ ] snapshotName)
void        deleteSnapshot(String snapshotName)
void        deleteTable(byte[ ] tableName)
void        deleteTable(String tableName)
HTableDescriptor[ ]    deleteTables(Pattern pattern)
HTableDescriptor[ ]    deleteTables(String regex)
//修改
void        addColumn(byte[ ] tableName, HColumnDescriptor column)
void        addColumn(String tableName, HColumnDescriptor column)
void        modifyColumn(byte[ ] tableName, HColumnDescriptor descriptor)
void        modifyColumn(String tableName, HColumnDescriptor descriptor)
void        modifyColumn(byte[ ] tableName, HTableDescriptor htd)
```

其中，每种类别的多种重载方法都有不同的参数数目和参数类型，以适用于多种情况。例如，createTable(HTableDescriptor desc)只有一个参数，需要创建一个 HTableDescriptor 实例，而 createTable(HTableDescriptor desc, byte[ ] startKey, byte[ ] endKeys, int numRegions)方法可以指定开始和结束的行键以及 Region 的数量，以更细粒度地操作创建表过程。

2）表的状态信息方法

表状态信息相关的方法更多，包括上线和下线表以及批量操作、获取集群信息、获取协处理器信息、获取表描述信息、获取表 Region 信息、表上线和下线状态信息等，详细如下：

```
void        disableTable(byte[ ] tableName)
void        disableTable(String tableName)
void        disableTableAsync(byte[ ] tableName)
void        disableTableAsync(String tableName)
HTableDescriptor     disableTables(Pattern pattern)
HTableDescriptor     disableTables(String regex)
```

```
void          enableTable(byte[ ] tableName)
void          enableTable(String tableName)
void          enableTableAsync(byte[ ] tableName)
void          enableTableAsync(String tableName)
HTableDescriptor    enableTables(Pattern pattern)
HTableDescriptor    enableTables(String regex)
ClusterStatus       getClusteStatus( )
String[ ]           getMasterCoprocessors( )
HTableDescriptor    getTableDescriptor(byte[ ] tableName)
HTableDescriptor    getTableDescriptor(List<String> tableNames)
List<HRegionInfo>   getTableRegions(byte[ ] tableName)
boolean       isMasterRunning( )
boolean       isTableDisabled(byte[ ] tableName)
boolean       isTableDisabled(String tableName)
boolean       isTableEnabled(byte[ ] tableName)
boolean       isTableEnabled(String tableName)
```

3) Region 相关操作方法

Region 相关的操作方法属于 HBase 的高级特性部分，包括 Region 分配、负载均衡、Region 的合并、分裂、移动等。其中，合并与分裂包含多种重载方法，并且合并方法分为小合并和大合并两类，每类有多种重载方法。详细代码如下：

```
void          assign(byte[ ] regionName)
void          unassign(byte[ ] regionName, Boolean force)
boolean       balancer( )
void          closerRegion(byte[ ] regionname, String serverName)
void          closerRegion(ServerName sn, HRegionInfo hri)
void          closeRegion(String regionname, String serverName)
boolean       closeRegionWithEncodeRegionName(String encodedRegionName, String serverName)
void          compact(byte[ ] tableNameOrRegionName)
void          compact(byte[ ] tableNameOrRegionName, byte[ ] columnFamily)
void          compact(String tableNameOrRegionName)
void          compact(String tableOrRegionName, String columnFamily)
void          majorCompact(byte[ ] tableNameOrRegionName)
void              majorCompact(byte[ ] tableNameOrRegionName, byte[ ] columnFamily)
void          majorCompact(String tableNameOrRegionName)
void          majorCompact(String tableNameOrRegionName, String columnFamily)
void          move(byte[ ] encodedRegions, byte[ ] destServerName)
void          split(byte[ ] tableNameOrRegionName)
void          split(byte[ ] tableNameOrRegionName, byte[ ] splitPoint)
void          split(String tableNameOrRegionName)
void          split(String tableNameOrRegionName, String splitPoint)
```

4) 快照相关操作方法

快照就是一份元信息的合集，允许管理员恢复到表的先前状态。快照不是表的复制而是一个文件名称列表，因而不会复制数据。完全快照恢复是指恢复到之前的"表结构"以及当时的数据，快照之后发生的数据不会恢复。

HBaseAdmin 中也提供了几种快照的操作方法，可以根据表名和快照名生成快照，方法 API 代码如下：

```
void        snapshot(byte[ ] snapshotName, byte[ ] tableName)
void        snapshot(HBaseProtos, SnapshotDescription snapshot)
void        snapshot(String snapshotName, String tableName)
void        snapshot(String snapshotName, String tableName, HBaseProtos.SnapshotDescription.
Type type)
```

5）其他操作方法

除了上面提到的表、Region 和快照相关的方法，HBaseAdmin 中还有两类方法：flush()和 listTables()，其中 flush()方法用于将表或者 Region 在内存中的数据序列化到硬盘上，listTables()方法用于罗列 HBase 集群上所有的表名，方法 API 细节如下所示：

```
void              flush(byte[ ] tableNameOrRegionName)
void              flush(String tableNameOrRegionName)
HTableDescriptor[ ]         listTables()
HTableDescriptor[ ]         listTables(Pattern pattern)
HTableDescriptor[ ]         listTables(String regex)
```

4. 连接池类 HTablePool

假如多线程访问 HBase，需要创建多个 HTable 对象，并且需要单独对每个 HTable 对象的创建、使用、消亡的整个过程进行维护，这样不但增加开发成本，也不利于资源的合理利用，同时 HTable 写入时不是线程安全的。HBase 官方提供了通过连接池的方式使用 HTable，即 HTablePool 类。该类支持多线程使用 HTable，并且多线程同时写入时是线程安全的，这非常类似于 MySQL 数据库连接池。HTablePool 的构造方法有四种，分别对应不同的情形，如图 7.12 所示。

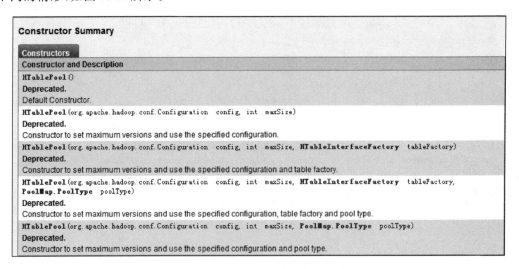

图 7.12　HTablePool 类的构造函数

从图 7.12 可以看出，第一个构造函数是空构造，剩余三个都有不同数量的参数，下面的内容将详细讲解有参数的三种构造函数。这三个构造函数的第一个参数都是配置类，这里的配置类与前面讲到的配置类一致，此处不再赘述，重点讲解其他几个参数。

1) 连接池容量

参数 maxsize 表示这个连接池的大小，默认值是 Integer 类型的最大值，即 Integer.MAX_VALUE。当使用 HTablePool 获取 HTable 时，HTablePool 会维护线程池容量。当容量大于或等于最大值时，将不再添加线程到线程池中。需要注意的是，如果线程个数大于最大值，会导致写入始终是自动 Flush。

其实，在 HTablePool 的使用过程中，通过 putTable()方法将新生成的 HTable 对象添加到线程池中，添加的同时进行最大值判断。通过 getTable()从线程池中直接获取 HTable 对象，如果线程池中没有，则创建封装类 PooledHTable。HTablePool 的使用代码示例如下：

```
Result result = mull;
HTable table = null;
try {
    table = ( HTable)pool.getTable(tableName ) ;
    if ( table == null ) throw new RuntimeException ( "this table is not exist!" ) ;
    result = table.get ( new get ( Bytes.toBytes ( "rk1" ) ) ) ;
} catch ( IOException e ) {
    Throw new RuntimeException ( e ) ;
} finally {
    if ( table != null ) {
        try {
            pool.putTable ( ( HTableInterface ) table ) ;
        } catch ( IOException e )
            e.printStackTrace ( ) ;
        }
    }
}
```

2) 连接池类型

HTable 类中提供了三种连接池类型：ReusablePool、RoundRobinPool 和 ThreadLocalPool，默认是 ReusablePool。这三种类型的底层数据结构不同，但总体实现都比较简单，三种类型的对比介绍如下：

- ReusablePool——底层使用 ConcurrentLinkedQueue 实现，实现比较简单。
- ThreadLocalPool——底层使用 ThreadLocal 实现，ThreadLocal 为每一个线程都维护了自己独有的变量副本。每个线程都拥有了自己独立的一个变量，竞争条件被彻底消除了，所以没有任何必要对这些线程进行同步，这样能最大限度地由 CPU 调度并发执行，是一种以空间来换取线程安全性的策略。访问的性能更高一些。
- RoundRobinPool——虽然将其罗列为一种类型，但是无法使用；详见 HTablePool 的构造函数中的代码实现。

3) 工厂类 HTableFactory

HTableFactory 用于创建 HTable 示例的工厂类，可以自定义 HTable 的配置属性，例如自动 Flush、写缓存大小等，可以为每个 HTable 定义不同的配置属性，其使用方法如下：

```
Configuration conf = HBaseConfiguration.create ( ) ;
HTableFactory factory = new HTableFactory ( ) ;
```

```
        HTablePool pool = new HTablePool ( conf, 30, factory, PoolType.ThreadLocal ) ;
        Result result = null ;
        HTable table = null ;
        Try {
            table = ( HTable ) pool.getTable ( tableName ) ;
            if ( table == null )
                throw new RuntimeException ( "This table is not exist ! " ) ;
            result = table.get ( new Get ( Bytes.toBytes ( "rk1" ) ) ) ;
        } catch ( IOExcption e ) {
            throw new RuntimeException (e);
        } finally {
            if ( table != null ) {
                try {
                    pool.putTable ( ( HTableInterface ) table ) ;
            } catch ( IOException e ) {
                    throw new RuntimeException ( e ) ;
            } finally {
             If ( table != null ) {
                try {
                    pool.putTable ( HTableInterface ) table ) ;
                } catch ( IOException e ) {
                    e.printStackTrace ( ) ;
                }
            }
        }
```

如果深入 HTablePool 代码实现,会发现其实 HTablePool 并不是常规意义上的线程池,而更类似于一个简单的计数器实现。HTablePool 是 HBase 连接池的老用法,该类在 0.94、0.95 和 0.96 版本中已经不建议使用,在 0.98.1 版本以后已经移除,对于 HTablePool 类的用法只要求了解即可。

7.2.2 创建表

下面尝试使用原生 Java 客户端的方式创建 HBase 表,下面代码将创建一个名为 test1 的表。该表拥有一个列族 cf1。

```
Configuration conf = HBaseConfiguration.create();
HBaseAdmin admin;
try {
    admin = new HBaseAdmin(conf);
    if(admin.tableExists("test1")){
        System.out.println("table Exists!");
    }else{
        HTableDescriptor tableDesc = new HTableDescriptor("test1");
        tableDesc.addFamily(new HColumnDescriptor("cf1"));
        admin.createTable(tableDesc);
        System.out.println("create table success!");
    }
} catch (IOException e) {
    // TODO Auto-generated catch block
```

```
            e.printStackTrace();
    }
```

7.2.3 删除表

使用原生 Java 客户端删除表与创建表的操作不同,删除一张表需要分两步进行:第一步,下线表;第二步,删除表。删除表使用的也是 HBaseAdmin 管理入口类,具体代码如下:

```
try{
            String tablename = "test1";
            @SuppressWarnings({ "deprecation", "resource" })
            HBaseAdmin admin = new HBaseAdmin(HBaseConfiguration.create());
            admin.disableTable(tablename);
            admin.deleteTable(tablename);
            System.out.println("delete table success!");
    }catch(Exception e){
            e.printStackTrace();
    }
```

其中,disableTable()方法用于下线表,deleteTable()用于删除表。与创建表操作相同,如果删除的表不存在,也会抛出异常信息,所以在删除之前最好判断表是否存在。

HBase 的删除并不像传统关系型数据库的删除,HBase 删除动作并不会立刻将 HBase 存储的数据进行删除,而是先在指定的 KeyValue 存储单元上打上删除标志。等到下一次 region 合并、分裂等操作时才会将所有的数据进行移除。

7.2.4 插入数据

前面的操作已经创建了一张 demo_test 表,接下来向该表中插入数据。向该表中插入一条行键是 row-1、列族是 cf1、列名是 row-id、时间戳是 100、值为 2015 的数据,代码如下:

```
HTable table = null;
        try {
                table = new HTable(HBaseConfiguration.create(),Bytes.toBytes("test1"));
                Put put = new Put(Bytes.toBytes("row-7"),100);
put.add(Bytes.toBytes("cf1"),Bytes.toBytes("row-id"),Bytes.toBytes("2017"));
                table.put(put);
                System.out.println("insert data success!");
        } catch(IOException e){
                e.printStackTrace();
        } finally {
                if(null != table){
                    try{
                        table.close();
                    }catch(IOException e){
                        e.printStackTrace();
                    }
                }
        }
```

从上面的代码可以看到,使用 HTable 类操作数据写入的整个过程分为三个步骤:第一步,初始化 HTable 类;第二步,构造 Put 实体类,该类封装写入的数据;第三步,执行写入操作。第二步中有很多细节需要讲解,重点介绍 Put 类构造函数和已实现方法。

Put 类的构造函数有六个,分别具备不同的参数类型,如图 7.13 所示。其中,第四个和第五个已经过期,第一个和第六个很少用到,下面的内容围绕第二个和第三个构造函数展开。

```
Constructor and Description
Put(byte[] row)
Create a Put operation for the specified row.
Put(byte[] rowArray, int rowOffset, int rowLength)
We make a copy of the passed in row key to keep local.
Put(byte[] rowArray, int rowOffset, int rowLength, long ts)
We make a copy of the passed in row key to keep local.
Put(byte[] row, long ts)
Create a Put operation for the specified row, using a given timestamp.
Put(ByteBuffer row)
Put(ByteBuffer row, long ts)
Put(Put putToCopy)
Copy constructor.
```

图 7.13 Put 类的构造函数

第二个和第三个构造函数的共性是:它们的第一个参数都是行键,类型是 byte[ ]。而第三个构造函数有两个参数,其中第二个参数 ts 表示写入单元格的时间戳,可见,在初始化 Put 实例时,可以直接赋值时间戳变量。该时间戳是长整型,并且该值可以根据实际需要定义,并不一定是实际的 UNIX 时间戳。

Put 类中的方法并不算多,可以将这些方法划分成三类:添加字段、获取键值对和判断是否存在,方法描述如下:

```
// 添加字段
Put         add ( byte[ ] family , byte[ ] qualifier , byte[ ] value )
Put         add ( byte[ ] family , byte[ ] qualifier , long ts , byte[ ] value )
Put         add ( KeyValue kv )
// 获取键值对
List<KeyValue>   get ( byte[ ] family , byte[ ] qualifier )
// 判断是否存在
boolean     has( byte[ ] family , byte[ ] qualifier )
boolean     has( byte[ ] family , byte[ ] qualifier , byte[ ] value )
boolean     has( byte[ ] family , byte[ ] qualifier , long ts )
boolean     has( byte[ ] family , byte[ ] qualifier , long ts , byte[ ] value )
```

其中,添加字段和判断是否存在两类都包含多种重载方法。添加字段的三种重载方法是开发过程中最常用的。add(byte[ ] family,byte[ ] qualifier,byte[ ] value)方法有三个参数:family(列族)、qualifier(列名)和 value(值),其时间戳默认使用系统当前时间(如果在初始化 Put 实例的时候已经对时间戳赋值)。add(byte[ ] family,byte[ ] qualifier,long ts,byte[ ]value)方法比之前一个方法多一个参数——时间戳 ts。该参数也是对单元格的时间戳字段赋值。这里的赋值比初始化 Put 实例时的赋值优先级更高,如果初始化 Put 实例和用此 add()方法同时赋值,则使用此方法的赋值。

add(KeyValue kv)方法的参数是 KeyValue 实例,需要首先构造 KeyValue 实例,构造

代码如下:

```
KeyValue kv = new KeyValue( row,family , qualifier , ts , KeyValue.Type.Put , Value) ;
```

7.2.5 查询数据

原生 Java 客户端有两种查询数据的方式:单行读和扫描读。其中,单行读使用 HTable 类的 get(Get)方法,参数是 Get 实体类;扫描读使用 HTable 类的 getScanner (Scan)方法,参数是 Scan 实体类。下面详细介绍这两种数据查询方式。

1. 单行读

单行读就是查询表中的某一行记录,可以是一行记录的全部字段,可以是某个列族的全部字段,或者某一个字段。单行读的示例代码如下:

```
try{
        HTable table = new HTable(HBaseConfiguration.create(),Bytes.toBytes("test1"));
        Get get = new Get(Bytes.toBytes("row-1"));
get.addColumn(Bytes.toBytes("cf1"),Bytes.toBytes("row-id"));
        Result dbResult = table.get(get);
        System.out.println("size = " + dbResult.size() + ",value = " + Bytes.toString
(dbResult.list().get(0).getValue()));
    } catch(Exception e){
        e.printStackTrace();
    }
```

从上面的示例代码中可以看到,整个单行读过程分为三步:第一步,初始化 HTable 实例;第二步,构造实体类 Get,Get 类封装所需的行键、列族、列名;第三步,执行查询并打印结果。其中,实体类 Get 是一个新的 API,接下来详细介绍该类的构造函数和主要实现方法。

Get 类的构造函数有两个,分别具备不同的参数类型,如图 7.14 所示。其中,第一个为 Writable 服务,第三个已经过期。所以需要关注的只有第二个构造函数。第二个钩造函数的参数是行键,类型是 byte[]数组。

Constructor and Description
Get(byte[] row)
Create a Get operation for the specified row.
Get(Get get)
Copy-constructor

图 7.14 Get 类的构造函数

Get 类的主要实现方法可以划分成三类:添加列或列族、设置查询属性和查看属性信息,方法描述如下:

```
// 添加列或列族
Get      addColumn(byte[] family,byte[] qualifier)
Get      addFamily(byte[] family)
// 设置查询属性
```

```
void        setCacheBlocks(Boolean cacheBlocks)
Get         setFilter(Filter filter)
Get         setMaxVersions()
Get         setMaxVersions(int maxVersions)
Get         setTimeRange(long timestamp,long maxStamp)
Get         setTimeStamp(long timestamp)
//查看属性信息
Set<byte[]> familySet()
boolean     getCacheBlocks()
Map<byte[],NavigableSet<byte[]>>    getFamilyMap()
Filter      getFilter()
Map<String,Object> getFingerprint()
long        getLockId()
int         getMaxVersions()
byte[] getRow()
RowLock     getRowLock()
TimeRange getTimeRange()
boolean     hasFamilies()
int         numFamilies()
```

其中，添加列或列族、设置查询属性是最常用的两类方法。addColumn()方法用于添加单个列，addFamily()方法用于添加单个列族。

设置查询属性的方法能够更细粒度地控制查询操作；setCacheBlocks()方法可以设置是否使用 BlockCache，用于提升查询性能。setFilter()方法用于设置过滤器，例如键值过滤器、列名过滤器等。setMaxVersion()方法用于控制查询返回的版本数量，该方法默认返回所有版本，即 Integer 类型的最大值。setMaxVersions(int)方法用于设置返回多少版本，int 类型参数表示返回版本数量。setTimeRange(long)方法用于设置返回哪个时间戳的版本，如果不命中，则返回小于等于参数值的最接近的版本。setTimeRange(long,long)方法用于设置返回版本的时间戳区间，第一个参数是开始时间戳，第二个参数是结束时间戳，第一个参数应该小于第二个参数。

查看属性信息方法用于查看 Get 类已经设置的一些属性信息，例如最大版本数量、行键、锁 ID、列族数量等。这些方法并不常用，实际应用中使用最多的还是通过设置相关查询条件获取符合条件的值。

2．扫描读

扫描读一般是在不确定行键的情况下，遍历全表或者表的某部分数据。当然，遍历过程中也可以进行细粒度控制，如时间戳、版本数量、列族和列名等。扫描读的示例代码如下：

```
HTable table;
        try {
            table = new HTable(HBaseConfiguration.create(),Bytes.toBytes("test1"));
            Scan scanner = new Scan();
            /* version */
            //Scanner.setTimeRange(startTime,endTime);
            /* batch and caching */
            scanner.setBatch(0);
            scanner.setCaching(100000);
```

```
                    ResultScanner rsScanner = table.getScanner(scanner);
                    for(Result res : rsScanner){
                        for (KeyValue kv : res.raw()) {       // 遍历每一行的各列
                            StringBuffer sb = new StringBuffer()
.append(Bytes.toString(kv.getRow())).append("\t")
                                .append(Bytes.toString(kv.getFamily()))
                                .append("\t")
                                .append(Bytes.toString(kv.getQualifier()))
.append("\t").append(Bytes.toString(kv.getValue()));
                            System.out.println(sb.toString());
                        }

                    }
                    rsScanner.close();
                } catch (IOException e) {
                    // TODO Auto-generated catch block
                    e.printStackTrace();
                }
```

  从上面的示例代码中可以看到，整个扫描读过程分为三步：第一步，初始化 HTable 实例；第二步，构造实体类 Scan，Scan 类封装所需的列族、列名和其他属性设置；第三步，执行查询并输出结果。其中，实体类 Scan 是一个新的 API，接下来详细介绍该类的构造函数和主要实现方法。

  Scan 类的构造函数有六个，分别具备不同的参数类型，如图 7.15 所示。其中，每个构造函数都很常用，多数情况下在不同的场景下都会用到。

Constructor and Description
Scan()   Create a Scan operation across all rows.
Scan(byte[] startRow)   Create a Scan operation starting at the specified row.
Scan(byte[] startRow, byte[] stopRow)   Create a Scan operation for the range of rows specified.
Scan(byte[] startRow, Filter filter)
Scan(Get get)   Builds a scan object with the same specs as get.
Scan(Scan scan)   Creates a new instance of this class while copying all values.

图 7.15　Scan 类的构造函数

  在展示的构造函数中，第一个构造函数是空构造，该方式使用很频繁。第二个构造函数中有一个参数：开始行键，也就是表示从表的哪一行开始扫描。第三个构造函数的参数既有开始行键，又有结束行键，表示扫描表的某一段区域。第四个构造函数的第一个参数表示开始行键，第二个参数表示过滤器，可以通过设置过滤器的方式进行数据过滤，提升访问性能。第五个构造函数参数是 Get 实例，该构造函数表示使用 Scan 实现 Get，实际上，Get 方法内部也是使用 Scan 去实现的。第六个构造函数可以将另一个 Scan 实例的所有属性都复制到本 Scan 实例中，参数是 Scan 实例。

  Scan 类中的主要实现方法可以划分成三类：添加列或列族、设置查询属性和查看属性信息，方法描述如下：

```
//添加列或列族
Scan       addColumn(byte[] family,byte[] qualifier)
Scan       addFamily(byte[] family)
//设置查询属性
void       setBatch( int batch)
void       setCacheBlocks(Boolean cacheBlocks)
void       setCaching(int caching)
Scan       setFamilyMap(Map < byte[],NavigableSet < byte[]>> familyMap)
Scan       setFilter(Filter filter)
Scan       setMaxVersions()
Scan       setMaxVersions(int maxVersions)
void         setRaw(boolean raw)
Scan       setStartRow(byte[] startRow)
Scan       setStopRow(byte[] stopRow)
Scan       setTimeRange(long minStamp, long maxStamp)
Scan       setTimeStamp(long timestamp)
//查看查询属性
int        getBatch()
boolean    getCacheBlocks()
int        getCaching()
byte[][]     getFamilies()
Map < byte[],NavigableSet < byte[]>>  getFamilyMap()
Filter     getFilter()
Map < String,Object >  getFingerprint()
byte[]     getStartRow()
byte[]     getStopRow()
TimeRange  getTimeRange()
boolean    hasFamilies()
boolean    hasFilter()
boolean    isGetScan()
boolean    isRaw()
int        numFamilies()
```

对比 Get 和 Scan 类的实现方法会发现，它们的很多方法都是一样的，对于相同的方法此处不再赘述，只介绍与 Get 类存在区别的方法。其中，setRaw()方法用于设置行键，如果设置了行键，则等同于 Get 类的使用；setstartRow()和 setStopRow()方法用于设置扫描的开始和结束行键。查看查询属性类别中也有一些有关设置开始和结束行健的方法，其他的方法与 Get 类都相同。值得注意的是，Scan 类也可以使用过滤器，对于扫描操作来讲，设置过滤器比单行读更重要。

7.2.6　删除数据

删除操作也是原生 Java 客户端所支持的 CRUD 操作之一，下面将介绍如何使用 HTable 类实现删除数据。原生 Java 客户端的删除操作可以删除整行、某个列族、某个列，也可以删除某个单元格。下面的代码示例中就包含删除某个单元格、某个列、某个列族的操作。

```
HTable table = null;
    try{
        table = new HTable(HBaseConfiguration.create(),Bytes.toBytes("test1"));
```

```
        } catch(IOException e) {
            e.printStackTrace();
        }
        Delete del = new Delete(Bytes.toBytes("row-4"));

    del.deleteColumn(Bytes.toBytes("cf1"),Bytes.toBytes("row-id"),1000);

    //del.deleteColumns(Bytes.toBytes("cf1"),Bytes.toBytes("row-id"));
        //del.deleteFamily(Bytes.toBytes("cf1"));
        try {
            table.delete(del);
            System.out.println("delete data success!");
        } catch(Exception e) {
            e.printStackTrace();
        }
```

从上面的示例代码中可以看到，整个删除过程分为三步：第一步，初始化 HTable 示例；第二步，构造实体类 Delete，Delete 类封装所需的行键、列族或者列名；第三步，执行删除。其中，实体类 Delete 是一个新的 API，接下来详细介绍该类的构造函数和主要实现方法。

Delete 类的构造函数有五个，分别具备不同的参数类型。如图 7.16 所示。其中，第一个为 Writable 服务，第四个已经过期，第五个一般用不到，所以最需要关注的是第二个和第三个构造函数。

Constructor and Description
Delete(byte[] row)
Create a Delete operation for the specified row.
Delete(byte[] rowArray, int rowOffset, int rowLength)
Create a Delete operation for the specified row and timestamp.
Delete(byte[] rowArray, int rowOffset, int rowLength, long ts)
Create a Delete operation for the specified row and timestamp.
Delete(byte[] row, long timestamp)
Create a Delete operation for the specified row and timestamp.
Delete(Delete d)

图 7.16 Delete 类的构造函数

第二个和第三个构造函数的第一个参数都是行键，由此可以得出结论：删除时必须指定某一行。第三个构造函数的第二个参数是时间戳，表示指定删除某个版本的单元格。

Delete 类中的主要实现方法比较少，主要是列族、列和时间戳相关的几种方法，方法描述如下所示：

```
Delete    deleteColumn(byte[] family, byte[] qualifier)
Delete    deleteColumn(byte[] family, byte[] qualifier, long timestamp)
Delete    deleteColumns(byte[] family, byte[] qualifier)
Delete    deleteColumns(byte[] family, byte[] qualifier, long timestamp)
Delete    deleteFamily(byte[] family)
Delete    deleteFamily(byte[] family, long timestamp)
Delete    setTimestamp(long timestamp)
```

其中，deleteColumn()方法用于删除列的某个版本；deleteColumns()方法用于删除某个列的所有版本；deleteFamily()方法用于删除某个列族；setTimestamp()方法用于设定删除操作的时间戳。

第 8 章

并行数据流处理引擎：Pig

8.1 Pig 概述

Pig 是一个 Apache 开源项目，Pig 提供了一个基于 Hadoop 的并行执行数据流处理的引擎，它包含了一种脚本语言，称为 Pig Latin，用来描述这些数据流。Pig Latin 本身提供了许多传统的数据操作（如 join、sort、filter 等），同时允许自己开发一些自定义函数用来读取、处理和写数据。

8.1.1 Pig 是什么

8.1.1.1 Pig 是基于 Hadoop 的

Pig 运行于 Hadoop 之上，它同时用到 Hadoop 分布式文件系统的 HDFS 和 Hadoop 处理系统 MapReduce。

HDFS 是一个分布式文件系统，它将文件存储到 Hadoop 集群的各个节点上。它负责将文件分割成许多数据块，然后分发到不同的节点机器上，其中包括对每个数据块进行多份冗余备份，这样可以避免因为某台机器宕掉而造成的数据丢失。HDFS 提供了一种类似 POSIX 的用户交互形式给用户。默认情况下，Pig 从 HDFS 中读取输入文件，使用 HDFS 来存放 MapReduce 任务所生成的中间数据，最终将输出写入 HDFS 中。

MapReduce 是一个简单而强大的并行数据处理算法。MapReduce 计算框架下的每个任务都由三个主要阶段组成：map 阶段、shuffle 阶段和 reduce 阶段。在 map 阶段，程序可以并行独立操作输入数据中的每一条记录。因为可以同时运行多个 map 任务，所以即使输入的数据量达到吉字节或者太字节级别，只要有足够多的机器，map 阶段通常可在 1 分钟内完成。

MapReduce 任务的一个特别之处在于需要确定数据是根据哪个键进行收集的。map 阶段后紧跟着就是 shuffle 阶段,在这个阶段数据已经根据用户指定的键收集起来并且分发到不同的机器上去了,这是为 reduce 阶段做准备。包含同一键的所有记录将会交由同一个 reducer 处理。

在 reduce 阶段,程序将提取每个键以及包含该键的所有记录。这个过程也是在多台机器上并行执行完成的。当处理完所有组时,reducer 就可以写输出了。下面通过一个简单的 MapReduce 程序进行演示。

> MapReduce 演示程序

假设现在有一个 MapReduce 程序对一个文本文件进行词频统计。该程序本身是 MapReduce 提供的演示过程。map 阶段会从文本文件中一次读取一行,然后分割出每个词作为一个字符串,之后对于分割出的每个单词,会输出单词本身以及数字 1,数字 1 表示这个单词出现过 1 次。在 shuffle 阶段,将使用单词作为键,哈希分布对应的记录到不同的 reducer 中去。在 reduce 阶段会将相同的单词对应的出现次数相加,并最终将求和后的数值和单词本身一起输出。以童谣"Mary Had a Little Lamb"为例,输入将是:

```
Mary had a little lamb
its fleece was white as snow
and everywhere that Mary went
the lamb was sure to go.
```

这里假设每一行都被发送到不同的 map 任务中去了。当然事实上,每个 map 任务处理的数据要远远大于这个数量,这里只是为了后面更好地去描述。MapReduce 整个过程的数据流如图 8.1 所示。

map 阶段一旦结束,shuffle 阶段将会把包含相同单词的所有记录提交到同一个 reducer 中。对于这个例子假设有两个 reducer:以 A~L 开头的单词提交到第一个 reducer 中,而以 M~Z 开头的单词提交到第二个 reducer 中。这两个 reducer 最终将会把每个单词的出现次数分别相加然后输出。

Pig 所有数据处理过程都是使用 MapReduce 来执行的。Pig 将所写的 Pig Latin 脚本编译成一个或者多个 MapReduce 任务,然后在 Hadoop 上执行。下面的例子展示了如何使用 Pig Latin 脚本来对童谣"Mary Had a Little Lamb"进行词频统计。

```
-- 加载文件名为 Mary 的文件,
-- 并将记录中的唯一字段命名为'line'.
input = load 'mary' as (line);

-- TOKENIZE 将 line 按单词分割成列
-- flatten 接受 TOKENIZE 操作后产生的记录集合然后分开成独立的列,
-- 这个独立的列称为 word
words = foreach input generate flatten(TOKENIZE(line)) as word;

-- 现在按照 word 进行分组
grpd = group words by word;
```

```
    -- 计数
cntd = foreach grpd generate group,COUNT(words);
    -- 打印结果
dump cntd;
```

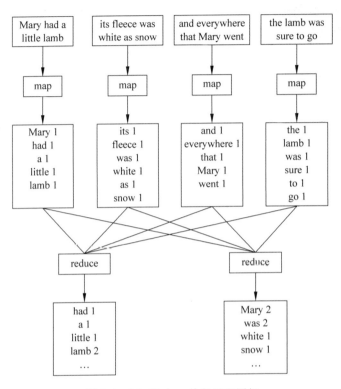

图 8.1　MapReduce 执行过程图解

在使用 Pig 时无须去过度关注 map、shuffle 和 reduce 阶段，因为 Pig 会将脚本中的操作解析成相应的 MapReduce 阶段。

8.1.1.2　Pig Latin

Pig Latin 是一种数据流语言，这意味着它允许去描述如何从一个或多个数据源并行读取数据，然后并行地进行处理，最后将处理结果并行地输出到一个或多个结果集中。这些数据流可以是个简单的线性流，也可以是复杂的工作流，其中可以包含一些加入多个输入的节点，也可以包含一些将输入数据分割为多个流的节点，这些节点都是通过不同的操作符来处理的。用数学语言来描述，Pig Latin 描述的是一个有向无环图（DAG），在这个图中，节点代表处理数据的操作符，节点间的向量代表数据流。

这意味着 Pig Latin 和之前见过的许多编程语言有所不同。在 Pig Latin 中没有 if 语句，也没有 for 循环操作。这是因为传统的过程语言和面向对象语言描述的是控制流，从而数据流只处于一个从属地位；而 Pig Latin 更专注于数据流。

1. 查询语言和数据流语言的比较

尽管 Pig Latin 和 SQL 有一定的相似性，但是其实两者具有非常多的差异。SQL 是一种查询语言,它关注于允许构造查询,它允许去描述想得到什么问题的答案,而不是如何给出问题的答案。然而在 Pig Latin 中,可以详细描述如何对输入的数据进行处理。

Pig Latin 和 SQL 的另一个主要区别是 SQL 面向的是回答一个问题,因此当想同时进行多个数据操作时,要么使用多个查询语句,这时需要将一些查询的中间数据存放到临时表中;要么写一个大的包含子查询的查询语句,将一些初始的处理过程由子查询来完成。然而,很多人都发现子查询令人困惑而且也并非那么容易去构建。同时,子查询使用的是由内而外的设计,也就是说,在数据管理通道最里面的子查询会最先执行。

Pig 被设计为实现知道将要进行的一系列的数据操作,因此不需要通过颠倒顺序的子查询的方式来写数据管道,也无须使用临时表来存放中间数据。

现在假设有个想先按某个键对表进行 group 分组操作,然后和第二张表进行 join 连接操作。在 SQL 查询中,因为 join 操作发生在 group 操作之后,所以要么使用子查询,要么写两个查询语句,同时将中间结果保存到临时表中。下面的例子用到了一个临时表,因为这样可读性更好。

SQL 中先进行分组,然后进行连接操作。

```
CREATE TEMP TABLE t1 AS
SELECT customer,sum(purchase)AS total_purchases
FROM transactions
GROUP BY customer;

SELECT customer,total_purchases,zipcode
FROM t1,customer_profile
WHERE t1.customer = customer_profile.customer;
```

在 Pig Latin 中,是另一种方式,如下例所示。

Pig Latin 中先进行分组,然后进行连接操作:

```
--加载汇报文件,按照 customer 字段进行分组,然后计算总购物金额
txns = load'transactions'as (customer,purchase);
grouped = group txnx by customer;
total = foreach grouped generate group,SUM(txns.purchase) as tp;
--加载 customer_profile 文件
profile = load'customer_profile'as (customer,zipcode);
--对已经分好组并进行了累加计算的汇报文件数据和 customer_profile 文件进行连接
answer = join total by group,profile by customer;
--将结果输出到控制台
dump answer;
```

此外,SQL 和 Pig Latin 各因不同的应用场景而生。SQL 的应用场景是 RDBMS,在这种场景下,数据是标准化的,并且加上了模式和其他一些特有的约束(例如,null 值也是不可以脱离约束单独存在的等)。Pig 是为 Hadoop 数据处理环境而设计的,在这种环境下,模式有时是未知的或不一致的,数据可能没有进行恰当的约束而且很少进行数据标准化。基于这些不同,Pig 不需要将数据事先导入表中,当数据导入 HDFS 中后,它就可以直接操作这些存放在 HDFS 的数据。

如果语言和文化类似,那么融入一个新的环境可能会更加容易些。在数据处理范畴里,

SQL 就是英语。它有个非常好的特点就是无论是人还是工具都认识它，也就是说，它的入门门槛很低。我们的目标是使 Pig 成为像 Hadoop 那样的并行数据处理系统范畴里的母语。尽管这可能要求需要进行一定的学习才能使用，但是它可以更加充分地利用 Hadoop 提供的计算能力。

2. Pig 和 MapReduce 的区别是什么

Pig 比直接使用 MapReduce 相比有几个优点。Pig Latin 提供了所有标准的数据处理操作，例如 join、filter、group by、order by、union 等。MapReduce 直接提供了 group by 操作（也就是 shuffle 和 reduce 两个阶段做的事情），同时通过实现分组操作间接地提供了 order by 操作。过滤器操作和推测执行操作可以在 map 阶段进行简单实现。但是其他的操作，特别是 join 操作无法提供，所以必须由用户自己进行代码实现。

Pig 提供了一些对这些标准的数据操作的复杂的、完备的实现。例如，因为每个键对应的记录的个数很少是均匀分布在集群中的，所以提交给 reducer 的数据经常会产生数据倾斜。也就是说，有的 reducer 需要比别的 reducer 处理多 10 倍或更多倍的数据。Pig 具有 join 和 order by 操作可以处理这种情况，而且（在一些情况下）可以重新均衡 reducer 负荷。这些需要 Pig 团队花费几个月的时间编写 MapReduce 程序，然后再重构代码，这确实耗费时间。

在 MapReduce 中，在 map 阶段和 reduce 阶段的内部的数据处理对于系统来说是不透明的。这意味着 MapReduce 没有机会优化或者检查的代码。另一方面，Pig 可以通过分析 Pig Latin 脚本来了解描述的数据流。这意味着 Pig 可以在早期进行错误检查（例如是否将一个 string 类型的字段放到一个 integer 类型的字段中？）和优化（例如这两个 group 操作是否可以合并？）。

MapReduce 没有一个类型系统，是有意这么设计的，因为这样可以提供更大的自由度去使用用户自己的数据类型和序列化框架。但这样就产生了一个不好的问题，就是限制了系统在运行前和运行时对代码进行检查的能力。

这几个方面都表明 Pig Latin 相对于 MapReduce Java 代码更容易编写和维护。下面做了一个实验，对于同一个操作分别使用 Pig Latin 和 MapReduce 进行实现。假设有个文件存有数据，另一个文件存放了对于某个网站的点击数据，下例所示的 Pig Latin 脚本将找到年龄为 18～25 岁的用户访问最多的五个页面。

查找访问次数最多的前五个 URL：

```
Users = load 'users' as (name,age);
Fltrd = filter Users by age >= 18 and age <= 25;
Pages = load 'pages' as (user,url);
Jnd   = join Fltrd by name, Pages by user;
Grpd  = group Jnd by url;
Smmd  = foreach Grpd generate group,COUNT(Jnd) as clicks;
Strd  = order Smmd by clicks desc;
Top5  = limit Strd 5;
Store Top5 into 'top5sites';
```

这段脚本的第 1 行表示加载文件名为 users 的文件，同时声明这份数据有两个字段：name 和 age，而且为这个输入取别名为 Users。第 2 行是个过滤器，将 Users 中 age 这个字段值大于或等于 18 而且小于等于 25 的记录过滤出来，不满足条件的数据将被忽略。经过过滤后，留下的数据就是在我们感兴趣的年龄范围内的了。我们将这个过滤器的结果取别名为 Fltrd。

第 3 行是第 2 个 load 加载数据语句，这个语句加载了文件 Pages，并取别名为 Pages，它声明了两个字段：user 和 url。

"Jnd=join"这一行以 Fltrd.name 和 Pages.user 为键，对 Fltrd 和 Pages 进行 join 连接操作。通过这次 join 操作，可以得到每个访问过的所有 URL 链接了。

"Grpd=group"这一行按照 URL 进行分组。紧跟着的下一行会统计每个 URL 对应的记录个数。在这一行后我们就知道了每个 URL 被年龄为 18~25 岁的用户访问了多少次。

之后的一件事就是按访问次数从访问最多到访问最少进行排序。"Strd=order"这一行就是根据前一行的统计结果进行 desc(降序)排列。因此，最大值将在第 1 行。因为最终还需要最前面的五条记录，所以最后一行将统计结果限制在前五行。最后的结果重新存放到 HDFS 中一个叫做 top5sites 的文件中。

在 Pig Latin 中整个处理过程需要写九行代码，耗时 15 分钟左右，其中包括写代码和对代码进行调试的时间。如果以 MapReduce 来写的话，需要差不多 170 行的代码而且花费了四个小时才调试成功。Pig Latin 同样便于维护，因为这段代码，对于后来的其他开发者同样是容易理解和方便修改的。

Pig 带来的这些便利同样是有代价的。通过 MapReduce 框架可以开发一些算法，在 Pig 中却很难实现。同时对于开发者，他们需要放弃一个层次的控制权。一名优秀的工程师，只要给予其足够的时间，总是可以将一个普通的系统做得足够好。因此对于不常见的算法或者是对于性能要求很高的话，这种情况下使用 MapReduce 仍然是正确的选择。基本上这种情况也和选择 Java 编码而不选择使用像 Python 这样的脚本语言是一样的。Java 功能强大，但是因为它是高级程序语言，所以使用它开发需要比脚本语言花费更多的时间。开发者需要根据实际情况选择合适的工具。

8.1.1.3 Pig 的用途

Pig Latin 的使用场景可以分为独立的三大类：传统的抽取转换加载(ETL)数据流、原生数据研究和迭代处理。

最大的使用场景就是数据流了。一个通常的例子就是网络公司从其 Web 服务器上收集到日志，进行数据清洗，之后进行简单的聚合预计算，然后导入数据仓库中。在这种情况下，数据被加载到计算网格中，之后使用 Pig 从数据泥潭中清理出有价值的数据。同时还可以使用 Pig 将网页操作数据和数据库信息进行 join 连接，这样可以将 cookie 和已知的信息关联起来。

另一个数据流应用的例子是使用 Pig 处理离线数据来建立行为预测模型。Pig 被用来扫描所有的和网站的交互数据，最终将分为各种各样的群组。然后，对于每个群组会生成一个数学模型，根据该模型可以预知这个群组的对各种类型的广告或者新闻文章的反应是怎样的。通过这种方式，网站可以知道展示什么样的广告可能获得更多的点击，或者发布什么样的新闻故事更有可能吸引用户的再次访问。

传统上，使用像 SQL 这样的语言执行点对点的查询可以快速地为问题准备好相应的数据。然而，对于原始数据的研究，一些还是偏向使用 Pig Latin 脚本。因为 Pig 可以在无模式、模式信息不全或者模式不一致的情况下进行操作，同时因为 Pig 可以很容易地控制封装的数据，因此对于那些期望在数据没有进行清洗也没有写入数据仓库的情况下，分析数据的研究人员经常更倾向于使用 Pig。经常处理大规模数据集的研究人员经常会使用像 Prel 或者 Python 这样的脚本语言进行处理。具有这些使用背景的人通常更喜欢使用 Pig 这样的数据流范式而非像 SQL 那样的声明式查询语言。

创建迭代处理模式的人也开始使用 Pig。假设有一个新闻门户网站，它保留了一个它跟踪的关于该网站的所有新闻故事的图。在这个图中每个新闻故事都是一个节点，节点间的连线表示的是相关故事间的关系。例如，所有关于即将来临的选举故事都是联系在一起的。每五分钟都有一组新的故事进来，这时数据处理引擎需要将这组故事增加到图中。这些故事中有一些是新的，有一些是对之前的故事进行的更新，还有一些是替代之前已经存储的一些故事的。这时需要对整个故事图做一些数据处理步骤。例如，对于建立行为目的模型的处理过程就需要将数据和整个故事图进行连接。每五分钟重新运行整个图是不可行的，因为对于适当数量的硬件资源来说，在五分钟内运行出结果是不可能的。但是模型创建者不想只是每天更新一次这些模型，因为那意味着会错过一整天的时间来提供机会。

为了解决这个问题，首先有必要定期地对整个图进行连接，例如可以按照天来进行连接。然后，每五分钟后一旦有数据进来，就可以立即对新进来的数据进行连接操作，同时这个结果是可以和对整个图做连接的结果整合在一起的。这个组合步骤并不容易，因为需要在五分钟内完成对整个图进行插入、更新和删除操作。使用 Pig Latin 来表达这种组合关系是可以的，并且是相当方便的。

目前所说的一切都隐含着一点：Pig（与 MapReduce 一样）是面向数据批处理的。如果需要处理的是 GB 或者 TB 数量级的数据，那么 Pig 是个不错的选择。但是因为它期望的是序列地读取一个文件中的所有记录然后序列地将输出写入存储中，因此对于需要写单条或者少量记录，或者查询随机序列下的多条不同记录的任务，Pig（与 MapReduce 一样）并非是个好选择。

8.1.2 Pig 的发展简史

Pig 最初是作为 Yahoo！的一个探索性的项目，Yahoo！的科学家们设计了 Pig 并且给出了一个原型实现。正如 2008 年发表在《数据管理专业委员会》（SIGMOD）杂志上的一篇论文所描述的，研究者认为 Hadoop 所描述的 MapReduce 框架"过于底层和严格，需要花费大量的时间编写代码，而且很难维护和重用。"同时他们注意到的 MapReduce 对像 SQL 这样的声明式语言并不熟悉。因此他们着手开发"一种叫做 Pig Latin 的新语言，这种语言被设计为在像 SQL 这样的声明式类型的语言和像 MapReduce 这种较底层的过程式的语言之间达到一个非常好的平衡点。"

最初 Yahoo！的 Hadoop 使用者开始采用 Pig。之后一个开发工程师团队开始接手 Pig 的最初原型并将 Pig 原型开发成一个达到产品级别的可用产品。在这个时间点，也就是 2007 年的秋天，Pig 通过 Apache 孵化器进行开源。一年后也就是 2008 年的 9 月，Pig 的第一个发布版本出现了。同年晚些时候，Pig 从孵化器中毕业，正式提升为 Apache Hadoop 项目的一个子项目。

2009 年年初，一些公司在其数据处理中开始使用 Pig。Amazon 也将 Pig 加入它的弹性 MapReduce 服务中的一部分。2009 年年末，Yahoo! 公司所运行的 Hadoop 任务有一半是 Pig 任务。在 2010 年，Pig 的发展持续增长，这一年 Pig 从 Hadoop 的子项目中脱离出来，成为了一个最高级别的 Apache 项目。

8.2 Pig 的安装和使用

8.2.1 下载和安装 Pig

首先需要下载和安装 Pig，才能在本地机器或者 Hadoop 集群上使用它。既可以直接下载 Pig 安装包，也可以先下载源代码然后自行进行编译。当然同样也可以以 Hadoop 分支的方式获得 Pig。

8.2.1.1 从 Apache 下载 Pig 软件包

这个是 Apache Pig 的官方版本。它是一个软件包，里面包含了运行 Pig 所需的所有 JAR 包。可以通过访问 Pig 发布页面进行下载。

Pig 不需要安装到 Hadoop 集群中去。它运行在提交 Hadoop 任务的那台机器上。尽管可以从个人笔记本电脑或者台式计算机运行 Pig，但是在实际操作中，大部分的集群管理员会配置好一到多台可以访问其 Hadoop 集群的服务器，虽然这些服务器可以不是集群的一部分。通过这种方式，管理员可以方便地升级 Pig 和将工具集成在一起，同时也可以很好地限制人员对于集群的访问。这些机器被称为网关机或者缝边机。本书中称之为网关机。

我们需要在这些网关机上安装 Pig。如果是通过个人台式计算机或者笔记本电脑访问 Hadoop 集群，那么同样需要在个人台式计算机或者笔记本电脑上安装 Pig。当然，如果想通过本地模式使用 Pig，那么也可以将 Pig 安装到个人的本地机器上。

Pig 的核心是用 Java 语言编写的，因此它是跨平台的。启动 Pig 的 shell 脚本是个 bash 脚本，所以它需要一个 UNIX 环境。Pig 所基于的 Hadoop，即使使用的是本地模式，也需要是 UNIX 环境的，因为 Hadoop 的文件操作是基于 UNIX 的。实际情况是，大部分的 Hadoop 集群采用的是 Linux 系统。很多 Pig 开发者是在 Mac OS X 系统上开发和测试 Pig 的。

如果想在本地模式下使用 Pig 或者想把它安装到一个没有安装 Hadoop 的网关机上，就不需要额外去下载 Hadoop。一旦下载完 Pig，就可以把它放在自己喜欢的任何位置，Pig 无须依赖一个特定位置。

安装第一步，将压缩包放在期望的目录下然后执行解压。

```
tar -zxvf filename
```

其中 filename 是用户所下载的 tar 压缩包文件。

安装第二步，是确定环境变量 JAVA_HOME 是否设置指向到包含一个 Java 发布版的目录，这一步是为执行 Pig 做环境准备。如果没有设置这个环境变量，那么 Pig 会立即执行失败。可以通过 shell 命令设置环境变量，也可以在调用 Pig 的时候通过命令行指定该环境变量,或者在位于刚才解压后的文件 bin 目录下的 Pig 脚本文件中显示的指定好 JAVA_HOME 环境变量。可以通过执行 which java 命令查看当前 java 所在路径,把这条命令返回的结果的 bin/java 去掉,就是当前环境中 JAVA_HOME 的值。

8.2.1.2 从 Cloudera 下载 Pig

除了官方的 Apache 版本之外,同样有一些其他公司会重新包装和分发 Hadoop 以及与其相关的工具。其中最受欢迎的是 Cloudera,它为 Red Hat 系列的系统开发了相应的 RPM 包并为 Debian 系统开发了相应的 APT 包。对于不可以使用这些包管理器的其他系统,Cloudera 同样提供了 tar 压缩包文件。使用像 Cloudera 这样的发行版的一个优点是与 Hadoop 相关的所有工具都是打包在一起并且在一起测试完成的。同样,如果需要专业的技术支持,它也是提供的。缺点是将受限于其使用的发行版的提供商的发行速度。Apache 推出一个新版本后,然后到各个分支出现不同的发行版,这中间会有一定的时间延迟。

如果想获得从 Cloudera 下载和安装 Hadoop 和 Pig 的详细说明,请访问 Cloudera 下载页面。应注意的是,需要单独下载 Pig,因为 Hadoop 包中没有包含 Pig。

8.2.1.3 Pig 安装步骤

1. 本地安装 pig

(1) 打开终端,解压软件包:pig-0.16.0.tar.gz。

```
tar - zxvf pig-0.16.0.tar.gz
```

(2) 在/usr 目录下创建 pig 文件夹,并将解压后的 pig-0.16.0 文件移至 pig 文件夹中。

```
mkdir /usr/pig
mv /root/下载/pig-0.16.0 /usr/pig
```

(3) 进入文件 pig-0.16.0,如图 8.2 所示。

```
cd /usr/pig/pig-0.16.0
ls (显示当前文件夹中的内容)
```

```
[root@master pig]# cd /usr/pig/pig-0.16.0
[root@master pig-0.16.0]# ls
bin          docs       lib-src             pig-0.16.0-core-h2.jar    src
build.xml    ivy        license             README.txt                test
CHANGES.txt  ivy.xml    LICENSE.txt         RELEASE_NOTES.txt         tutorial
conf         legacy     NOTICE.txt          scripts
contrib      lib        pig-0.16.0-core-h1.jar   shims
```

图 8.2 pig-0.16.0 目录下内容图

(4) 配置环境:编辑文件 vim~/.bashrc,在文本的最后添加下面的语句,保存文件退出并使用 source~/.bashrc 编译文件,如图 8.3 所示。

注:进入 vim 后,输入字母"i"进入可写状态(后面将不再具体指出此操作)。

```
export PATH = /usr/java/jdk1.8.0_60/bin:/usr/hadoop/hadoop-2.7.1/bin:/usr/pig/pig-0.16.
0/bin: $ PATH
export PIG_HOME = /usr/pig/pig-0.16.0
export PATH = $ PIG_HOME/bin: $ HADOOP_HOME/bin: $ JAVA_HOME/bin: $ PATH
Esc + shift + : wq!
(保存退出,后面将不再具体指出此操作)
source ~/.bashrc
```

```
[root@master pig-0.16.0]# vim ~/.bashrc
[root@master pig-0.16.0]# source ~/.bashrc
[root@master pig-0.16.0]#
```

图 8.3 编辑及编译图

.bashrc 文件的配置内容如图 8.4 所示。

```
# User specific aliases and functions
alias rm='rm -i'
alias cp='cp -i'
alias mv='mv -i'
# Source global definitions
if [ -f /etc/bashrc ]; then
    . /etc/bashrc
fi

export JAVA_HOME=/usr/java/jdk1.8.0_60
export CLASSPATH=/usr/java/jdk1.8.0_60/lib/dt.jar:/usr/java/jdk1.8.0_60/lib/tool
.jar
export HADOOP_HOME=/usr/hadoop/hadoop-2.7.1
export HADOOP_MAPRED_HOME=$HADOOP_HOME
export HADOOP_COMMON_HOME=$HADOOP_HOME
export HADOOP_HDFS_HOME=$HADOOP_HOME
export YARN_HOME=$HADOOP_HOME
export HADOOP_COMMON_LIB_NATIVE_DIR=$HADOOP_HOME/lib/native
export HADOOP_OPTS="-Djava.library.path-$HADOOP_HOME/lib"
export PATH=/usr/java/jdk1.8.0_60/bin:/usr/hadoop/hadoop-2.7.1/bin:/usr/pig/pi
g-0.16.0/bin:$PATH
export PIG_HOME=/usr/pig/pig-0.16.0
export PATH=$PIG_HOME/bin:$PATH
```

图 8.4 配置内容图

(5) 输入下面命令,进入 grunt,如图 8.5 所示。

```
pig - x local
quit (使用 quit 命令退出 pig)
```

2. Hadoop 集群下安装 Pig

(1) 在 Hadoop 的配置中,设置 PIG_CLASSPATH,启动的命令也是不同的。我们需要设置 PIG_CLASSPATH 这个环境变量到那个目录下。要注意的是,需要指向的是那些 XML 配置文件所在的目录,而不是这些配置文件本身。Pig 会自动加载读取那个目录下的所有 XML 和 properties 类型的文件。

(2) 配置环境:编辑文件 vim~/.bashrc,在文本的最后添加下面的语句,保存文件退出并编译,如图 8.6 所示。

图 8.5　本地模式 Pig 配置完成图

```
export PIG_CLASSPATH = /usr/hadoop/hadoop-2.7.1/conf
source ~/.bashrc
```

```
[root@master pig-0.16.0]# vim ~/.bashrc
[root@master pig-0.16.0]# source ~/.bashrc
[root@master pig-0.16.0]#
```

图 8.6　编辑及编译图

.bashrc 的配置内容如图 8.7 所示。

```
# User specific aliases and functions
alias rm='rm -i'
alias cp='cp -i'
alias mv='mv -i'
# Source global definitions
if [ -f /etc/bashrc ]; then
        . /etc/bashrc
fi

export PATH=/usr/java/jdk1.8.0_60/bin:/usr/hadoop/hadoop-2.7.1/bin:/usr/pig/pig
-0.16.0/bin:$PATH
export JAVA_HOME=/usr/java/jdk1.8.0_60
export PIG_HOME=/usr/pig/pig-0.16.0
export CLASSPATH=/usr/java/jdk1.8.0_60/lib/dt.jar:/usr/java/jdk1.8.0_60/lib/too
l.jar
export HADOOP_HOME=/usr/hadoop/hadoop-2.7.1
export HADOOP_MAPRED_HOME=$HADOOP_HOME
export HADOOP_COMMON_HOME=$HADOOP_HOME
export HADOOP_HDFS_HOME=$HADOOP_HOME
export YARN_HOME=$HADOOP_HOME
export HADOOP_COMMON_LIB_NATIVE_DIR=$HADOOP_HOME/lib/native
export HADOOP_OPTS="-Djava.library.path=$HADOOP_HOME/lib"
export PIG_CLASSPATH=/usr/hadoop/hadoop-2.7.1/conf
```

图 8.7　配置内容图

(3) 进入 /usr/hadoop/hadoop-2.7.1 目录，执行 sbin/stop-all.sh 命令，然后启动 Hadoop。

```
cd /usr/hadoop/hadoop-2.7.1
sbin/stop-all.sh
sbin/start-all.sh
jps
```

启动 Hadoop 的日志如图 8.8 所示，可以执行 jps 命令用于查看是否成功启动 Hadoop，出现如图 8.9 所示的结果，表示 Hadoop 启动完成，接下来启动 pig。若 jps 命令查看时没有 dataNode 节点，表示 Hadoop 未能成功启动，原因是 namenode 和 datanode 的 ID 不一致，此时需要进入 /hdfs/namenode/current 和 /hdfs/datanode/current，编辑 VERSION 文件，保证 namenode 的 ID 和 datanode 的 ID 一致，修改好后重新启动 Hadoop。

```
[root@master hadoop-2.7.1]# sbin/start-all.sh
This script is Deprecated. Instead use start-dfs.sh and start-yarn.sh
16/08/09 17:21:24 WARN util.NativeCodeLoader: Unable to load native-hadoop library for your platform... using builtin-java classes where applicable
Starting namenodes on [localhost]
localhost: namenode running as process 9917. Stop it first.
localhost: datanode running as process 10019. Stop it first.
Starting secondary namenodes [0.0.0.0]
0.0.0.0: secondarynamenode running as process 10209. Stop it first.
16/08/09 17:21:31 WARN util.NativeCodeLoader: Unable to load native-hadoop library for your platform... using builtin-java classes where applicable
starting yarn daemons
resourcemanager running as process 10361. Stop it first.
localhost: nodemanager running as process 10463. Stop it first.
```

图 8.8　启动 Hadoop

```
[root@master hadoop-2.7.1]# jps
6017 NameNode
1987 VmServer.jar
6115 DataNode
6453 ResourceManager
7512 Jps
6553 NodeManager
6303 SecondaryNameNode
```

图 8.9　Hadoop 启动成功图

(4) 输入 pig，进入 grunt，出现如图 8.10 所示内容，表示 Pig 启动成功。

```
pig
quit (使用 quit 命令退出 pig)
```

8.2.2　命令行使用以及配置选项介绍

Pig 具有许多可以使用的命令行选项。可以通过输入 pig-h 命令查看完整的选项列表。在本节将讨论如下几个不同的命令选项：

```
-e 或者 -execute
```

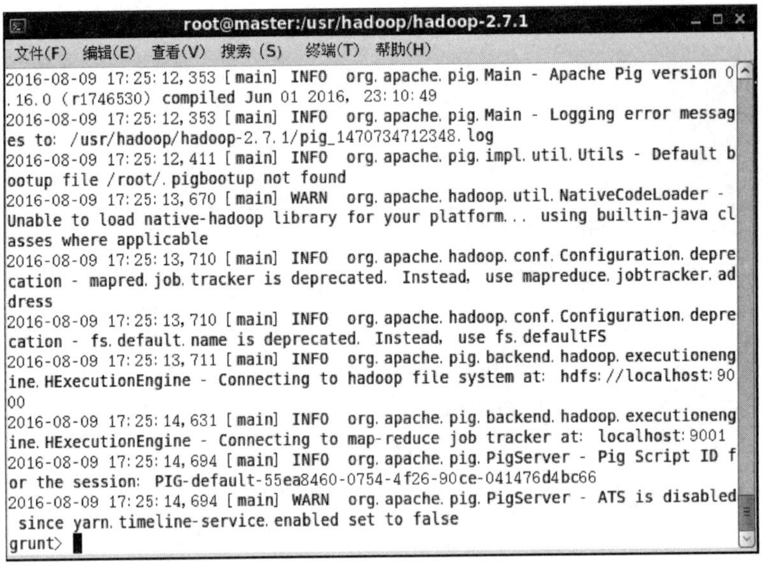

图 8.10 Hadoop 模式下 Pig 配置完成图

在 Pig 中单独执行一条命令。例如，pig -e fs -ls 将会列出用户根目录下的文件。

-h 或者 -help

列举出可用的命令行选项，如图 8.11 所示。

```
[root@master work]# pig -h

Apache Pig version 0.16.0 (r1746530)
compiled Jun 01 2016, 23:10:49

USAGE: Pig [options] [-] : Run interactively in grunt shell.
       Pig [options] -e[xecute] cmd [cmd ...] : Run cmd(s).
       Pig [options] [-f[ile]] file : Run cmds found in file.
  options include:
    -4, -log4jconf - Log4j configuration file, overrides log conf
    -b, -brief - Brief logging (no timestamps)
    -c, -check - Syntax check
    -d, -debug - Debug level, INFO is default
    -e, -execute - Commands to execute (within quotes)
    -f, -file - Path to the script to execute
    -g, -embedded - ScriptEngine classname or keyword for the ScriptEngine
    -h, -help - Display this message. You can specify topic to get help for that
 topic.
        properties is the only topic currently supported: -h properties.
    -i, -version - Display version information
    -l, -logfile - Path to client side log file; default is current working dire
ctory.
    -m, -param_file - Path to the parameter file
    -p, -param - Key value pair of the form param=val
```

图 8.11 可用的命令行选项图

-h properties

列举出 Pig 将要使用的属性值(如果设置了这些属性值)。

```
-P 或者-propertyFile
```

指定一个 Pig 应该读取的属性值配置文件。

```
-version
```

打印出 Pig 的版本信息，如图 8.12 所示。

```
[root@master work]# pig -version
Apache Pig version 0.16.0 (r1746530)
compiled Jun 01 2016, 23:10:49
```

图 8.12　pig 版本信息图

Pig 同样会使用很多的 Java 属性值。属性值信息的完整列表可以通过执行 pig-h propertie 命令打印出来。

Hadoop 同样具有许多用于决定其行为的 Java 属性。在 Pig 0.8 版本和之后版本中，这些 Java 属性也可以传送给 Pig，Pig 然后会在调用到 Hadoop 的时候再将这些参数传送给 Hadoop。对于 0.8 版本之前的其他版本，这些属性值必须要在 hadoop-site.xml 配置文件中配置才可以，这样 Hadoop 客户端会自己加载这些配置信息。

这些属性值可以通过命令行选项-D 传送给 Pig，传送的格式和普通的 Java 属性一样。当使用命令行的方式加载这些属性时，必须要在 Pig 专有的命令行选项前定义这些属性。也可以在所使用的 Pig 发行版目录下的 conf/pig.properties 文件中定义属性值。最后，还可以通过-P 选项指定另一个不同的配置文件。如果同时使用了命令行的方式和属性文件的方式设置属性值，优先以命令行设置的值为准。

8.2.3　返回码

Pig 使用如表 8.1 所示的返回码来传达是运行成功还是失败。

表 8.1　Pig 的返回码

值	含　义	备　注
0	成功	
1	失败，但还可以重试	
2	失败	
3	部分失败	Mulitquery 下使用
4	传递给 Pig 是非法参数	
5	抛出 IOException 异常	通常是由 UDF 抛出
6	抛出 PigException	通常意味着有 Python UDF 抛出了异常
7	抛出 ParseException 异常（参数代入完成后进行解析时可能会抛出这样的异常）	
8	抛出 Throwable 异常（一种未知异常）	

8.3 命令行交互工具

8.3.1 Grunt 概述

Grunt 是 Pig 的命令行交互工具。它允许交互地输入 Pig Latin 脚本以及以交互的方式操作 HDFS。

不需要执行任何脚本和命令直接调用 Pig，就可以进入 Grunt。命令如下：

```
pig - x local
```

立刻就会出现如下信息：

```
grunt >
```

显示结果如图 8.13 所示。

图 8.13　本地模式运行 Pig 图

这样就开启了一个访问本地文件系统的 Grunt shell 界面。如果在上述命令中省略 -x local 而且同时在其 PIG_CLASSPATH 中包含了集群的配置信息，那么将会开启一个访问集群的 HDFS 文件系统的 Grunt shell 界面。

```
pig
```

立刻就会出现如图 8.14 所示的信息。

Grunt 与其他 shell 界面一样，也会提供命令行操作界面进行编辑，同时也通过 Tab 键

```
[root@master pig-0.16.0]# pig
16/08/10 17:11:35 INFO pig.ExecTypeProvider: Trying ExecType : LOCAL
16/08/10 17:11:35 INFO pig.ExecTypeProvider: Trying ExecType : MAPREDUCE
16/08/10 17:11:35 INFO pig.ExecTypeProvider: Picked MAPREDUCE as the ExecType
2016-08-10 17:11:35,488 [main] INFO  org.apache.pig.Main - Apache Pig version 0.
16.0 (r1746530) compiled Jun 01 2016, 23:10:49
2016-08-10 17:11:35,490 [main] INFO  org.apache.pig.Main - Logging error message
s to: /usr/pig/pig-0.16.0/pig_1470820295486.log
2016-08-10 17:11:35,550 [main] INFO  org.apache.pig.impl.util.Utils - Default bo
otup file /root/.pigbootup not found
2016-08-10 17:11:36,890 [main] WARN  org.apache.hadoop.util.NativeCodeLoader - U
nable to load native-hadoop library for your platform... using builtin-java clas
ses where applicable
2016-08-10 17:11:36,964 [main] INFO  org.apache.hadoop.conf.Configuration.deprec
ation - mapred.job.tracker is deprecated. Instead, use mapreduce.jobtracker.addr
ess
2016-08-10 17:11:36,964 [main] INFO  org.apache.hadoop.conf.Configuration.deprec
ation - fs.default.name is deprecated. Instead, use fs.defaultFS
2016-08-10 17:11:36,964 [main] INFO  org.apache.pig.backend.hadoop.executionengi
ne.HExecutionEngine - Connecting to hadoop file system at: hdfs://master:9000
2016-08-10 17:11:38,061 [main] INFO  org.apache.pig.PigServer - Pig Script ID fo
r the session: PIG-default-786a4ea6-397d-4773-b0d1-32781365e461
2016-08-10 17:11:38,061 [main] WARN  org.apache.pig.PigServer - ATS is disabled
since yarn.timeline-service.enabled set to false
grunt>
```

图 8.14　集群模式下运行 Pig

自动补全命令,但是没有提供通过 Tab 键补全文件名的功能。也就是说,如果输入 kil 然后再按 Tab 键,那么就会自动补全命令为 kill;但是如果在的本地文件系统中有一个文件名称为 foo,输入 ls fo,之后再按 Tab 键,不会自动补全命令为 ls foo。这是因为通过 HDFS 去连接然后再确认文件是否存在耗时太久,以至于根本没有使用的必要。

尽管 Grunt 是个很有用的 shell 工具,但它并非包含了普通 shell 具有的所有功能。它没有提供标准 UNIX shell 所提供的很多功能,例如管道、重定向和后台执行等功能。可以通过输入 quit 或者按 Ctrl+D 组合键退出 Grunt。

8.3.2　在 Grunt 中输入 Pig Latin 脚本

Grunt 的主要用途之一就是以交互式会话的方式输入 Pig Latin 脚本。这对于快速地对数据进行抽样以及原型设计更好的 Pig Latin 脚本是非常有用的。

可以在 Grunt 中直接输入 Pig Latin 脚本。Pig 不会执行输入的 Pig Latin 脚本,直到它发现输入 store 或者 dump 命令。不过,它会做一些基本的语法和语义检查,这样可以方便快速地捕获到错误。如果在 Grunt 中有一行 Pig Latin 脚本输入错误了,那么可以采用相同的别名重新输入那条脚本,Pig 会以最后那行输入的实例为准。例如:

```
pig -x local
(严格注意空格以及标点的正确输入,否则会出现错误)
grunt> dividends = load 'NYSE_dividends' as (exchange,symbol,date,dividend);
grunt> symbols = foreach dividends generate symbl;
…Error…
grunt> symbols = foreach dividends generate symbol;
```

结果如图 8.15 所示。

```
grunt> dividends = load 'NYSE_dividends' as (exchange, symbol, date, dividend);
2016-08-10 17:17:42,878 [main] INFO  org.apache.hadoop.conf.Configuration.deprec
ation - io.bytes.per.checksum is deprecated. Instead, use dfs.bytes-per-checksum
2016-08-10 17:17:42,879 [main] INFO  org.apache.hadoop.conf.Configuration.deprec
ation - fs.default.name is deprecated. Instead, use fs.defaultFS
grunt> symbols = foreach dividends generate symbl;
2016-08-10 17:18:27,332 [main] ERROR org.apache.pig.tools.grunt.Grunt - ERROR 10
25:
<line 2, column 37> Invalid field projection. Projected field [symbl] does not e
xist in schema: exchange:bytearray,symbol:bytearray,date:bytearray,dividend:byte
array.
Details at logfile: /usr/pig/pig-0.16.0/pig_1470820641079.log
grunt> symbols = foreach dividends generate symbol;
grunt>
```

图 8.15 基本语法语义检查图

8.3.3 在 Grunt 中使用 HDFS 命令

除了可以交互式地输入 Pig Latin 脚本，Grunt 的另一个主要用途是作为访问 HDFS 的一个 shell 端口。在 Pig 0.5 版本以及之后的版本中，所有的 hadoop fs shell 命令都可以在 Grunt 中使用。这些命令可以通过关键字 fs 访问。hadoop fs 命令后面所跟的短横线（-）在 Grunt 中同样是需要的：

```
hadoop fs -ls
```

执行结果如图 8.16 所示。

```
[root@master work]# hadoop fs -ls
16/08/10 17:21:19 WARN util.NativeCodeLoader: Unable to load native-hadoop libra
ry for your platform... using builtin-java classes where applicable
Found 2 items
-rw-r--r--   1 root supergroup      17027 2016-08-10 16:52 NYSE_dividends
drwxr-xr-x   - root supergroup          0 2016-08-10 16:53 average_dividend
```

图 8.16 列出 Hadoop 集群中各个文件信息图

```
pig -- 进入 grunt
grunt> fs -ls
```

执行结果如图 8.17 所示。

```
grunt> fs -ls
Found 2 items
-rw-r--r--   1 root supergroup      17027 2016-08-10 16:52 NYSE_dividends
drwxr-xr-x   - root supergroup          0 2016-08-10 16:53 average_dividend
```

图 8.17 列出 Hadoop 集群中各个文件信息图

许多命令与 UNIX shell 中的相同而且功能也是类似的，如 chgrp、chmod、chown、cp、du、ls、mkdir、mv、rm 和 stat。其他的一些命令看起来和 UNIX 命令相似但是功能有点不同，包括：

```
cat filename
```

将一个文件的内容打印到标准输出 stdout，结果如图 8.18 所示（截图只选取了显示的部分内容）。

```
grunt > cat NYSE_dividends          -- 此处为熟悉命令，查看的文件可以为任意
```

```
NYSE    CHW    2009-05-07    0.08
NYSE    CHW    2009-04-08    0.08
NYSE    CHW    2009-03-09    0.08
NYSE    CHW    2009-02-06    0.08
NYSE    CRP    2009-12-08    0.494
NYSE    CRP    2009-09-08    0.494
NYSE    CRP    2009-06-08    0.494
NYSE    CRP    2009-03-09    0.494
NYSE    CHL    2009-09-10    0.868
NYSE    CHL    2009-05-11    0.906
NYSE    CYN    2009-11-02    0.1
NYSE    CYN    2009-08-03    0.1
NYSE    CYN    2009-05-04    0.1
NYSE    CYN    2009-02-02    0.25
NYSE    COP    2009-10-28    0.5
NYSE    COP    2009-07-29    0.47
NYSE    COP    2009-05-21    0.47
NYSE    COP    2009-02-19    0.47
NYSE    CEO    2009-09-04    2.58
NYSE    CEO    2009-05-14    2.581
NYSE    CS     2009-04-27    0.096
NYSE    CMC    2009-09-30    0.12
NYSE    CMC    2009-06-30    0.12
NYSE    CMC    2009-04-01    0.12
NYSE    CMC    2009-01-06    0.12
grunt> █
```

图 8.18 输出文件内容图

```
copyFromLocal localfile hdfsfile
```

上面的代码从本地磁盘中复制一个文件到 HDFS 中，该命令是以串行而非并行的方式处理的。结果如图 8.19 所示。

```
grunt > copyFromLocal average_dividend.pig average_dividend.pig
-- 此处为熟悉命令，复制的文件可以为任意，若文件 ABC 不在进入 pig 的目录中，使用该命令时，第一个文件需要指明文件的路径
```

```
grunt> copyFromLocal average_dividend.pig average_dividend.pig
grunt> fs -ls
Found 3 items
-rw-r--r--   1 root supergroup      17027 2016-08-10 16:52 NYSE_dividends
drwxr-xr-x   - root supergroup          0 2016-08-10 16:53 average_dividend
-rw-r--r--   1 root supergroup        206 2016-08-10 17:26 average_dividend.pig
```

图 8.19 本地文件复制到 HDFS 文件系统图

```
copyToLocal hdfsfile localfile
```

上面的代码从 HDFS 文件系统中复制一个文件到的本地磁盘，该命令是以串行而非并行的方式处理的。结果如图 8.20 所示。

```
grunt> copyToLocal average_dividend /root/average_dividend    -- 此处为熟悉命令,复制的文件
                                                                 可以为任意
```

```
[root@master ~]# ls
anaconda-ks.cfg    install.log           work   模板  图片  下载  桌面
average_dividend   install.log.syslog    公共的 视频  文档  音乐
```

图 8.20　HDFS 文件系统文件复制到本地图

```
rm filename
```

上面的代码递归地移除文件,该命令相当于 UNIX 中的 rm-r 命令,要慎重使用此命令,如图 8.21 所示。

```
grunt> rm ABC        -- 此处为熟悉命令,移除的文件可以为任意
```

```
grunt> rm average_dividend.pig
2016-08-10 17:31:06,156 [main] INFO  org.apache.pig.tools.grunt.GruntParser - Wa
ited 0ms to delete file
grunt> fs -ls
Found 2 items
-rw-r--r--   1 root supergroup      17027 2016-08-10 16:52 NYSE_dividends
drwxr-xr-x   - root supergroup          0 2016-08-10 16:53 average_dividend
```

图 8.21　移除文件图

在 Pig 0.5 版本前的其他版本中,还没有提供 hadoop fs 系列命令,而是 Grunt 本身实现了如下一些命令:cat、cd、copyFromLocal、copyToLocal、cp、ls、mkdir、mv、pwd、rm(该命令与 Hadoop 中的 rmr 命令效果一样,而不是 Hadoop 中的 rm)和 rmf。在 Pig 0.8 版本中(本书使用 pig-0.16.0 版本),所有这些命令都是存在的。但是除了 cd 和 pwd 两个命令外,其他都不推荐使用,而且它们在未来可能会被移除,所以推荐使用 hadoop fs。

在 0.8 版本,Grunt 中新增了一个命令:sh。该命令允许使用本地的 shell 命令,就像 fs 提供给访问 HDFS 的功能一样。没有使用管道或者重定向的简单的 shell 命令都可以被执行。因为 sh 并非总是能正确地获取到当前的工作目录,所以最好还是使用绝对路径。

8.3.4　在 Grunt 中控制 Pig

Grunt 同样提供了控制 Pig 和 MapReduce 的命令:

```
Kill jobid
```

终止指定 jobid 的 MapReduce 任务。产生任务的 Pig 命令在输出信息中会列举出它多产生的所有任务的 ID。也可以通过 Hadoop 的 JobTracker 的 GUI(图形用户界面)查看到所有任务的 ID 号。在 GUI 中会列出当前集群中正在运行的所有任务信息。需要注意的是,该命令只会终止特定的那个 MapReduce 任务。如果 Pig 任务包含了不依赖所终止的那个任务的其他 MapReduce 任务时,那么这些任务还会继续执行下去。如果想终止一个特定的 Pig 任务所触发的所有 MapReduce 任务,那么最好的办法是终止这个 Pig 进程,然后再

使用那个命令终止所有真正执行的 MapReduce 任务。请确定使用 Ctrl+C 或者 UNIX 中的 kill 命令来终止 Pig 进程,而不要使用 UNIX 中的 kill-9 命令,因为这个命令不会触发 Pig 进行临时文件清理,这样会在集群中留下垃圾文件。

```
Exec [[ - param param_name = param_value]] [[ - param_file filename]] script
```

执行 Pig Latin 脚本文件 script。script 这个文件中的别名不会被传入到 Grunt 中。当在一个 Grunt 会话中测试的 Pig Latin 脚本时,这个命令非常有用。

```
run [[ - param param_name = param_value]] [[ - param_file filename]] script
```

在当前 Grunt shell 中执行 Pig Latin 脚本 script。因此 script 脚本文件中使用的所有别名在 Grunt 中都是有效的,同时 script 文件中的命令也是通过 shell 历史记录可以查看到的。这是另一种方式在一个 Grunt 会话中测试的 Pig Latin 脚本。

第 9 章

Pig Latin 的使用

是时候开始研究 Pig Latin 了。本章将介绍 Pig Latin 的基础知识，以帮助读者编写第一个可以使用的脚本。

9.1 Pig Latin 概述

9.1.1 基础知识

Pig Latin 是一种数据流语言，每个处理步骤都会产生一个新的数据集，或者产生一个新的关系。在 input=load 'data' 这句脚本中，input 是加载数据集 data 后所得结果的关系名称。这里所说的关系名称也就是通常所说的别名。关系名称和变量的概念相似，但是它们不是变量。一旦声明了变量，那么这个分配就是不变的了。关系名称是可以被重用的，例如，如下的用法是被允许的：

```
A = load 'NYSE_dividends' (exchange,symbol,date,dividends);
A = filter A by dividends > 0;
A = foreach A generate UPPER(symbol);
```

但是，这种做法是不被推荐的。这里看上去好像是在重定义 A，其实是创建了一个新的关系并命名为 A，这也就丢掉了和之前那个叫做 A 的关系的联系了。虽然 Pig 会尽可能地保持关系，但这仍然不是一个好的做法。这会导致在阅读脚本程序和查看错误日志的时候，让人感到困惑，不知道指定的到底是哪个 A。

除了关系名称外，Pig Latin 还有一个概念就是字段名称。它代表的是一个关系所包含的字段(或者称为列)的名称。例如,在前面提到的 Pig Latin 脚本片段中,dividends 和 symbol 就是字段名称。这与变量在数据管道中传送时,对于不同的记录包含不同的值这一点有些相像,但是无法对它们进行赋值。

无论是关系名称还是字段名称都必须以字母字符开头,之后可以跟上零个或多个字母、数字或者下画线(_)。名称中的所有字符必须都是 ASCII 码。

9.1.1.1 区分大小写

不幸的是,Pig Latin 无法明确说是否是区分字母大小写的。Pig Latin 中的关键字是不区分大小写的,例如,LOAD 和 load 是等价的。但是关系名称和字段名称是区分大小写的。因此"A=load'foo';"和"a = load 'foo';"是不等价的。用户自定义函数 UDF 的名称也是区分大小写的,因此 COUNT 和 count 所指的并非是同一个 UDF。

9.1.1.2 注释

Pig Latin 具有两种注释方式:SQL 样式的单行注释(--)和 Java 样式的多行注释(/**/)。

例如:

```
A = load 'foo';          -- 这个是单行注释
/*
* 这个是多行注释
*/
B = load/* 这个是中间的注释*/'bar';
```

9.1.2 输入和输出

在可以做一些自身感兴趣的事情之前,需要知道如何为数据流增加输入和输出。

9.1.2.1 加载

任何一种数据流的第一步都是要指定输入。这在 Pig Latin 中是通过 load 语句来完成的。默认情况下,load 使用默认加载函数 PigStorage 加载存放在 HDFS 中并且以制表键进行分隔的文件。例如,语句"divs=load'data/examples/NYSE_dividends';"会在文件夹目录/data/examples 下查找文件名为 NYSE_dividends 的文件。当然也可以写相对路径。默认情况下,Pig 任务会在当前的 HDFS 中的目录下执行,例如/user/yourlogin。除非改变了这个目录,否则所有的相对路径都以那个目录为基准。也可以指定一个完整的 URL 路径,例如 hdfs://nn.acme.com/data/examples/NYSE_dividends,这可以从 NameNode 为 nn.acme.com 的 HDFS 实例中读取文件。

实际上,大部分数据并非是使用制表键作分隔符的文本文件,也有可能需要从其他非HDFS 的存储系统中加载数据。Pig 允许在加载数据时通过 using 句式指定其他加载函数。

例如，如果想从 HBase 中加载数据，那么需要使用为 HBase 准备的加载函数：

```
divs = load 'NYSE_dividends' using HBaseStorage ();
```

如果没有指定加载函数，那么会使用内置的加载函数 PigStorage。同样可以通过 using 句式为使用的加载函数指定参数。例如，如果想读取以逗号分隔的文本文件数据，那么 PigStorage 会接受一个指定分隔符的参数：

```
divs = load 'NYSE_dividends' using PigStorage (',');
```

load 语句中也可以有 as 句式，这个句式可以为加载的数据指定模式。

```
divs = load 'NYSE_dividends' as (exchange,symbol,date,dividends);
```

当从 HDFS 访问指定"文件"的时候，也可以指定文件夹。在这种情况下，Pig 会遍历指定的文件夹下的所有文件并将它们作为 load 语句的输入。因此，如果有一个名称为 input 的文件夹，其中包含两个文件 today 和 yesterday，那么可以将文件夹 input 作为加载的输入文件，Pig 会将 today 和 yesterday 两个文件同时作为输入进行处理。如果指定的文件夹下还有其他文件夹，那么所有这些文件夹下的文件也都是会被包括在内的。

PigStorage 和 TextLoader 这两个内置的可以操作 HDFS 文件的 Pig 加载函数是支持模式匹配的。通过模式匹配，可以读取不在同一文件夹下的多个文件或者读取一个文件夹下的部分文件。表 9.1 描述的是在 Hadoop 0.20 中所提供的合法的模式匹配语法。需要注意的是，这些正则字符的含义是由 Pig 下面的 HDFS 决定的，因此这些正则字符是否可以工作取决于使用的是什么版本的 HDFS。同样，如果是从 UNIX shell 命令行运行 Pig Latin 命令，那么还需要将这些正则字符拆分为多个部分。

表 9.1 Hadoop 0.20 中提供的正则匹配语法

正 则 字 符	含 义
?	匹配任何单字符
*	匹配零个或多个字符
[abc]	匹配字符集合{a,b,c}所包含的任意一个字符
[a-z]	匹配指定范围内的任意字符
[^abc]	匹配未包含的任意字符，其中^符号匹配输入字符串的开始位置
[^a-z]	匹配任何不在指定范围内的任意字符
\c	移除(转义)字符 c 的多表达的特殊含义
[ab,cd]	配置字符串集合{ab,cd}中的任一字符串

9.1.2.2 存储

当处理完数据之后，需要把结果写到某个地方。Pig 提供了 store 语句来进行写数据操作。在许多方面，它是 load 语法的镜像。默认情况下，Pig 会用 PigStorage 将结果数据以制表键作为分隔符存储到 HDFS 中：

```
store processed into'/data/examples/processed';
```

执行上面的语句，Pig 会将处理完的结果数据存储到/data/examples 路径下的 processed 文件夹下。当然可以指定相对路径，同样也可以指定完整的 URL 路径，例如：

```
hdfs://nn.name.com/data/examples/processed
```

如果并没有显式地指定存储函数，那么将会默认使用 PigStorge。可以使用 using 语句指定一个不同的存储函数：

```
store processed into'processed'using HBaseStorage();
```

也可以传参数给其使用的存储函数。例如，如果想将数据存储为以逗号分隔的文本数据，PigStorage 会接受一个指定分隔符的参数：

```
store processed into'processed'using PigStorage(',');
```

当写到文件系统中后，processed 是一个包含多个部分文件的文件夹，而不是一个文件。但是至于会生成多少个部分文件，这要取决于执行 store 前的最后一个任务的并行数，该值由为这个任务所设置的并行级别所决定。如果这是一个 map-only 的任务，那么生成的文件个数取决于 map 任务的个数，这是由 Hadoop 而非 Pig 控制的。

9.1.2.3 输出

在大多数情况下，当处理完数据后会希望将结果存储到某个地方，但是偶尔也会想在屏幕上看看这些结果数据。这对于调试阶段和原型研究阶段是特别有用的。对于一些特殊的运行很快的任务也是有用的。dump 语句可以将脚本的输出打印到屏幕上。

```
dump processed;
```

一直到 Pig 0.7 版本，dump 的输出数据都是符合 Pig Latin 中定义的常量格式的。因此，long 类型的值会以 L 结尾，float 类型的数据会以 F 结尾，map 用[]（方括号）分隔，tuple 使用()（圆括号）分隔，bag 使用{}（花括号）分隔。从 0.8 版本开始，long 值的 L 和 float 类型值的 F 被移除了，然而复杂数据类型的表示方式被保留了下来。null 表示数据缺失值，字段是以逗号分隔的。因为输出中的每条记录都是一个 tuple，所以它是被()（圆括号）包围的。

9.2 关系操作

关系操作符是 Pig Latin 提供给用于操作数据的主要工具。使用关系操作符可以对数据进行排序、分组、连接、推测和过滤等转换。本节将介绍一些基本的关系操作符，掌握了这些内容,已基本能够使用 Pig Latin 进行编程了。

9.2.1 foreach

foreach 语句接受的是一组表达式,然后在数据管道中将它们应用到每条记录中,因此命名为 foreach。通过这些表达式,它将会产生出新的数据并传送给下一个操作符。对于那些熟悉数据库术语的人来说,这个就是 Pig 的推测操作符。

如下代码加载了完整的记录,然后对于每条记录只保留 user 和 id 两个字段。

(1) 在目录/root/work 下,创建如图 9.1 所示的数据文件 input(数据之间使用 Tab 键分隔),并将数据文件 input 导入 HDFS 中。

```
cd /root/work
vim input
pig                (进入 pig)
grunt > copyFromLocal input input
```

```
P1    001    beijing     13811110001    teacher
P2    002    shanghai    13811110002    teacher
P3    003    shanghai    13811110003    doctor
P4    004    sichuang    13811110004    student
P5    005    anhui       13811110005    student
```

图 9.1 数据文件 input 内容图

(2) 输入下面的命令,然后对于每条记录只保留 user 和 id 两个字段。

```
grunt > A = load 'input' as (user:chararray, id:long, address:chararray, phone:chararray,
preferences:map[]);          -- 注意' = '前后需要有空格
grunt > B = foreach A generate user,id;
grunt > dump B;              -- 输出 B
```

运行结果如图 9.2 所示。

```
(P1,1)
(P2,2)
(P3,3)
(P4,4)
(P5,5)
```

图 9.2 保留 user 和 id 两个字段结果图

1. foreach 语句中的表达式

foreach 支持大量的表达式,最简单的就是常量和字段引用。字段引用可以通过别名进行引用或者通过位置引用。位置引用是通过 $(美元符号)加上从 0 开始的整数构成的。

(1) 将上例中所需的数据文件 NYSE_daily 复制到文件夹/root/work 中。

```
cp /root/下载/NYSE_daily /root/work/    (此处 NYSE_daily 数据文件已存于模板 2 的本地文件夹
                                        【下载】中)
cd /root/work/                          (回到工作目录/root/work/下)
```

(2)进入 Pig,输入下面的命令,查看结果。

```
grunt> copyFromLocal NYSE_daily NYSE_daily
grunt> prices = load 'NYSE_daily' as (exchange,symbol,date,open,high,low,close,volume,adj_close);
grunt> gain = foreach prices generate close - open;
grunt> gain2 = foreach prices generate $6 - $3;
grunt> dump gain              --输出结果
grunt> dump gain2             --gain 和 gain2 的输出结果相同,如图 9.3 部分所示(截图只选取了显
                                示的部分内容)
```

```
(0.26000000000000156)
(0.4399999999999977)
(0.00999999999999801)
(-0.019999999999999574)
(0.7100000000000009)
(-0.38000000000000256)
(0.21999999999999886)
(-0.39000000000000057)
(-0.2900000000000027)
(-0.36999999999999744)
(0.07000000000000028)
(-0.33999999999999986)
(-0.2699999999999996)
(-1.4400000000000013)
(0.16999999999999815)
(1.0300000000000011)
(-0.05000000000000071)
(-0.46000000000000085)
(-0.22000000000000242)
(-0.3299999999999983)
(-0.2699999999999996)
(-1.0199999999999996)
(-0.350000000000014)
(0.8900000000000006)
(-0.05000000000000071)
```

图 9.3 位置引用结果图

在上面的脚本中,关系名称 gain 和 gain2 会存储相同的值。位置引用这种方式在不知道模式或者没有声明模式的时候会有用。

除了使用别名和位置,也可以使用..(两个点)来指定字段区间。当字段很多而且不想在 foreach 命令中重写一遍时,这是非常好用的。

(3)继续输入下面的命令,查看结果。

```
grunt> prices = load 'NYSE_daily' as (exchange,symbol,date,open,high,low,close,volume,adj_close);
grunt> beginning = foreach prices generate .. open;    --produces exchange,symbol,
                                                         date,open
grunt> middle = foreach prices generate open .. close; --produces open,high,low,close
grunt> end = foreach prices generate volume ..;        --produces volume,adj_close
grunt> dump beginning;                                 --输出 exchange,symbol,date,
                                                         open 的内容,如图 9.4 所示
                                                         (截图只选取了显示的部分
                                                         内容)
```

```
(NYSE, CVA, 2009-02-06, 18.63)
(NYSE, CVA, 2009-02-05, 18.19)
(NYSE, CVA, 2009-02-04, 18.42)
(NYSE, CVA, 2009-02-03, 18.61)
(NYSE, CVA, 2009-02-02, 17.74)
(NYSE, CVA, 2009-01-30, 17.67)
(NYSE, CVA, 2009-01-29, 17.59)
(NYSE, CVA, 2009-01-28, 18.22)
(NYSE, CVA, 2009-01-27, 18.12)
(NYSE, CVA, 2009-01-26, 18.49)
(NYSE, CVA, 2009-01-23, 18.29)
(NYSE, CVA, 2009-01-22, 18.88)
(NYSE, CVA, 2009-01-21, 19.49)
(NYSE, CVA, 2009-01-20, 20.92)
(NYSE, CVA, 2009-01-16, 20.76)
(NYSE, CVA, 2009-01-15, 19.40)
(NYSE, CVA, 2009-01-14, 19.69)
(NYSE, CVA, 2009-01-13, 20.11)
(NYSE, CVA, 2009-01-12, 20.46)
(NYSE, CVA, 2009-01-09, 21.08)
(NYSE, CVA, 2009-01-08, 21.08)
(NYSE, CVA, 2009-01-07, 22.34)
(NYSE, CVA, 2009-01-06, 22.85)
(NYSE, CVA, 2009-01-05, 21.79)
(NYSE, CVA, 2009-01-02, 21.76)
```

图 9.4 beginning 字段区间结果图

```
grunt > dump middle;       -- 输出 open,hign,low,close 的内容,如图 9.5 所示(截图只选取了显示
                              的部分内容)
```

```
(18.42, 18.81, 18.38, 18.43)
(18.61, 18.82, 18.32, 18.59)
(17.74, 18.71, 17.64, 18.45)
(17.67, 17.96, 17.10, 17.29)
(17.59, 18.04, 17.51, 17.81)
(18.22, 18.34, 17.53, 17.83)
(18.12, 18.25, 17.52, 17.83)
(18.49, 18.84, 17.95, 18.12)
(18.29, 18.55, 17.87, 18.36)
(18.88, 18.90, 18.29, 18.54)
(19.49, 19.83, 18.75, 19.22)
(20.92, 21.04, 19.31, 19.48)
(20.76, 21.06, 20.19, 20.93)
(19.40, 20.75, 19.30, 20.43)
(19.69, 19.95, 19.00, 19.64)
(20.11, 20.11, 19.31, 19.65)
(20.46, 20.88, 19.91, 20.24)
(21.08, 21.63, 20.30, 20.75)
(21.08, 21.25, 20.43, 20.81)
(22.34, 22.50, 21.22, 21.32)
(22.85, 22.92, 22.25, 22.50)
(21.79, 22.89, 21.49, 22.68)
(21.76, 22.80, 21.46, 21.71)
```

图 9.5 middle 字段区间结果图

```
grunt > dump end;          -- 输出 volume,adj_close 的内容,如图 9.6 所示(截图只选取了显示的
                              部分内容)
```

```
(1141500,18.43)
(1192600,18.59)
(2453600,18.45)
(1672500,17.29)
(1228000,17.81)
(1739300,17.83)
(1360900,17.83)
(1340900,18.12)
(1258500,18.36)
(1623300,18.54)
(1598400,19.22)
(1435400,19.48)
(966600,20.93)
(1063900,20.43)
(1744100,19.64)
(1261200,19.65)
(700800,20.24)
(723800,20.75)
(1633700,20.81)
(1169500,21.32)
(1058700,22.50)
(981100,22.68)
(1344600,21.71)
```

图 9.6　end 字段区间结果图

对于整数和浮点数，标准的算术操作符也是支持的："＋"表示加法，"－"表示减法，"＊"表示乘法，"/"表示除法。这些操作符会根据数值本身的类型确定返回值的类型，因此 5/2 是 2，5.0/2.0 的值是 2.0。除此之外，对于整数，还支持取模操作符％。对于整数和浮点数，同样支持一元负值操作符(－)。Pig Latin 遵循标准的数学执行优先级规则。

null 值对于所有的算术操作符都是抵消的。也就是说，x＋null＝null，这里不管 x 是什么类型的值都是成立的。

Pig 同样提供了一个三元条件操作符，经常被称为 bincond。它以一个布尔表达式开始，后面跟着一个？（问号），然后跟着的是如果前面布尔表达式为真时返回的值，后面再跟着：（冒号），最后跟着如果前面布尔表达式为假时应返回的值。如果前面的布尔表达式返回 null，那么整个表达式返回 null。表达式中的？后的两个值应该是相同的数据类型：

```
2 == 2?1:4              -- 返回 1
2 == 3?1:4              -- 返回 4
null == 2?1:'fred'      -- 类型错误；冒号两边的值应该是相同的数据类型
2 == 2?1:'fred'         -- 类型错误；冒号两边的值应该是相同的数据类型
```

为了从复杂的数据类型中提取出值，需要使用投射运算符。对于 map 是使用♯（也就是散列表），然后后面跟着一个字符串类型的键的名称。需要注意的是，一个键对应的值可以是任意数据类型的。如果使用的键在 map 中不存在，那么结果是 null。

在 shell 命令行中输入下面的命令：

```
cp /root/下载/baseball /root/work/ (移动数据文件 baseball 至/root/work/中,此处 NYSE_daily 数据文件已存于模板的本地文件夹【下载】中)
cd /root/work/
```

进入 Pig,输入下面的命令：

```
grunt> copyFromLocal baseball baseball
grunt> bball = load 'baseball' as (name:chararray, team:chararray, position:bag{(p:
chararray)},bat:map[]);
grunt> avg = foreach bball generate bat# 'batting_average';
grunt> dump avg;        --结果如图9.7所示(截图只选取了显示的部分内容)
```

```
(0.181)
(0.192)
(0.171)
(0.5)
(0.5)
()
(0.25)
()
()
(0.083)
(0.292)
(0.141)
(0.29)
(0.266)
(0.302)
()
(0.236)
(0.251)
()
()
(0.284)
(0.104)
(0.26)
```

图 9.7　运行结果图

tuple 的映射是用.(点操作符)符号完成的。最外层的数据记录可以通过字段名指定一个字段(如果对于这个 tuple 定义了一个模式)。对 tuple 通过位置进行引用,如果该位置不存在,则会返回 null。如果使用字段名进行引用,而 tuple 中没有这个字段名,那么将会产生一个错误：

```
A = load 'input' as(t:tuple(x:int,y:int));
B = foreach A generate t.x t.$1;
```

bag 的映射并不像 map 和 tuple 映射那么简单明了。bag 并不会保证它内部存放的 tuple 是有序的,因此对一个 bag 中的 tuple 进行映射是没有意义的。反之,当需要映射一个 bag 中的字段时,可以通过创建一个包含需要的字段的 bag：

```
A = load'input'as(b:bag{t:(x:int,y:int)});
B = foreach A generate b.x;
```

这会创建一个新的 bag,其中只有字段 x 在里面。可以通过使用圆括号中标明逗号分隔的字段名称的方式映射多个字段：

```
A = load'input'as(b:bag{t:(x:int,y:int)});
B = foreach A generate b.(x,y);
```

这一点区别是 b.x 是一个 bag,而不是一个可以进行计算的数量值。

请思考如下的 Pig Latin 脚本,第一个脚本是错误的,第二个脚本是正确的。

(1) 在本地目录/root/work 下创建数据文件 foo 和 Pig Latin 脚本 foo.pig。

foo 文件内容如下:

```
vim foo
1  2  3
```

vim foo.pig 内容如下:

```
-- foo.pig
-- 添加内容如下:
A = load 'foo' as (x:chararray,y:int,z:int);
B = group A by x; --产生包含对于 x 给定的值对应的都有记录的 bag A
C = foreach B generate SUM(A.y + A.z);
```

(2) 在当前目录下进入 pig,将两文件导入 HDFS 中,运行脚本,出现错误,如图 9.8 所示。

```
grunt> copyFromLocal foo foo
grunt> copyFromLocal foo.pig foo.pig
grunt> run foo.pig
2016-08-11 10:03:46,747 [main] INFO  org.apache.hadoop.conf.Configuration.deprec
ation - fs.default.name is deprecated. Instead, use fs.defaultFS
2016-08-11 10:03:46,858 [main] INFO  org.apache.hadoop.conf.Configuration.deprec
ation - fs.default.name is deprecated. Instead, use fs.defaultFS
grunt> A = load 'foo' as (x:chararray, y:int, z:int);
2016-08-11 10:03:47,823 [main] INFO  org.apache.hadoop.conf.Configuration.deprec
ation - fs.default.name is deprecated. Instead, use fs.defaultFS
grunt> B = group A by x;    --产生包含对于x给定的值对应的都有记录的bag A
grunt> C = foreach B generate SUM(A.y+A.z);
2016-08-11 10:03:48,326 [main] ERROR org.apache.pig.tools.grunt.Grunt - ERROR 10
39:
<line 3, column 30> (Name: Add Type: null Uid: null)incompatible types in Add Op
erator left hand side:bag : tuple(y:int)    right hand side: bag : tuple(z:int)
Details at logfile: /root/work/pig_1470880989005.log
```

图 9.8 脚本运行出现错误图

这里我们很清楚编程人员在想什么。但是因为 A.y 和 B.y 是 bag,而 bag 类型间是没有加法操作的,因此会报错。在 Pig Latin 中正确地进行这个计算的方式如下:

(1) 在本地目录/root/work 下创建 Pig Latin 脚本 foo2.pig。

foo2.pig 内容如下:

```
-- foo2.pig 创建脚本 foo2.pig
-- 添加内容如下:
A = load 'foo'as(x:chararray,y:int,z:int);
A1 = foreach A generate x,y + z as yz;
B = group A1 by x;
C = foreach B generate SUM(A1.yz);
dump C;
```

(2) 进入 Pig,将 foo2.pig 脚本导入 HDFS 中,并成功运行脚本,求和结果如图 9.9 所示。

(5)

图 9.9　脚本运行正确结果图

2. foreach 语句中的 UDF

在 foreach 句式中可以使用用户自定义函数（UDF），这些 UDF 被称为求值函数。因为它们是作为 foreach 语句的组成部分，所以这些 UDF 每次读取一条记录然后产生一条输出。需要记住的是，这里不管是输入还是输出都可以是一个 bag，因此这里的一条记录可以包含一个 bag 单元的记录。

在 /root/work 目录下，编写 udf_in_foreach.pig 脚本如下：

```
-- udf_in_foreach.pig
 divs = load 'NYSE_dividends' as (exchange,symbol,date,dividends);
  -- 确保所有的字符串都是大写的
upped = load 'NYSE_dividends' as (exchange,symbol,date,dividends);
grpd = group upped by symbol;           -- 为每个 symbol 值输出一个 upped bag
  -- 接受一个 bag 的整数，为每个分组产生一个结果
sums = foreach grpd generate group,SUM(upped.dividends);
dump sums;                              -- 每个分组产生一个结果
```

编写完成后，直接进入 Pig，将 udf_in_foreach.pig 导入 HDFS 中，输入下面的命令，运行脚本 udf_in_foreach.pig，结果如图 9.10 所示（截图只选取了显示的部分内容）。

```
grunt> run udf_in_foreach.pig          -- 运行
```

```
(CUZ,0.68)
(CVB,0.776)
(CVC,0.4)
(CVE,23.749999)
(CVS,0.304)
(CVX,2.66)
(CWF,0.408000000000000014)
(CWT,1.18)
(CWZ,2.218)
(CXE,0.3650000000000001)
(CXH,0.6130000000000001)
(CYD,0.1)
(CYE,0.6800000000000002)
(CYN,0.5499999999999999)
(CYS,0.9)
(CYT,0.164)
(CASC,0.12000000000000001)
(CATO,0.66)
(CLNY,0.07)
```

图 9.10　udf_in_foreach.pig 脚本结果图

除此之外，评估函数可以接受 * 作为参数，这个符号表示整个记录传送给该函数。即使函数本身是无参数的，它们也是可以被引用的。

3. foreach 语句中的字段命名

每一个 foreach 语句的结果都是一个新的 tuple，这种情况下输出结果的模式与 foreach

的输入数据的模式是不同的。Pig 可以从 foreach 语句中推断出输出结果的各个字段的数据类型，但是它并非总能够推断出这些字段的名称。对于那些没有使用其他操作符的简单的映射的字段来说，Pig 会保留它们之前使用的名称。

进入 Pig，输入下面的命令：

（注：因为之前操作已将数据文件 NYSE_dividends 导入过 HDFS 中，所以这里可以直接加载该数据文件。）

```
grunt > divs = load 'NYSE_dividends' as (exchange:chararray,symbol:chararray,date:chararray,
dividends:float);
grunt > sym = foreach divs generate symbol;
grunt > describe sym;
sym: {symbol: chararray}              -- 如图 9.11 所示
```

```
grunt> divs = load 'NYSE_dividends' as (exchange: chararray, symbol: chararray, dat
e: chararray, dividends: float);
2016-08-11 10:37:59,563 [main] INFO  org.apache.hadoop.conf.Configuration.deprec
ation - fs.default.name is deprecated. Instead, use fs.defaultFS
grunt> sym = foreach divs generate symbol;
grunt> describe sym;
sym: {symbol: chararray}
```

图 9.11 各个字段的数据类型图

一旦其中有一个使用的不是简单的映射操作，Pig 就不会对这个字段重新赋一个新的名称。如果不是显式地给这个字段命个名称，那么该字段就没有名称，也就只能通过位置参数来获得，例如 $0。可以用 as 句式为字段命名。

进入 Pig，输入下面的命令：

```
grunt > divs = load 'NYSE_dividends' as (exchange:chararray,symbol:chararray,date:chararray,
dividends:float);
grunt > int_cents = foreach divs generate dividends * 100.0 as dividend,dividends * 100.0;
grunt > describe int_cents;
int_cents: {dividend: double,double}              -- 结果如图 9.12 所示
```

```
grunt> divs = load 'NYSE_dividends' as (exchange: chararray, symbol: chararray, date
: chararray, dividends: float);
2016-08-11 10:40:11,661 [main] INFO  org.apache.hadoop.conf.Configuration.deprec
ation - fs.default.name is deprecated. Instead, use fs.defaultFS
grunt> int_cents = foreach divs generate dividends *100.0 as dividend, dividends*
100.0;
2016-08-11 10:40:25,121 [main] WARN  org.apache.pig.newplan.BaseOperatorPlan - E
ncountered Warning IMPLICIT_CAST_TO_DOUBLE 2 time(s).
grunt> describe int_cents;
2016-08-11 10:40:37,576 [main] WARN  org.apache.pig.newplan.BaseOperatorPlan - E
ncountered Warning IMPLICIT_CAST_TO_DOUBLE 2 time(s).
int_cents: {dividend: double, double}
```

图 9.12 int_cents 字段数据类型图

需要注意的是，在 foreach 中 as 关键字是跟在每个表达式之后的，这点与 load 句式指定字段名的方式不同。在 load 句式中字段定义语句是跟在整个语句之后的。

9.2.2 Filter

通过 filter 语句可以选择将哪些数据保留在数据流中。filter 中包含了一个断言。对于一条记录，断言如果为 true，那么这条记录就会在数据流中传下去，否则就不会向下传。

断言式可以包含等值比较操作符，包括判断是否相等的＝＝，以及！＝、＞、＞＝和＜＝。这些比较操作符都可以用于基本数据类型。＝＝和！＝也可以用于 map 和 tuple 这样复杂数据类型的比较。如果在两个 tuple 间进行比较，那么还要求这两个 tuple 要么有相同的模式，要么都没有模式。所有上述的等值判断操作符都不可以用于 bag。

Pig Latin 遵循大多数程序语言都支持的标准运算符操作优先级规则，其中有一项就是算术计算操作符的优先级要大于等值比较操作符的优先级。因此，x＋y＝＝a＋b 和（x＋y）＝＝（a＋b）是等价的。

对于 chararray 数据类型，使用者可以判断这个 chararray 是否符合指定的正则表达式。在/root/work 目录下，编写 filter_matches.pig 脚本如下：

```
-- filter_matches.pig
divs = load 'NYSE_dividends' as (exchange:chararray, symbol:chararray, date:chararray, dividends:float);
startswithcm = filter divs by symbol matches 'CM.*';
dump startswithcm; -- 筛选出 symbol 为 CM 开头的所有记录
```

编写完成后，直接进入 Pig，将 filter_matches.pig 导入 HDFS 中，输入下面的命令，运行脚本 filter_matches.pig，结果如图 9.13 所示（截图只选取了显示的部分内容）：

```
grunt > run filter_matches.pig          -- 运行
```

```
(NYSE, CMK, 2009-04-13, 0.041)
(NYSE, CMK, 2009-03-09, 0.041)
(NYSE, CMK, 2009-02-09, 0.041)
(NYSE, CMK, 2009-01-12, 0.041)
(NYSE, CMI, 2009-11-18, 0.175)
(NYSE, CMI, 2009-08-19, 0.175)
(NYSE, CMI, 2009-05-20, 0.175)
(NYSE, CMI, 2009-02-18, 0.175)
(NYSE, CMS, 2009-11-04, 0.125)
(NYSE, CMS, 2009-08-06, 0.125)
(NYSE, CMS, 2009-05-06, 0.125)
(NYSE, CMS, 2009-02-04, 0.125)
(NYSE, CMO, 2009-12-29, 0.54)
(NYSE, CMO, 2009-09-28, 0.56)
(NYSE, CMO, 2009-06-26, 0.58)
(NYSE, CMO, 2009-03-27, 0.56)
(NYSE, CMA, 2009-12-11, 0.05)
(NYSE, CMA, 2009-09-11, 0.05)
(NYSE, CMA, 2009-06-11, 0.05)
(NYSE, CMA, 2009-03-11, 0.05)
(NYSE, CMC, 2009-09-30, 0.12)
(NYSE, CMC, 2009-06-30, 0.12)
(NYSE, CMC, 2009-04-01, 0.12)
(NYSE, CMC, 2009-01-06, 0.12)
```

图 9.13　filter_matches.pig 脚本结果图

可以通过在语句前面加上关键字 no 来查找那些不满足一个指定正则表达式的 chararray 类型数据。

继续编写 filter_not_matches.pig 脚本如下：

```
-- filter_not_matches.pig
divs = load 'NYSE_dividends' as (exchange:chararray, symbol:chararray, date:chararray, dividends:float);
notstartswitchcm = filter divs by not symbol matches 'CM.*';
dump notstartswitchcm;                    -- 筛选出 symbol 为不以 CM 开头的所有记录
```

编写完成后，进入 Pig，将 filter_not_matches.pig 导入 HDFS 中，输入下面的命令，运行脚本 filter_not_matches.pig，结果如图 9.14 所示（截图只选取了显示的部分内容）：

```
grunt> run filter_not_matches.pig        -- 运行
```

```
(NYSE, CHW, 2009-08-06, 0.06)
(NYSE, CHW, 2009-07-08, 0.06)
(NYSE, CHW, 2009-06-08, 0.08)
(NYSE, CHW, 2009-05-07, 0.08)
(NYSE, CHW, 2009-04-08, 0.08)
(NYSE, CHW, 2009-03-09, 0.08)
(NYSE, CHW, 2009-02-06, 0.08)
(NYSE, CRP, 2009-12-08, 0.494)
(NYSE, CRP, 2009-09-08, 0.494)
(NYSE, CRP, 2009-06-08, 0.494)
(NYSE, CRP, 2009-03-09, 0.494)
(NYSE, CHL, 2009-09-10, 0.868)
(NYSE, CHL, 2009-05-11, 0.906)
(NYSE, CYN, 2009-11-02, 0.1)
(NYSE, CYN, 2009-08-03, 0.1)
(NYSE, CYN, 2009-05-04, 0.1)
(NYSE, CYN, 2009-02-02, 0.25)
(NYSE, COP, 2009-10-28, 0.5)
(NYSE, COP, 2009-07-29, 0.47)
(NYSE, COP, 2009-05-21, 0.47)
(NYSE, COP, 2009-02-19, 0.47)
(NYSE, CEO, 2009-09-04, 2.58)
(NYSE, CEO, 2009-05-14, 2.581)
(NYSE, CS, 2009-04-27, 0.096)
```

图 9.14 filter_not_matches.pig 脚本结果图

可以将多个断言表达式通过布尔操作符 and 和 or，以及用于取相反的布尔操作符 not 联合在一起使用。正如通常的标准规则，布尔操作符的执行优先级从高到低依次是：not、and、or。因此 a and b or not c 和 (a and b) or (not c) 是等价的。

Pig 会在一些情况下对缩短布尔操作符的判断路径。如果对于 and 操作符左边第一个表达式结果 false，and 后面的判断就会被忽略。因此对于 1==2 and udf(x) 这个表达式，该 UDF 函数永远也不会被调用。同样，如果一个 or 的左边表达式结果是 true，那么 or 的右边也不会被调用。因此对于 1==1 or udf(x) 这个表达式，该 udf 函数也同样永远不会被调用。

对于布尔操作符，null 值遵循 SQL 的三元判断逻辑。也就是说，x==null 的结果是一个 null 值，而不是 true（即使当 x 也是 null 时）或者 false。过滤器 filter 只会允许布尔值判断为 true 的值通过。因此如果一个字段包含了 2、null 和 4 共 3 个值,如果的过滤条件是 x==2,那么只有第一个值也就是 2 可以通过这个过滤器。然而,对于过滤条件 x!=2 将只会返回最后一条记录,也就是值 4。而查找 null 值的正确方式是表达式使用 is null 操作符,当值为 null 时会返回 true。如果想得到值不为 null 的记录,则应使用 is not null。

同样,对于任何正则表达式,null 既不会匹配上也不会失败。

正如求值函数中使用到的一些 UDF 一样,也有一些专门处理过滤数据的 UDF,我们称之为过滤函数。这些函数可以返回一个布尔值并且可以在 filter 语句中调用。注意过滤函数不可以在 foreach 语句中使用。

9.2.3　Group

group 语句可以把具有相同键值的数据聚合在一起,这是本书中遇到的第一个与 SQL 语法相同的操作符,但是很重要的一点是我们意识到 Pig Latin 中的 group 操作与 SQL 中的 group 操作有着本质的区别。在 SQL 中,group by 子句创建的组必须直接注入一个或多个聚合函数中。在 Pig Latin 中的 group 和聚合函数之间没有直接的关系。然而,group 关键字正如它字面所表达的:将包含了特定的键对应的值的所有记录封装到一个 bag 中。之后,可以将这个结果传递给一个聚合函数或者使用它做其他一些处理。

在 /root/work 目录下,编写 count.pig 脚本如下:

```
-- count.pig
daily = load 'NYSE_daily' as (exchange,stock);
grpd = group daily by stock;
cnt = foreach grpd generate group,COUNT(daily);
dump cnt;              -- 输出分组结果
```

编写完成后,直接进入 Pig,并将 count.pig 导入 HDFS 中,运行脚本 count.pig,结果如图 9.15 所示(截图只选取了显示的部分内容):

```
grunt > run count.pig         -- 运行
```

上面的例子展示了对以 stock 为键的值进行分组然后做计数统计。在进行 group 分组操作后直接存储到文件系统中用于后期的再处理,在 Pig 中,这也是合法的操作:

```
-- group.pig
daily = load 'NYSE_daily' as(exchange,stock);
grpd = group daily by stock;
store grpd into 'by_group';
```

进入 Pig,将 group.pig 导入 HDFS 中,运行脚本 group.pig,结果如图 9.16 所示(截图只选取了显示的部分内容):

```
grunt > run group.pig         -- 运行
grunt > cat by_group          -- 查看分组结果
```

```
(CVI, 252)
(CVO, 252)
(CVS, 252)
(CVX, 256)
(CWF, 252)
(CWT, 252)
(CWZ, 252)
(CXE, 252)
(CXG, 252)
(CXH, 252)
(CXO, 252)
(CXS, 74)
(CXW, 252)
(CYD, 252)
(CYE, 252)
(CYH, 252)
(CYN, 255)
(CYS, 139)
(CYT, 252)
(CZZ, 252)
(CACI, 252)
(CASC, 252)
(CATO, 252)
(CLNY, 69)
```

图 9.15 count.pig 脚本结果图

```
SE, CATO), (NYSE, CATO), (NYSE, CATO), (NYSE, CATO), (NYSE, CATO), (NYSE, CATO), (NYSE, CATO)
, (NYSE, CATO), (NYSE, CATO), (NYSE, CATO), (NYSE, CATO), (NYSE, CATO), (NYSE, CATO), (NYSE, C
ATO), (NYSE, CATO), (NYSE, CATO), (NYSE, CATO), (NYSE, CATO), (NYSE, CATO), (NYSE, CATO), (NY
SE, CATO), (NYSE, CATO), (NYSE, CATO), (NYSE, CATO), (NYSE, CATO), (NYSE, CATO), (NYSE, CATO)
, (NYSE, CATO), (NYSE, CATO), (NYSE, CATO), (NYSE, CATO), (NYSE, CATO), (NYSE, CATO), (NYSE, C
ATO), (NYSE, CATO), (NYSE, CATO), (NYSE, CATO), (NYSE, CATO), (NYSE, CATO), (NYSE, CATO), (NY
SE, CATO), (NYSE, CATO), (NYSE, CATO), (NYSE, CATO), (NYSE, CATO), (NYSE, CATO), (NYSE, CATO)
, (NYSE, CATO), (NYSE, CATO), (NYSE, CATO), (NYSE, CATO), (NYSE, CATO), (NYSE, CATO), (NYSE, C
ATO), (NYSE, CATO), (NYSE, CATO), (NYSE, CATO), (NYSE, CATO), (NYSE, CATO), (NYSE, CATO), (NY
SE, CATO), (NYSE, CATO), (NYSE, CATO), (NYSE, CATO), (NYSE, CATO), (NYSE, CATO), (NYSE, CATO)
, (NYSE, CATO), (NYSE, CATO), (NYSE, CATO), (NYSE, CATO), (NYSE, CATO), (NYSE, CATO), (NYSE, C
ATO), (NYSE, CATO), (NYSE, CATO), (NYSE, CATO), (NYSE, CATO), (NYSE, CATO), (NYSE, CATO), (NY
SE, CATO), (NYSE, CATO), (NYSE, CATO), (NYSE, CATO), (NYSE, CATO), (NYSE, CATO), (NYSE, CATO)
, (NYSE, CATO), (NYSE, CATO), (NYSE, CATO), (NYSE, CATO), (NYSE, CATO), (NYSE, CATO)}
CLNY    {(NYSE, CLNY), (NYSE, CLNY), (NYSE, CLNY), (NYSE, CLNY), (NYSE, CLNY), (NYSE, CLNY)
, (NYSE, CLNY), (NYSE, CLNY), (NYSE, CLNY), (NYSE, CLNY), (NYSE, CLNY), (NYSE, CLNY), (NYSE, C
LNY), (NYSE, CLNY), (NYSE, CLNY), (NYSE, CLNY), (NYSE, CLNY), (NYSE, CLNY), (NYSE, CLNY), (NY
SE, CLNY), (NYSE, CLNY), (NYSE, CLNY), (NYSE, CLNY), (NYSE, CLNY), (NYSE, CLNY), (NYSE, CLNY)
, (NYSE, CLNY), (NYSE, CLNY), (NYSE, CLNY), (NYSE, CLNY), (NYSE, CLNY), (NYSE, CLNY), (NYSE, C
LNY), (NYSE, CLNY), (NYSE, CLNY), (NYSE, CLNY), (NYSE, CLNY), (NYSE, CLNY), (NYSE, CLNY), (NY
SE, CLNY), (NYSE, CLNY), (NYSE, CLNY), (NYSE, CLNY), (NYSE, CLNY), (NYSE, CLNY), (NYSE, CLNY)
, (NYSE, CLNY), (NYSE, CLNY), (NYSE, CLNY), (NYSE, CLNY), (NYSE, CLNY), (NYSE, CLNY), (NYSE, C
LNY), (NYSE, CLNY), (NYSE, CLNY), (NYSE, CLNY), (NYSE, CLNY), (NYSE, CLNY), (NYSE, CLNY), (NY
SE, CLNY), (NYSE, CLNY), (NYSE, CLNY), (NYSE, CLNY), (NYSE, CLNY), (NYSE, CLNY), (NYSE, CLNY)
, (NYSE, CLNY), (NYSE, CLNY), (NYSE, CLNY)}
```

图 9.16 group.pig 脚本结果图

group by 语句的输出结果包含两个字段：一个是键，另一个是包含了聚集的记录的 bag。存放键的字段别名为 group。而 bag 的别名和被分组的那条语句的别名相同，因此在前面的例子中别名应该为 daily，同时和关系 daily 具有相同的模式。如果关系 daily 没有模式，那么数据包 daily 也将是没有模式的。对于分组中的每一条记录，整个记录（包括那个键）都在这个数据包 bag 中。将上述脚本最后一行的 store grpd…改为“describe grpd；”，将会可以查看这个关系 grpd 的模式如图 9.17 所示。

```
grunt> describe grpd
grpd: {group: bytearray, daily: {(exchange: bytearray, stock: bytearray)}}
```

图 9.17 关系 grpd 的模式图

也可以对多个键进行分组,但是这些键必须包含在一组圆括号内。结果记录同样包含两个字段。在这种情况下,group 字段将是包含每个键值的 tuple。

在/root/work 目录下,编写 twokey.pig 脚本如下:

```
-- twokey.pig
daily = load 'NYSE_daily' as (exchange,stock,date,dividends);
grpd = group daily by (exchange,stock);
avg = foreach grpd generate group,AVG(daily.dividends);
-- dump avg;
describe grpd;
```

编写完成后,直接进入 Pig,将 twokey.pig 导入 HDFS 中,运行脚本 twokey.pig,运行结果如图 9.18 所示,其中显示出关系 grpd 的模式。

```
grunt> describe grpd;
grpd: {group: (exchange: bytearray,stock: bytearray),daily: {(exchange: bytearray,stock: bytearray,date: bytearray,dividends: bytearray)}}
```

图 9.18 twokey.pig 脚本结果图

也可以使用关键字 all 对数据流中的所有字段进行分组。

在/root/work 目录下,编写 countall.pig 脚本如下:

```
-- countall.pig
daily = load'NYSE_daily'as (exchange,stock);
grpd = group daily all;
cnt = foreach grpd generate COUNT(daily);
dump cnt;
```

编写完成后,直接进入 Pig,将 countall.pig 导入 HDFS 中,运行脚本 countall.pig,运行结果如图 9.19 所示。

```
grunt> run countall.pig
```

```
(57391)
```

图 9.19 countall.pig 脚本结果图

group all 的输出记录是以字符串文字 all 为键的值。一般情况下,这没有什么问题。因为通常会把数据包 bag 直接传送给像 COUNT 这样的聚集函数。但是如果计划把记录存储起来或者用于其他用途,那么可能需要事先知道全部的字段。

group 是我们到目前为止遇到的第一个通常会触发一个 reduce 过程的操作符。分组就意味着收集所有键中包含相同的值的记录。如果数据流处于一个 map 阶段,那么就会迫使它先进行 shuffle,然后再进行 reduce。如果数据流处于一个 reduce 阶段,那么就会迫使它先进行 map，然后再进行 shuffle，最后进入 reduce 阶段。

因为分组收集了所有键中包含相同的值的记录，所以经常会获得数据倾斜的结果。也就是说，若是已经制定了的任务具有 100 个 reducer，那么也就没有理由期望每个键对应的值会按照个数均匀地分发出去。它们可能是一个高斯分布或幂次法则分布。例如，假设有一些网页索引数据，同时按照根 URL 进行分组。一些特定的值，例如 Yahoo.com 会比大多数的其他根 URL 具有更多的记录，这也就意味着一些 reducer 会比其他的 reducer 获得更多的数据。因为 MapReduce 任务是直到所有的 reducer 都完成时才会结束，那么这个数据倾斜会在相当大的程度上拖慢处理过程。在一些情况下，这也并非是一个 reducer，可以处理的。

Pig 具有一系列的处理方法来控制数据倾斜，以达到 reducer 间的负荷均衡。Hadoop 的组合器是适用于分组的。这不会移除所有的数据倾斜，但是会在上面加上限制。同时因为对于大多数的任务，mapper 的个数都是数以万计的，即使 reducer 获得了一些倾斜的数据，每个 reducer 所处理的实际记录也会非常小，足以使这些 reducer 能够快速处理好。

不幸的是，并非所有的计算都是可以通过这个组合完成的。对于可以分解成任意数量的处理步骤的计算，也就是所谓的可分布处理的计算可以很好地使用组合器运算。而那些可以分解为一个初始化过程，任意多个中间处理过程，最后会有一个结束处理过程的计算，我们称之为代数运算。

可分布处理的计算是代数计算的一种特殊情况：初始过程、中间处理过程和结束处理过程都是相同的操作。追踪在网站上的行为的会话分析是一个非代数运算的例子。在开始分析和网站的交互行为之前必须对所有的记录按照时间戳进行排序处理。

Pig 的操作符和内置的 UDF 在需要的时候都会使用这个组合器，因为它可以降低数据倾斜的特性，并且在处理早期进行聚合处理的操作大大降低了网络间的数据传输和写磁盘的数据量，因此可以显著提高处理效率。用户自定义函数 UDF 可以通过实现代数计算接口来决定在什么时候使用这个组合器。

需要记住的是，当使用 group all 的时候，有必要序列化数据流。也就是说，这个步骤和其后的其他所有步骤，在没有将当前包含了所有记录的数据包分离出来之前都是无法以并行的方式运行的。

最后需要说明的是，group 处理 null 值的方式与 SQL 处理 null 值的方式是一样的：将以 null 作为键的所有记录汇聚到相同的组中。需要注意的是，这与表达式处理 null 值的方式以及 join 处理 null 值的方式是截然不同的。

9.2.4 Order by

order 语句是对数据进行排序，产生一个全排序的输出结果。全排序意味着不仅是将每个部分的数据进行排序，同时也会保证对于 n 个部分文件进行排序，那么第 n 个文件中的记录序号要比第 $n-1$ 个部分文件中的记录序号少。当数据存放在 HDFS 中的时候，每个部分都是一个部分文件，这意味着使用 cat 命令将会使数据全部以有序的方式输出。

order 语句的语法与 group 语句的语法相似。需要指定想按照哪一个或者多个键对数据进行排序。一个比较明显的区别是在 order 语句中指定键不需要使用圆括号。

在/root/work 目录下，编写 order.pig 脚本如下：

```
-- order.pig
daily = load 'NYSE_daily' as (exchange:chararray,symbol:chararray,date:chararray,open:float,
high:float,low:float,close:float,volume:int,adj_close:float);
bydate = order daily by date;
dump bydate;
```

编写完成后,直接进入 Pig,将 order.pig 导入 HDFS 中,运行脚本 order.pig,结果如图 9.20 所示(截图只选取了显示的部分内容)。

```
grunt > run order.pig                 -- 运行
```

```
(NYSE, CDE, 2009-12-31, 18.33, 18.48, 18.05, 18.06, 1055300, 18.06)
(NYSE, CSR, 2009-12-31, 7.64, 7.75, 7.61, 7.64, 650600, 7.64)
(NYSE, CIX, 2009-12-31, 7.56, 7.57, 7.56, 7.57, 200, 7.57)
(NYSE, CBC, 2009-12-31, 1.94, 1.97, 1.91, 1.96, 146600, 1.96)
(NYSE, CYH, 2009-12-31, 36.39, 36.45, 35.57, 35.6, 551500, 35.6)
(NYSE, CEE, 2009-12-31, 32.57, 33.25, 32.57, 32.99, 46800, 32.99)
(NYSE, CSQ, 2009-12-31, 8.69, 8.76, 8.67, 8.76, 464000, 8.71)
(NYSE, CVA, 2009-12-31, 18.18, 18.25, 18.05, 18.09, 475600, 18.09)
(NYSE, CWZ, 2009-12-31, 23.4, 23.4, 23.18, 23.4, 6300, 23.4)
(NYSE, CVH, 2009-12-31, 24.76, 24.88, 24.23, 24.29, 627600, 24.29)
(NYSE, CF, 2009-12-31, 92.63, 92.63, 90.76, 90.78, 1105600, 90.78)
(NYSE, CIB, 2009-12-31, 45.18, 45.67, 44.91, 45.51, 56200, 45.51)
(NYSE, CAH, 2009-12-31, 32.63, 32.8, 32.22, 32.24, 1624300, 32.24)
(NYSE, CPT, 2009-12-31, 43.39, 43.84, 42.02, 42.37, 869200, 42.37)
(NYSE, CSV, 2009-12-31, 3.94, 3.94, 3.9, 3.93, 6700, 3.93)
(NYSE, CRY, 2009-12-31, 6.41, 6.58, 6.36, 6.42, 186900, 6.42)
(NYSE, CVI, 2009-12-31, 7.1, 7.13, 6.86, 6.86, 417200, 6.86)
(NYSE, CPO, 2009-12-31, 29.9, 30.0, 29.21, 29.23, 225300, 29.23)
(NYSE, CAG, 2009-12-31, 23.39, 23.45, 23.04, 23.05, 1960900, 22.85)
(NYSE, CBU, 2009-12-31, 19.71, 19.82, 19.31, 19.31, 79600, 19.31)
(NYSE, CMK, 2009-12-31, 8.02, 8.02, 7.96, 8.0, 2100, 7.92)
(NYSE, CCK, 2009-12-31, 25.64, 25.72, 25.5, 25.58, 793600, 25.58)
(NYSE, CVC, 2010-02-01, 25.87, 26.48, 25.81, 26.38, 2009000, 26.38)
```

图 9.20 order.pig 脚本结果图

在/root/work 目录下,编写 order2key.pig 脚本如下:

```
-- order2key.pig
daily = load 'NYSE_daily' as (exchange:chararray,symbol:chararray,date:chararray,open:float,
high:float,low:float,close:float,volume:int,adj_close:float);
bydatensymbol = order daily by date,symbol;         -- 先按 date 排序,再按 symbol
dump bydatensymbol;
```

编写完成后,直接进入 Pig,将 order2key.pig 导入 HDFS 中,运行脚本 order2key.pig,结果如图 9.21 所示(截图只选取了显示的部分内容):

```
grunt > run order2key.pig              -- 运行
```

在指定的键后面加上 desc 关键字可以使结果按照降序排列。如果 order 语句中指定了多个键,那么 desc 关键字之后对紧靠着它的那个键起作用。其他的键还会按照升序排列。

```
(NYSE, CVO, 2009-12-31, 8.94, 9.1, 8.71, 8.75, 224500, 8.75)
(NYSE, CVS, 2009-12-31, 32.61, 32.76, 32.19, 32.21, 4962600, 32.13)
(NYSE, CVX, 2009-12-31, 77.72, 77.78, 76.93, 76.99, 4246600, 76.26)
(NYSE, CW, 2009-12-31, 31.74, 31.95, 31.32, 31.32, 163000, 31.32)
(NYSE, CWF, 2009-12-31, 3.8, 3.8, 3.76, 3.77, 22500, 3.74)
(NYSE, CWT, 2009-12-31, 37.06, 37.39, 36.75, 36.82, 61100, 36.52)
(NYSE, CWZ, 2009-12-31, 23.4, 23.4, 23.18, 23.4, 6300, 23.4)
(NYSE, CX, 2009-12-31, 11.91, 12.05, 11.82, 11.82, 2583900, 11.82)
(NYSE, CXE, 2009-12-31, 4.83, 4.83, 4.76, 4.83, 92500, 4.77)
(NYSE, CXG, 2009-12-31, 29.93, 30.0, 29.38, 29.52, 87000, 29.52)
(NYSE, CXH, 2009-12-31, 9.35, 9.35, 9.26, 9.26, 8100, 9.15)
(NYSE, CXO, 2009-12-31, 45.25, 45.59, 44.84, 44.9, 191400, 44.9)
(NYSE, CXS, 2009-12-31, 13.97, 14.1, 13.87, 13.96, 37300, 13.96)
(NYSE, CXW, 2009-12-31, 24.96, 25.01, 24.55, 24.55, 462600, 24.55)
(NYSE, CY, 2009-12-31, 10.74, 10.81, 10.54, 10.56, 2349100, 10.56)
(NYSE, CYD, 2009-12-31, 14.87, 15.0, 14.7, 14.74, 294700, 14.74)
(NYSE, CYE, 2009-12-31, 6.36, 6.37, 6.28, 6.35, 86600, 6.3)
(NYSE, CYH, 2009-12-31, 36.39, 36.45, 35.57, 35.6, 551500, 35.6)
(NYSE, CYN, 2009-12-31, 46.22, 46.22, 45.6, 45.6, 473900, 45.5)
(NYSE, CYS, 2009-12-31, 13.51, 13.55, 13.49, 13.51, 25400, 13.51)
(NYSE, CYT, 2009-12-31, 36.9, 37.06, 36.33, 36.42, 220300, 36.41)
(NYSE, CZZ, 2009-12-31, 8.77, 8.77, 8.67, 8.7, 694200, 8.7)
(NYSE, CVC, 2010-02-01, 25.87, 26.48, 25.81, 26.38, 2009000, 26.38)
```

图 9.21 order2key.pig 脚本结果图

在/root/work 目录下,编写 orderdesc.pig 脚本如下:

```
-- orderdesc.pig
daily = load 'NYSE_daily' as (exchange:chararray, symbol:chararray, date:chararray, open:
float, high:float, low:float, close:float, volume:int, adj_close:float);
byclose = order daily by close desc, open;        -- 按照 close 降序排序
dump byclose;                                      -- 字段 open 还是按照升序排序
```

编写完成后,直接进入 Pig,将 orderdesc.pig 导入 HDFS 中,运行脚本 orderdesc.pig,结果如图 9.22 所示(截图只选取了显示的部分内容):

```
grunt> run orderdesc.pig                          -- 运行
```

```
(NYSE, CHB, 2009-12-22, 0.2, 0.2, 0.2, 0.2, 0, 0.2)
(NYSE, CHB, 2009-12-23, 0.2, 0.2, 0.2, 0.2, 0, 0.2)
(NYSE, CHB, 2009-12-24, 0.2, 0.2, 0.2, 0.2, 0, 0.2)
(NYSE, CHB, 2009-12-28, 0.2, 0.2, 0.2, 0.2, 0, 0.2)
(NYSE, CHB, 2009-12-29, 0.2, 0.2, 0.2, 0.2, 0, 0.2)
(NYSE, CHB, 2009-12-30, 0.2, 0.2, 0.2, 0.2, 0, 0.2)
(NYSE, CHB, 2009-12-31, 0.2, 0.2, 0.2, 0.2, 0, 0.2)
(NYSE, CHB, 2009-11-16, 0.2, 0.2, 0.2, 0.2, 0, 0.2)
(NYSE, CHB, 2009-11-17, 0.2, 0.2, 0.2, 0.2, 0, 0.2)
(NYSE, CHB, 2009-11-18, 0.2, 0.2, 0.2, 0.2, 0, 0.2)
(NYSE, CHB, 2009-11-30, 0.2, 0.2, 0.2, 0.2, 0, 0.2)
(NYSE, CHB, 2009-03-06, 0.21, 0.23, 0.15, 0.2, 671800, 0.2)
(NYSE, CHB, 2009-11-09, 0.22, 0.23, 0.19, 0.2, 2790100, 0.2)
(NYSE, CHB, 2009-11-13, 0.22, 0.22, 0.2, 0.2, 1253800, 0.2)
(NYSE, CHB, 2009-03-05, 0.28, 0.3, 0.2, 0.2, 1001800, 0.2)
(NYSE, CHB, 2009-03-09, 0.2, 0.22, 0.19, 0.19, 463300, 0.19)
(NYSE, CHB, 2009-10-28, 0.28, 0.28, 0.18, 0.19, 6253100, 0.19)
(NYSE, CHB, 2009-03-18, 0.17, 0.2, 0.15, 0.17, 2281900, 0.17)
(NYSE, CHB, 2009-03-17, 0.17, 0.17, 0.15, 0.17, 500100, 0.17)
(NYSE, CHB, 2009-03-16, 0.19, 0.19, 0.15, 0.17, 1660800, 0.17)
(NYSE, CHB, 2009-03-19, 0.2, 0.2, 0.16, 0.17, 845700, 0.17)
(NYSE, CHB, 2009-03-13, 0.15, 0.18, 0.13, 0.16, 2309600, 0.16)
(NYSE, CHB, 2009-03-12, 0.22, 0.22, 0.12, 0.12, 6520500, 0.12)
```

图 9.22 orderdesc.pig 脚本结果图

数据的排序方式取决于所指定的字段的类型：数值会按照数字顺序排序，chararray 类型的字符串字段会按照字典顺序排序，bytearray 类型的字段也按照字典顺序排序，但是按照字节值而不是字符值排序的。对 map、tuple 和 bag 类型的字段进行排序会报错。对于所有的数据类型，null 都是最小的，因此常常是最先显示的或者在使用 desc 排序时是最后显示的。

数据倾斜在数据中很常见。只有在使用了 group 的时候才会影响到 order，这会导致一些 reducer 比其他的 reducer 花费明显长的时间。为了解决这个问题，Pig 会在 reducer 间均衡输出。它的处理方法是对于 order 语句的输入先进行取样以获得键值分布情况的预算。基于这个样本，Pig 会创建一个分割器用于产生一个均衡的全排序。

Pig 为减少数据倾斜而采用的这种数据分布方式所产生的弊端之一，是它打破了同一个指定键所对应的所有实例都传送到同一个部分文件中这个 MapReduce 规则。如果需要其他基于这一规则的处理过程，那么就不要使用 Pig 的 order 语句去对数据进行排序。

order 总是会引起数据流通过一个 reduce 阶段，因此将所有相等的记录收集到一起是有必要的。同时，为了做抽样，Pig 也会在数据轮流中增加一个额外的 MapReduce 任务。因为这个抽样是非常轻量的，通常只会占用少于整个任务的 5% 的时间。

9.2.5 distinct

distinct 语句非常简单，它会将重复值去掉，只会对整个记录进行处理，而不是对字段级别进行计算。

在 /root/work 目录下，编写 distinct.pig 脚本如下：

```
-- distinct.pig
-- 为每笔交易生成一个去掉重复后的股票代码列表
-- 这个加载过程截断记录，只会选择前面两个字段
daily = load 'NYSE_daily' as (exchange:chararray,symbol:chararray);
uniq = distinct daily;
dump uniq;
```

编写完成后，直接进入 Pig，将 distinct.pig 导入 HDFS 中，运行脚本 distinct.pig，结果如图 9.23 所示（截图只选取了显示的部分内容）：

```
grunt > run distinct.pig
```

因此它需要将相似的记录收集到一起来判断它们是否有重复，所有 distinct 都会触发一个 reduce 处理过程。在 map 阶段也确实可以通过那个组合器将一些重复的数值去除掉。这里所展示的 distinct 用法与 SQL 中的 select distinct x 用法是相同的。

9.2.6 Join

join 是数据处理中非常重要的操作之一，同时是最有可能在很多的 Pig Latin 脚本中使用到的操作。join 可以将一个输入中的记录和另一个输入中的数据放在一起。通过指定每个输入的键可以达到中介目的。当这些键值相等的时候，数据就会被连接在一起，没有被匹配到的数据会被去掉。

```
(NYSE, CVO)
(NYSE, CVS)
(NYSE, CVX)
(NYSE, CWF)
(NYSE, CWT)
(NYSE, CWZ)
(NYSE, CXE)
(NYSE, CXG)
(NYSE, CXH)
(NYSE, CXO)
(NYSE, CXS)
(NYSE, CXW)
(NYSE, CYD)
(NYSE, CYE)
(NYSE, CYH)
(NYSE, CYN)
(NYSE, CYS)
(NYSE, CYT)
(NYSE, CZZ)
(NYSE, CACI)
(NYSE, CASC)
(NYSE, CATO)
(NYSE, CLNY)
```

图 9.23 distinct.pig 脚本结果图

在/root/work 目录下，编写 join.pig 脚本如下：

```
-- join.pig
daily = load 'NYSE_daily' as (exchange,symbol,date,open,high,low,close,volume,adj_close);
divs = load 'NYSE_dividends' as (exchange,symbol,date,dividends);
jnd = join daily by symbol, divs by symbol;          -- 以 symbol 为键值
dump jnd;
```

编写完成后，直接进入 Pig，将 join.pig 导入 HDFS 中，运行脚本 join.pig，结果如图 9.24 所示（截图只选取了显示的部分内容）：

```
grunt > run join.pig;                                -- 运行
```

```
(NYSE, CLNY, 2009-11-13, 19.42, 19.42, 19.35, 19.36, 38400, 19.29, NYSE, CLNY, 2009-12-29, 0.07)
(NYSE, CLNY, 2009-11-12, 19.39, 19.45, 19.35, 19.41, 18500, 19.34, NYSE, CLNY, 2009-12-29, 0.07)
(NYSE, CLNY, 2009-11-11, 19.50, 19.50, 19.31, 19.43, 40500, 19.36, NYSE, CLNY, 2009-12-29, 0.07)
(NYSE, CLNY, 2009-11-10, 19.35, 19.45, 19.35, 19.40, 19700, 19.33, NYSE, CLNY, 2009-12-29, 0.07)
(NYSE, CLNY, 2009-11-09, 19.42, 19.45, 19.33, 19.37, 9100, 19.30, NYSE, CLNY, 2009-12-29, 0.07)
(NYSE, CLNY, 2009-11-06, 19.40, 19.47, 19.31, 19.37, 42000, 19.30, NYSE, CLNY, 2009-12-29, 0.07)
(NYSE, CLNY, 2009-11-05, 19.39, 19.58, 19.39, 19.46, 44400, 19.39, NYSE, CLNY, 2009-12-29, 0.07)
(NYSE, CLNY, 2009-11-04, 19.49, 19.51, 19.16, 19.50, 50900, 19.43, NYSE, CLNY, 2009-12-29, 0.07)
(NYSE, CLNY, 2009-11-03, 19.42, 19.54, 19.42, 19.49, 52400, 19.42, NYSE, CLNY, 2009-12-29, 0.07)
(NYSE, CLNY, 2009-11-02, 19.50, 19.54, 19.41, 19.50, 80400, 19.43, NYSE, CLNY, 2009-12-29, 0.07)
(NYSE, CLNY, 2009-10-30, 19.33, 19.49, 19.33, 19.45, 34200, 19.38, NYSE, CLNY, 2009-12-29, 0.07)
```

图 9.24 join.pig 脚本结果图

也可以指定多个键来使用 join。每个语句中都有相同个数的键,同时它们必须是相同的或者是同一类的数据类型(同一类的数据类型是指可以通过隐式的类型转换得到的)。

在/root/work 目录下,编写 join2key.pig 脚本如下:

```
-- join2key.pig
daily = load 'NYSE_daily' as (exchange,symbol,date,open,high,low,close,volume,adj_close);
divs = load 'NYSE_dividends' as (exchange,symbol,date,dividends);
jnd = join daily by (symbol,date),divs by (symbol,date);
dump jnd;
```

编写完成后,直接进入 Pig,将 join2key.pig 导入 HDFS 中,运行文件 join2key.pig,结果如图 9.25 所示(截图只选取了显示的部分内容):

```
grunt> run join2key.pig                                          --运行
```

```
(NYSE, CYT, 2009-08-06, 26.06, 26.19, 25.37, 25.77, 363100, 25.75, NYSE, CYT, 2009-08-06, 0.013)
(NYSE, CYT, 2009-11-06, 34.45, 35.86, 34.33, 35.45, 478000, 35.44, NYSE, CYT, 2009-11-06, 0.013)
(NYSE, CASC, 2009-04-27, 22.60, 23.00, 21.51, 21.84, 48100, 21.76, NYSE, CASC, 2009-04-27, 0.05)
(NYSE, CASC, 2009-06-29, 17.44, 17.70, 16.61, 16.75, 97800, 16.74, NYSE, CASC, 2009-06-29, 0.05)
(NYSE, CASC, 2009-09-30, 27.22, 27.32, 25.81, 26.74, 46100, 26.73, NYSE, CASC, 2009-09-30, 0.01)
(NYSE, CASC, 2009-12-31, 27.47, 27.89, 27.31, 27.49, 8700, 27.49, NYSE, CASC, 2009-12-31, 0.01)
(NYSE, CATO, 2009-03-05, 13.20, 13.59, 12.85, 13.21, 212500, 12.87, NYSE, CATO, 2009-03-05, 0.165)
(NYSE, CATO, 2009-06-04, 19.96, 20.51, 19.45, 20.33, 119700, 19.98, NYSE, CATO, 2009-06-04, 0.165)
(NYSE, CATO, 2009-09-10, 18.20, 19.28, 18.18, 18.76, 475000, 18.60, NYSE, CATO, 2009-09-10, 0.165)
(NYSE, CATO, 2009-12-17, 19.41, 19.49, 18.81, 19.29, 78600, 19.29, NYSE, CATO, 2009-12-17, 0.165)
(NYSE, CLNY, 2009-12-29, 20.59, 20.75, 20.30, 20.44, 117500, 20.44, NYSE, CLNY, 2009-12-29, 0.07)
```

图 9.25 join2key.pig 脚本结果图

与 foreach 一样,join 也会保留传递给它的输入数据的字段的别名,同时也会保留这个字段所来自的关系的别名,可以通过::符号指定。将上面那个例子的最后一行"dump jnd;"替换成"describe jnd;"语句,会产生如图 9.26 所示的结果。

```
grunt> describe jnd;
jnd: {daily::exchange: bytearray, daily::symbol: bytearray, daily::date: bytearray
, daily::open: bytearray, daily::high: bytearray, daily::low: bytearray, daily::clos
e: bytearray, daily::volume: bytearray, daily::adj_close: bytearray, divs::exchange
: bytearray, divs::symbol: bytearray, divs::date: bytearray, divs::dividends: bytea
rray}
```

图 9.26 jnd 各个字段数据类型图

只有当字段名和记录中的字段不再相同时才需要使用 daily:: 这个前缀。在这个例子中,如果想在 join 操作之后引用这两个 date 字段中的任一个,需要使用 daily::date 和 divs::date 方式指定。但是像 open 和 divs 这样的字段就不需要使用那个前缀了,因为它们不会引起混淆。

Pig 同时也支持 outer join。在 outer join 中,两个表中不匹配的字段值也会被保留下来,同时使用 null 值填补缺失的字段。outer join 分为 left、right 和 full 3 种形式。left outer join 意味着左边的数据会被全部保留下来,即使在右边没有匹配的值。同样,right outer join 意味着右边的数据被全部保留下来,即使在左边没有匹配的值。full out join 意味着即使两边都没有匹配的值两边的值也全部保留下来。

在/root/work 目录下,编写 leftjoin.pig 脚本如下:

```
-- leftjoin.pig
daily = load 'NYSE_daily' as (exchange,symbol,date,open,high,low,close,volume,adj_close);
divs = load 'NYSE_dividends' as (exchange,symbol,date,dividends);
jnd = join daily by (symbol,date) left outer, divs by (symbol,date);
dump jnd;
```

编写完成后,直接进入 Pig,将 leftjoin.pig 导入 HDFS 中,运行脚本 leftjoin.pig,结果如图 9.27 所示(截图只选取了显示的部分内容):

```
grunt> run leftjoin.pig
```

```
(NYSE, CLNY, 2009-12-01, 18.85, 19.09, 18.60, 19.00, 104100, 18.94, , , )
(NYSE, CLNY, 2009-12-02, 18.91, 19.21, 18.86, 19.05, 196900, 18.98, , , )
(NYSE, CLNY, 2009-12-03, 19.00, 19.47, 19.00, 19.33, 72100, 19.26, , , )
(NYSE, CLNY, 2009-12-04, 19.32, 19.45, 19.16, 19.36, 81900, 19.29, , , )
(NYSE, CLNY, 2009-12-07, 19.28, 19.47, 18.99, 19.31, 28400, 19.24, , , )
(NYSE, CLNY, 2009-12-08, 19.17, 19.50, 19.17, 19.49, 108800, 19.42, , , )
(NYSE, CLNY, 2009-12-09, 19.48, 19.71, 19.40, 19.53, 121900, 19.46, , , )
(NYSE, CLNY, 2009-12-10, 19.47, 19.75, 19.36, 19.72, 47600, 19.65, , , )
(NYSE, CLNY, 2009-12-11, 19.71, 19.95, 19.61, 19.87, 105900, 19.80, , , )
(NYSE, CLNY, 2009-12-14, 19.80, 20.35, 19.75, 20.28, 220700, 20.21, , , )
(NYSE, CLNY, 2009-12-15, 20.22, 20.99, 20.00, 20.75, 378400, 20.68, , , )
(NYSE, CLNY, 2009-12-16, 20.47, 20.74, 20.13, 20.31, 158000, 20.24, , , )
(NYSE, CLNY, 2009-12-17, 20.13, 20.90, 19.87, 20.15, 374200, 20.08, , , )
(NYSE, CLNY, 2009-12-18, 20.05, 20.75, 20.02, 20.51, 1493800, 20.44, , , )
(NYSE, CLNY, 2009-12-21, 20.51, 20.79, 20.17, 20.53, 180700, 20.46, , , )
(NYSE, CLNY, 2009-12-22, 20.62, 20.63, 20.20, 20.48, 213400, 20.41, , , )
(NYSE, CLNY, 2009-12-23, 20.50, 20.56, 20.37, 20.41, 175200, 20.34, , , )
(NYSE, CLNY, 2009-12-24, 20.52, 20.58, 20.43, 20.51, 83100, 20.44, , , )
(NYSE, CLNY, 2009-12-28, 20.50, 20.50, 20.20, 20.50, 123400, 20.43, , , )
(NYSE, CLNY, 2009-12-29, 20.59, 20.75, 20.30, 20.44, 117500, 20.44, NYSE, CLNY, 2009-12-29, 0.07)
(NYSE, CLNY, 2009-12-30, 20.37, 20.58, 20.03, 20.52, 80800, 20.52, , , )
(NYSE, CLNY, 2009-12-31, 20.51, 20.53, 20.34, 20.37, 51900, 20.37, , , )
```

图 9.27 leftjoin.pig 脚本结果图

outer 是一个多余的关键字,是可以省略的。与一些 SQL 实现不同的是,full 关键字是不可以省略的。"C=join A by x outer, B by u;"将会产生一个语法错误,而不是一个 full outer join。

只有当 Pig 知道数据需要填补 null 值的哪边或哪两边的模式的时候才支持使用 outer join。因此对于 left outer join,Pig 必须知道右边的模式;对于 right outer join,Pig 必须要知道两边的模式。这是因为如果没有模式,Pig 就不会知道需要填补多少个 null 值。

与 SQL 一样,键所对应的 null 值和什么都不匹配,即使和另一边输入中的 null 值也无法匹配。因此,对于内部连接 inner join,所有包含 null 键值的记录都会被去除。对于外部连接 outer join,它们会被保留下来,但是不会和另一边输入的任何记录相匹配。

Pig 也支持在一个操作下做多个 join 操作,只要它们全部使用相同的键进行连接即可。以下只有做内部连接 inner join 时才能使用:

```
A = load 'input1' as (x,y);
B = load 'input2' as (u,v);
C = load 'input3' as (e,f);
alpha = join A by x, B by u, C by e;
```

自连接也是支持的,尽管这时数据需要被加载两次。

在 /root/work 目录下,编写 selfjoin.pig 脚本如下:

```
-- selfjoin.pig
-- 对于每支股票,找出在两个日期范围内股息都是增加的记录
divs1 = load 'NYSE_dividends' as (exchange:chararray, symbol:chararray, date:chararray, dividends);
divs2 = load 'NYSE_dividends' as (exchange:chararray, symbol:chararray, date:chararray, dividends);
jnd = join divs1 by symbol, divs2 by symbol;
increased = filter jnd by divs1::date < divs2::date and divs1::dividends < divs2::dividends;
dump increased;
```

编写完成后,直接进入 Pig,将 selfjoin.pig 导入 HDFS 中,运行文件 selfjoin.pig,结果如图 9.28 所示(截图只选取了显示的部分内容):

```
grunt > run selfjoin.pig            -- 运行
```

```
(NYSE, CXH, 2009-04-13, 0.051, NYSE, CXH, 2009-12-09, 0.054)
(NYSE, CXH, 2009-04-13, 0.051, NYSE, CXH, 2009-11-10, 0.054)
(NYSE, CXH, 2009-04-13, 0.051, NYSE, CXH, 2009-10-13, 0.054)
(NYSE, CXH, 2009-04-13, 0.051, NYSE, CXH, 2009-09-10, 0.053)
(NYSE, CXH, 2009-04-13, 0.051, NYSE, CXH, 2009-05-11, 0.052)
(NYSE, CXH, 2009-04-13, 0.051, NYSE, CXH, 2009-08-10, 0.053)
(NYSE, CXH, 2009-04-13, 0.051, NYSE, CXH, 2009-06-10, 0.052)
(NYSE, CXH, 2009-05-11, 0.052, NYSE, CXH, 2009-07-13, 0.053)
(NYSE, CXH, 2009-05-11, 0.052, NYSE, CXH, 2009-12-09, 0.054)
(NYSE, CXH, 2009-05-11, 0.052, NYSE, CXH, 2009-11-10, 0.054)
(NYSE, CXH, 2009-05-11, 0.052, NYSE, CXH, 2009-10-13, 0.054)
(NYSE, CXH, 2009-05-11, 0.052, NYSE, CXH, 2009-09-10, 0.053)
(NYSE, CXH, 2009-05-11, 0.052, NYSE, CXH, 2009-08-10, 0.053)
(NYSE, CXH, 2009-08-10, 0.053, NYSE, CXH, 2009-12-09, 0.054)
(NYSE, CXH, 2009-08-10, 0.053, NYSE, CXH, 2009-11-10, 0.054)
(NYSE, CXH, 2009-08-10, 0.053, NYSE, CXH, 2009-10-13, 0.054)
(NYSE, CXH, 2009-06-10, 0.052, NYSE, CXH, 2009-07-13, 0.053)
(NYSE, CXH, 2009-06-10, 0.052, NYSE, CXH, 2009-12-09, 0.054)
(NYSE, CXH, 2009-06-10, 0.052, NYSE, CXH, 2009-11-10, 0.054)
(NYSE, CXH, 2009-06-10, 0.052, NYSE, CXH, 2009-10-13, 0.054)
(NYSE, CXH, 2009-06-10, 0.052, NYSE, CXH, 2009-09-10, 0.053)
(NYSE, CXH, 2009-06-10, 0.052, NYSE, CXH, 2009-08-10, 0.053)
(NYSE, CYS, 2009-10-01, 0.35, NYSE, CYS, 2009-12-29, 0.55)
```

图 9.28 selfjoin.pig 脚本结果图

如果将上面的代码按照如下方式进行更改,则会执行失败:

```
-- selfjoin.pig
-- 对于每支股票,找出在两个日期范围内股息都是增加的记录
divs1 = load 'NYSE_dividends' as (exchange:chararray, symbol:chararray, date:chararray, dividends);
jnd = join divs1 by symbol,divs1 by symbol;
increased = filter jnd by divs1::date < divs2::date and divs1::dividends < divs2::dividends;
dump increased;
```

看上去这段代码应该是可以执行的,虽然 Pig 可以将 divs1 数据集进行分割后传送给 join 两次,但是问题在于连接操作后字段名会变得模糊不清,因此 load 语句必须要执行两次。下一个要做的事情是执行这两个 load 语句,因为输入数据是相同的,所以只需要执行一个 load 语句,但是目前来说还没有做这个优化。

Pig 通过在 map 处理阶段标注出每条记录是来自哪一个输入的 MapReduce 方式进行这些 join 操作,之后它再使用 join 的键作为 shuffle 阶段的键。因此 join 会触发一个新的 reduce 处理过程。当指定键具有相同值的所有记录被收集到一起的时候,Pig 会对两条输入的记录进行一次交互乘积运算。为了减少内存使用,它使用 map 处理阶段增加的输入标注对进入 reducer 的记录进行 MapReduce 排序。因此,所有记录中来自左边输入的记录会先到达,Pig 会将这些记录缓存在内存中,而右边的输入数据会在之后紧跟而来。这些数据每到来一条就会和左边的数据进行一次交叉运算,然后产生一条输出记录。在一个多路连接操作中,左边 $n-1$ 个输入都会被放入内存中,同时第 n 条记录直接通过。如果知道输入记录中对于选定的键每个值对应着多条记录,那么在 Pig 查询中写 join 语句时牢记这一点将会非常重要。将这样的输入记录放到的 join 语句的右边将会降低内存的使用率,也许能够提高脚本的执行效率。

9.2.7 Limit

有时只是想从结果中拿出几条数据看看,limit 语句就可以满足此需求。
在/root/work 目录下,编写 limit.pig 脚本如下:

```
-- limit.pig
divs = load 'NYSE_dividends';
first10 = limit divs 10;
dump first10;
```

编写完成后,直接进入 Pig,将 limit.pig 导入 HDFS 中,运行脚本 limit.pig,结果如图 9.29 所示(截图只选取了显示的部分内容):

```
grunt> run limit.pig                    -- 运行
```

这个例子最多只会返回 10 行记录,如果输入小于 10 行,那么会返回所有的记录。需要注意的是,除了 order 外的其他所有操作符,Pig 都不会保证产生的数据是按照一定次序排序的。因为 NYSE_dividends 文件记录大于 10 条,所以上述样例脚本每次运行可能返回不

```
(NYSE,CCS,2009-01-28,0.414)
(NYSE,CCS,2009-04-29,0.414)
(NYSE,CCS,2009-07-29,0.414)
(NYSE,CCS,2009-10-28,0.414)
(NYSE,CIF,2009-12-09,0.029)
(NYSE,CPO,2009-01-06,0.14)
(NYSE,CPO,2009-03-27,0.14)
(NYSE,CPO,2009-06-26,0.14)
(NYSE,CPO,2009-09-28,0.14)
(NYSE,CPO,2009-12-30,0.14)
```

图 9.29　limit.pig 脚本结果图

同的结果。在 limit 前如果加上 order 语句就可以保证每次执行返回的结果是一样的。

limit 会产生一个额外的 reduce 过程，因为它需要将记录收集起来以计算它应该返回多少记录。Pig 通过在每个 map 过程限制输出条数，然后再以对 reducer 限制输出条数的方式来优化这一执行过程。当 limit 和 order 一起使用的时候，在 map 和 reduce 阶段这两个处理过程会同时进行。也就是说，在 map 这一边，使用 MapReduce 对记录进行排序，然后在组合器中限制输出条数。在 shuffle 阶段会有一部分用 MapReduce 再次进行排序，Pig 这时对 reduce 阶段再次进行输出条数限制。

这里 Pig 没有提供一个可能的优化，即在早期读取输入数据时，一旦达到 limit 指定的条数时就停止读入。因此，在这个例子中，如果期望使用此优化的方式去读取一小片输入数据，那么将会失望，因为 pig 仍然会读取全部数据。

9.2.8　Sample

sample 语法提供了一种方式用于抽取样本数据。它会读取所有的数据，然后返回一定百分比的行数的数据。可以通过一个 0 和 1 间的 double 值指定百分比。因此，在如下的例子中，0.1 表示 10%。

在 /root/work 目录下，编写 sample.pig 脚本如下：

```
-- sample.pig
divs = load 'NYSE_dividends';
some = sample divs 0.1;
dump some;
```

编写完成后，直接进入 Pig，将 sample.pig 导入 HDFS 中，运行脚本 sample.pig，结果如图 9.30 所示（截图只选取了显示的部分内容）：

```
grunt> run sample.pig                              -- 运行
```

当前这个抽样函数非常简单。sample A by 0.1 其实被重写为 filter A by random()<=0.1。很明显这个也并非是精确的，因此使用 sample 的脚本每次执行的结果都是不同的。同时，百分比也并非是精确的，但一定是近似的。目前对于增加一些高级的抽样技术是有讨论的，但还没有得出结论。

```
(NYSE, CPV, 2009-09-09, 0.422)
(NYSE, CSX, 2009-11-25, 0.22)
(NYSE, CJR, 2009-09-11, 0.046)
(NYSE, CJR, 2009-08-13, 0.045)
(NYSE, CCU, 2009-12-29, 0.596)
(NYSE, CAT, 2009-07-16, 0.42)
(NYSE, CHY, 2009-05-07, 0.085)
(NYSE, CXE, 2009-07-13, 0.031)
(NYSE, CXE, 2009-04-13, 0.031)
(NYSE, CNQ, 2009-09-09, 0.098)
(NYSE, CW, 2009-11-23, 0.08)
(NYSE, CYT, 2009-05-07, 0.013)
(NYSE, CIM, 2009-09-29, 0.12)
(NYSE, CRE, 2009-09-09, 0.17)
(NYSE, CTS, 2009-03-25, 0.03)
(NYSE, CNX, 2009-11-02, 0.1)
(NYSE, CEL, 2009-03-12, 0.66)
(NYSE, CCZ, 2009-07-15, 0.392)
(NYSE, CII, 2009-03-12, 0.485)
(NYSE, CBS, 2009-06-08, 0.05)
(NYSE, CHW, 2009-09-08, 0.06)
(NYSE, CHW, 2009-05-07, 0.08)
(NYSE, CYN, 2009-11-02, 0.1)
```

图 9.30　sample.pig 脚本结果图

9.2.9　Parallel

Pig 的核心声明之一就是它提供一种并行数据处理语言。为了做这件事，Pig 提供了 parallel 语句。

parallel 语句可以附加到 Pig Latin 中任一个关系操作符后面。然后，它只会控制 reduce 阶段的并行，因此只有对于可以触发 reduce 过程的操作符使用才有意义。可以触发 reduce 过程的操作符有 group *、order、distinct、join *、cogroup * 和 cross。使用星号标记的操作符具有多个不同的实现，这些实现中有的会触发一个 reduce 过程，而有的则不会。

在本地模式下 parallel 会被忽略，原因是本地模式下所有的操作符都是串行执行的。

在 /root/work 目录下，编写 parallel.pig 脚本如下：

```
-- parallel.pig
daily = load 'NYSE_daily' as (exchange,symbol,date,open,high,low,close,volume,adj_close);
bysymbl = group daily by symbol parallel 10;
dump bysymbl;
```

编写完成后，直接进入 Pig，运将 parallel.pig 导入 HDFS 中，行脚本 parallel.pig，结果如图 9.31 所示（截图只选取了显示的部分内容）：

```
grunt> run parallel.pig            -- 运行
```

在这个例子中，parallel 会使 Pig 触发的 MapReduce 任务具有 10 个 reducer。parallel 只会对后面附加上它的语句起作用，它们并非是在整个脚本中一直起作用的。因此如果这里的 group 后面跟的是一个 order 语句，则需要单独为这个 order 语句设置 parallel 值。因为 group 操作最有可能显著地降低数据量大小，所以可能需要更改并行参数的值。

```
12564000,63.66),(NYSE,CVX,2009-02-17,67.23,67.93,66.11,66.18,18834600,63.72),(NY
SE,CVX,2009-02-13,69.71,70.74,63.00,69.73,10852400,67.14),(NYSE,CVX,2009-02-12,6
9.95,70.06,67.94,69.86,17237200,67.26),(NYSE,CVX,2009-02-11,71.47,72.10,70.28,71
.26,15533700,67.99),(NYSE,CVX,2009-02-10,74.36,75.24,70.56,71.12,20150600,67.85)
,(NYSE,CVX,2009-02-09,74.41,75.56,73.59,74.42,14894400,71.00),(NYSE,CVX,2009-02-
06,72.58,75.00,72.29,74.90,13345400,71.46),(NYSE,CVX,2009-02-05,71.48,73.42,70.9
7,73.25,14933800,69.88),(NYSE,CVX,2009-02-04,72.29,72.93,71.01,71.60,12058000,68
.31),(NYSE,CVX,2009-02-03,57.07,71.79,69.94,71.64,12509600,68.35),(NYSE,CVX,2009
-02-02,69.52,70.85,69.12,70.29,12445000,67.06),(NYSE,CVX,2009-01-30,72.30,72.70,
70.18,70.52,17880000,67.28),(NYSE,CVX,2009-01-29,72.51,72.69,70.61,70.62,1578050
0,67.38),(NYSE,CVX,2009-01-28,72.99,74.02,71.96,73.79,14649100,70.40),(NYSE,CVX,
2009-01-27,71.28,72.63,70.82,72.09,13632700,68.78),(NYSE,CVX,2009-01-26,71.43,72
.39,70.13,71.29,14023100,68.01),(NYSE,CVX,2009-01-23,68.40,71.11,68.18,70.82,134
22100,67.57),(NYSE,CVX,2009-01-22,69.92,70.97,68.28,69.95,14320100,66.74),(NYSE,
CVX,2009-01-21,68.43,71.49,68.41,71.23,15889900,67.96),(NYSE,CVX,2009-01-20,70.9
3,72.20,68.00,68.31,16084200,65.17),(NYSE,CVX,2009-01-16,72.08,72.94,70.38,71.74
,15594800,68.44),(NYSE,CVX,2009-01-15,69.70,71.25,68.25,70.77,19293100,67.52),(N
YSE,CVX,2009-01-14,70.81,71.18,68.83,69.69,15142400,66.49),(NYSE,CVX,2009-01-13,
70.58,72.70,70.58,71.82,14378800,68.52),(NYSE,CVX,2009-01-12,72.23,72.61,70.44,7
0.82,14195500,67.57),(NYSE,CVX,2009-01-09,73.88,74.00,72.20,72.82,14182600,69.47
),(NYSE,CVX,2009-01-08,73.61,74.84,73.22,74.24,11386700,70.83),(NYSE,CVX,2009-01
-07,76.41,77.07,73.33,73.96,12804300,70.56),(NYSE,CVX,2009-01-06,77.56,78.45,76.
33,77.35,15894500,73.80),(NYSE,CVX,2009-01-05,76.16,78.37,75.74,76.66,16954500,7
3.14)})
```

图 9.31 parallel.pig 脚本结果图(一)

修改 parallel.pig 脚本如下：

```
-- parallel.pig
daily = load 'NYSE_daily' as (exchange,symbol,date,open,high,low,close,volume,adj_close);
bysymbl = group daily by symbol parallel 10;
average = foreach bysymbl generate group,AVG(daily.close) as avg;
sorted = order average by avg desc parallel 2;
dump sorted;
```

修改完成后，直接进入 Pig，将 parallel.pig 导入 HDFS 中，运行脚本 parallel.pig，结果如图 9.32 所示(截图只选取了显示的部分内容)：

```
grunt> run parallel.pig                           -- 运行
```

然而，如果不想为脚本中每一个会触发 reduce 过程的操作符单独设置并行数值，那么可以通过 set 命令设置一个脚本范围内有效的参数值。

在 /root/work 目录下，编写 defaultparallel.pig 脚本如下：

```
-- defaultparallel.pig
daily = load 'NYSE_daily' as (exchange,symbol,date,open,high,low,close,volume,adj_close);
bysymbl = group daily by symbol;
average = foreach bysymbl generate group,AVG(daily.close) as avg;
sorted = order average by avg desc;
dump sorted;
```

编写完成后，直接进入 Pig，将 defaultparallel.pig 导入 HDFS 中，运行文件 defaultparallel.pig，结果如图 9.33 所示(截图只选取了显示的部分内容)：

```
grunt > run defaultparallel.pig                          -- 运行
```

```
(CYE, 5.057936507936504)
(COT, 5.046850393700784)
(CFI, 4.79980158730159)
(CBM, 4.62775590551181)
(CSU, 4.271746031746031)
(CXE, 4.182698412698414)
(CPF, 4.078214285714285)
(CMU, 3.8274603174603166)
(CBC, 3.783690476190475)
(CSE, 3.5607936507936513)
(CIM, 3.5448809523809506)
(CBR, 3.4740476190476186)
(CMM, 3.4425793650793657)
(CWF, 3.2880158730158717)
(CNO, 3.1945238095238095)
(CSV, 3.178968253968254)
(CBB, 2.7912992125984237)
(CIF, 2.08468253968254)
(CHP, 1.977420634920637)
(CT, 1.9655158730158724)
(CPE, 1.8293650793650797)
(CMZ, 0.9682142857142858)
(CHB, 0.436442687747036)
```

图 9.32 parallel.pig 脚本结果图（二）

```
(CYE, 5.057936507936504)
(COT, 5.046850393700784)
(CFI, 4.79980158730159)
(CBM, 4.62775590551181)
(CSU, 4.271746031746031)
(CXE, 4.182698412698414)
(CPF, 4.078214285714285)
(CMU, 3.8274603174603166)
(CBC, 3.783690476190475)
(CSE, 3.5607936507936513)
(CIM, 3.5448809523809506)
(CBR, 3.4740476190476186)
(CMM, 3.4425793650793657)
(CWF, 3.2880158730158717)
(CNO, 3.1945238095238095)
(CSV, 3.178968253968254)
(CBB, 2.7912992125984237)
(CIF, 2.08468253968254)
(CHP, 1.977420634920637)
(CT, 1.9655158730158724)
(CPE, 1.8293650793650797)
(CMZ, 0.9682142857142858)
(CHB, 0.436442687747036)
```

图 9.33 defaultparallel.pig 脚本结果图

在这个脚本中，所有的 MapReduce 都会在 10 个 reducer 中完成。当设置了一个默认的并行值的时候，仍然可以为脚本中的语句增加一个 parallel 语句来重写这个默认值。因此可以设置一个大多数情况使用的默认值，然后对于需要其他值的操作符通过 parallel 关键字显式地重新设置一个并行值。

所有这些都是静态的，如果对于不同特征的输入数据执行同一个脚本会怎么样？或者如果输入数据有时差异性非常大又会出现什么情况？我们当然不希望每次都去修改脚本。

使用参数替换,可以在写并行语句的时候使用变量,在运行时再为这些变量赋值。

如果没有设置并行值,在 0.8 版本之前,Pig 是让 MapReduce 来设置并行值的。MapReduce 的默认并行值是通过集群的配置文件控制的。安装的默认值是 1,大多数人都不会更改这个值。也就是说,很有可能只会以一个 reducer 来运行,这绝对不是期望的。

为了避免这种情况,Pig 在 0.8 版本中增加了一个试探性的做法,就是如果没有设置这个值,就通过一个粗略的估算给出一个并行值。它根据初始的输入值,减少的数据量大小不会改变,然后每 1GB 的数据就分配一个 reducer。需要强调的是,这并非是一个好算法,这只是为了防止因为一些设置错误导致脚本运行很慢,同时,在一些极端的情况下,因一些错误导致 MapReduce 本身就有问题。这是一个安全网,而不是一个优化。

9.3 用户自定义函数 UDF

Pig 的一大特色在于它允许用户通过 UDF 将 Pig 和操作符以及用户的代码或其他人提供的代码合并在一起使用。在 0.7 版本之前,所有的 UDF 必须是使用 Java 语言编写并实现 Java 类。这使得很容易通过写 Java 类向 Pig 中增加新 UDF 并且告诉 Pig 用户编写的 JAR 文件在什么地方。

在 0.8 版本中,也可以使用 Python 编写 UDF。Pig 使用 Jython 来执行 Python 编写的 UDF,因此这些 UDF 必须使用 Python 2.5 进行编译,这样也就不能使用 Python 3 中的功能了。

Pig 本身包含了一些 UDF。0.8 版本之前,本身只带了少量的 UDF,包括标准的 SQL 聚合函数和其他一些 UDF。在 0.8 版本中,增加了大量的标准字符串处理函数、数学函数和复杂类型 UDF。

Piggybank 是和 Pig 一起发布的,其包含了用户贡献的 UDF 集合。Piggybank 中的 UDF 并不包含在 Pig 的 JAR 中,因此如果想使用这些 UDF,需要在脚本中进行注册。

可以自己写 UDF,当然也可以使用其他用户提供的 UDF,一些 Java 静态函数也是可以作为 UDF 的。

9.3.1 注册 UDF

当需要使用一个非 Pig 内置的 UDF 时,需要告诉 Pig 到哪里去查找那个 UDF。可以通过 register 命令完成这个过程。

在 /root/work 目录下,编写 register.pig 脚本如下:

```
-- register.pig
register /usr/pig/pig-0.16.0/contrib/piggybank/java/piggybank.jar;   --此处为 piggybank.
                                                                      jar 的绝对路径
divs = load 'NYSE_dividends' as (exchange:chararray, symbol:chararray, date:chararray,
dividends:float);
backwards = foreach divs generate org.apache.pig.piggybank.evaluation.string.Reverse
(symbol);
dump backwards;
```

直接在 shell 命令行输入下面的命令,运行 register.pig 脚本:

```
pig register.pig                    --运行
```

运行结果如图 9.34 所示。

```
(WHC)
(WHC)
(PRC)
(PRC)
(PRC)
(PRC)
(LHC)
(LHC)
(NYC)
(NYC)
(NYC)
(NYC)
(POC)
(POC)
(POC)
(OEC)
(OEC)
(SC)
(CMC)
(CMC)
(CMC)
(CMC)
```

图 9.34　register.pig 脚本结果图

这个例子告诉我们 Pig 在生成 JAR 传送给 Hadoop 的时候需要包含 your_path_to_piggybank/ piggybank.jar 中的内容。Pig 会打开所有这些注册进来的 JAR,取出所有文件,然后把它们全部加入那个需要传送给 Hadoop 执行的任务的 JAR 中。

在这个例子中,需要告诉 Pig 完整的包名和 UDF 函数的类名。有两种方式可以简化这种冗长的使用方式。第一种方式是使用 define 命令,直接在 shell 命令行输入"pig register.pig;"命令的方式调用 Pig。第二种方式是在命令行为 Pig 指定一组路径用于查找需要的 UDF,需要在 shell 命令行中输入"pig-Dudf.import.list = org.apache.pig.piggybank.evaluation.string register.pig;"命令,并且将脚本修改成如下形式:

```
register /usr/pig/pig-0.16.0/contrib/piggybank/java/piggybank.jar;    --此处为 piggybank.
                                                                       jar 的绝对路径
divs = load 'NYSE_dividends' as (exchange:chararray, symbol:chararray, date:chararray,
dividends:float);
backwards = foreach divs generate Reverse(symbol);
dump backwards;
```

使用另一种属性,也可以不用先注册命令,即在 register.pig 脚本文件中不需要写"register …",但是需要增加"-Dpig.additional.jars =/usr/pig/pig-0.16.0/contrib/piggybank/java/piggyb bank.jar"语句来指明 JAR 包的路径,在 shell 命令行输入如下命令:PIG

```
-Dpig.additional.jars=/usr/pig/pig-0.16.0/contrib/p
iggybank/java/piggybank.jar
-Dudf.import.list=org.apache.pig.piggybank.evaluation.string register.pig;
```

很多情况下,在脚本中使用 register 和 define 命令显式地注册和定义 UDF 要比通过设置属性值的方式加载好。否则,使用脚本的也必须知道在命令行如何配置这些属性。然而,在一些情况下的脚本总是使用相同的一组 JAR,同时总是在一组相同的位置去查找它们。例如,用户可能有一组全公司都使用的 JAR。在这种情况下,把这些属性信息放到一个共享的属性配置文件中,然后使用这个配置运行 Pig 脚本将会使得共享这些 UDF 变得容易,同时也利于保证大家都在使用相同的正确版本的 UDF。

register 也可以在 Pig Latin 脚本中加载用 Python 编写的 UDF 资源。在这种情况下,不是注册一个 JAR 文件,而是注册一个包含的 UDF 的 Python 脚本文件。这个 Python 脚本必须在当前目录下。如下面的例子。

在/root/work 目录下,编写 production.py,内容如下:

```
-- vim production.py
@outputSchema("production:float")
def production(slugging_pct, onbase_pct):
return slugging_pct + onbase_pct
-- 最后一行需要注意:return 的前面需要按一下 Tab 键
```

在/root/work 目录下,继续编写 batting_production.pig 脚本如下:

```
-- batting_production.pig
register 'production.py' using jython as bballudfs;
players = load 'baseball' as (name:chararray, team:chararray,pos:bag{t:(p:chararray)}, bat:map[]);
nonnull = filter players by bat#'slugging_percentage' is not null and bat#'on_base_percentage' is not null;
calcprod = foreach nonnull generate name, bballudfs.production((float)bat#'slugging_percentage',(float)bat#'on_base_percentage');
dump calcprod;
```

编写完成后,直接进入 Pig,将 batting_production.pig 导入 HDFS 中,运行脚本 batting_production.pig,结果如图 9.35 所示(截图只选取了显示的部分内容):

```
grunt>run batting_production.pig                    -- 运行
```

这里一个重要的不同点是 register 语句中的 using jython 和 as bballudfs 部分。

using jython 告诉 Pig 这个 UDF 是使用 Python 编写,而不是使用 Java 编写的,需要使用 Jython 对这个 UDF 进行编译。Pig 并不知道系统中 Jython 解释器所在的路径,因此当调用 Pig 时需要保证 jython.jar 已经放到类路径中了。可以通过设置 PIG_CLASSPATH 环境变量来指定该路径。

as bballudfs 为从这个文件中加载的 UDF 定义了一个命名空间。所有这个文件定义的 UDF 目前都需要通过 bballudfs.udfname 的方式进行调用。加载的每一个 Python 文件都

```
(Josh Willingham, 0.8399999737739563)
(Dontrelle Willis, 0.6370000243186951)
(Reggie Willits, 0.6739999949932098)
(Bobby Wilson, 0.5230000019073486)
(Jack Wilson, 0.6840000152587891)
(Dewayne Wise, 0.6399999856948853)
(Randy Wolf, 0.4880000054836273)
(Brandon Wood, 0.5349999964237213)
(Kerry Wood, 0.4449999928474426)
(Tim Wood, 1.0)
(Mark Worrell, 2.5)
(Wesley Wright, 0.5)
(Tyler Yates, 0.16599999368190765)
(Kevin Youkilis, 0.8779999911785126)
(Chris Young, 0.38600000739097595)
(Delmon Young, 0.7380000054836273)
(Delwyn Young, 0.7119999825954437)
(Michael Young, 0.79800000786781311)
(Carlos Zambrano, 0.6389999985694885)
(Gregg Zaun, 0.7320000231266022)
(Ryan Zimmerman, 0.824999988079071)
(Barry Zito, 0.25200000405311584)
(Ben Zobrist, 0.804999977350235)
```

图 9.35 batting_production.pig 脚本结果图

需要给予一个不同的命名空间,这样可以避免注册两个 Python 脚本中相同函数名所产生的冲突。

这里有个警告：Pig 不会追踪 Python 脚本中的依赖关系而会向 Hadoop 集群中传送需要的 Python 模块。需要确保所使用到的 Python 模块在集群的任务节点中是存在的,而且在这些节点上已经设置好了 PYTHONPATH 这个环境变量,只有这样才能保证的 Python UDF 可以正确地把这些模块加载进去。

9.3.2 define 命令和 UDF

define 命令用于为 Java UDF 定义一个别名,那样就不需要写冗长的包名全路径了。它也可以为 UDF 的构造函数提供参数。define 命令同样可用于定义 streaming 命令。以下是一个使用 define 为 org.apache.piggybank.evluation.string.Reverse 定义一个别名的例子。

在/root/work 目录下,编写 define.pig 脚本如下：

```
-- define.pig
register '/usr/pig/pig-0.16.0/contrib/piggybank/java/piggybank.jar';
                                    -- 此处为 piggybank.jar 的绝对路径
define reverse org.apache.pig.piggybank.evaluation.string.Reverse();
divs = load 'NYSE_dividends' as (exchange:chararray, symbol:chararray, date:chararray,
dividends:float);
backwards = foreach divs generate reverse(symbol);
dump backwards;
```

编写完成后,直接进入 Pig,将 define.pig 导入 HDFS 中,运行脚本 define.pig,结果如图 9.36 所示(截图只选取了显示的部分内容)：

```
grunt > run define.pig                    -- 运行
```

```
(WHC)
(WHC)
(PRC)
(PRC)
(PRC)
(PRC)
(LHC)
(LHC)
(NYC)
(NYC)
(NYC)
(NYC)
(POC)
(POC)
(POC)
(POC)
(OEC)
(OEC)
(SC)
(CMC)
(CMC)
(CMC)
(CMC)
```

图 9.36　define.pig 脚本结果图

数学函数和过滤函数也是可以有一个或多个字符串类型的构造函数参数。如果使用的是一个接收构造函数参数的 UDF，define 命令就可以放置这些参数。

假设一个方法 CurrencyConverter 的构造函数需要两个参数：第一个参数是需要被转换的货币类型，第二个参数是需要转换成的货币类型。

（1）在目录 /root/work 下，创建 myudfs 文件夹。

```
mkdir myudfs
```

（2）通过 cd myudfs 进入文件夹 myudfs，编写货币类型转换的 Java 函数，内容如下：

```
-- vim CurrencyConverter.java
package myudfs;                    //当前目录为/root/work/myudfs
import org.apache.hadoop.io.WritableComparable;
import java.io.IOException;
import java.util.Map;
import java.util.HashMap;

import org.apache.pig.EvalFunc;
import org.apache.pig.backend.executionengine.ExecException;
import org.apache.pig.data.DataType;
import org.apache.pig.data.Tuple;
import org.apache.pig.impl.logicalLayer.schema.Schema;

public class CurrencyConverter extends EvalFunc<Float> {

    String from, to;
```

```java
        public CurrencyConverter(String from, String to) {
            super();
            this.from = from;
            this.to = to;
        }

    @Override
        public Float exec(Tuple input) throws IOException {
                if (input == null || input.size() != 1) return null;
                return (Float)input.get(0) * 1.5f;
        }

        @Override
        public Schema outputSchema(Schema input) {
            return new Schema(new Schema.FieldSchema(null, DataType.FLOAT));
        }

}
```

（3）修改配置文件~/.bashrc，在底端加入如下代码：

```
vim ~/.bashrc
export CLASSPATH = /usr/pig/pig-0.16.0/build/pig-0.16.0-SNAPSHOT.jar:/usr/hadoop/hadoop-2.7.1/share/hadoop/common/hadoop-common-2.7.1.jar:$CLASSPATH
```

（4）将本地目录/root/下载/中的 pig-0.16.0-src.tar.gz 解压，得到 pig-0.16.0-src，并将其移至/usr/pig 中。然后，进入/usr/pig/pig-0.16.0/目录，输入下面的命令并编译。

```
ant jar    （此处 ant 工具在模板 2 中已安装好，可以直接使用）
（注：编译的过程有点长，需等待）
```

（5）回到目录/root/work/myudfs 中，编译并打包成 acme.jar。

```
cd /root/work/myudfs
javac CurrencyConverter.java
cd .. （回到上级目录）
jar - cf acme.jar myudfs/CurrencyConverter.class          --打包
```

（6）创建脚本 define_constructor_args.pig，内容如下：

```
-- vim define_constructor_args.pig
register 'acme.jar';
define convert myudfs.CurrencyConverter('dollar', 'euro');
divs = load 'NYSE_dividends' as (exchange: chararray, symbol: chararray, date: chararray, dividends:float);
backwards = foreach divs generate convert(dividends);
dump backwards;
```

（7）进入 Pig，执行下面的命令，运行脚本文件 define_constructor_args.pig，输出结果如图 9.37 所示。

```
grunt> copyFromLocal acme.jar acme.jar
grunt> copyFromLocal define_constructor_args.pig define_constructor_args.pig
grunt> run define_constructor_args.pig
```

```
(0.12)
(0.12)
(0.741)
(0.741)
(0.741)
(0.741)
(1.3019999)
(1.359)
(0.15)
(0.15)
(0.15)
(0.375)
(0.75)
(0.705)
(0.705)
(0.705)
(3.87)
(3.8715)
(0.144)
(0.17999999)
(0.17999999)
(0.17999999)
(0.17999999)
```

图 9.37　define_constructor_args.pig 脚本结果图

9.3.3　调用静态 Java 函数

Java 具有丰富的工具集和函数库，因为 Pig 是使用 Java 实现的，所以 Java 中的一些函数也是可以暴露给 Pig 用户的。从 0.8 版本开始，Pig 提供了 invoker 方法，允许像使用 Pig UDF 一样使用一些特定的静态 Java 函数。

所有没有参数或者具有 int、long、float、double、String 或者 array 类型的参数，同时有 int、long、float、double 或 String 类型返回值的静态 Java 函数，都是可以通过这种方式进行调用的。

因为 Pig Latin 不支持对返回值的数据类型进行重载，因此对于每一种类型有一个对应的调用方法：InvokeForInt、InvokeForLong、InvokeForFloat、InvokeForDouble 和 InvokeForString。需要根据期望的返回值的数据类型调用适当的调用方法。该方法有两个构造参数：第一个参数是完整的包名、类名和方法名；第二个参数是以空格作为分隔符的参数列表，这些参数将传送给这个 Java 函数，只包含参数的类型。如果这个参数是数组，那么[]（方括号）会跟在这个类型名称后面。如果这个方法不需要参数，那么第二个构造参数会省略。

例如，如果想使用 Java 的 Integer 类将十进制的数值转换为十六进制的值，需要按如下方式使用。

在 /root/work 目录下，编写 invoker.pig 脚本如下：

```
-- invoker.pig
define hex InvokeForString('java.lang.Integer.toHexString', 'int');
divs = load 'NYSE_daily' as (exchange, symbol, date, open, high, low, close, volume, adj_close);
nonnull = filter divs by volume is not null;
inhex = foreach nonnull generate symbol, hex((int)volume);
dump inhex;
```

编写完成后,直接进入 pig,将 invoker.pig 导入 HDFS 中,运行脚本 invoker.pig,结果如图 9.38 所示(截图只选取了显示的部分内容):

```
grunt> run invoker.pig;              -- 运行
```

```
(CVA,123298)
(CVA,257060)
(CVA,198534)
(CVA,12bce0)
(CVA,1a8a24)
(CVA,14c404)
(CVA,1475e4)
(CVA,133404)
(CVA,18c504)
(CVA,1863c0)
(CVA,15e708)
(CVA,ebfc8)
(CVA,103bdc)
(CVA,1a9ce4)
(CVA,133e90)
(CVA,ab180)
(CVA,b0b58)
(CVA,18eda4)
(CVA,11d85c)
(CVA,10278c)
(CVA,ef86c)
(CVA,148458)
```

图 9.38 invoker.pig 脚本结果图

如果方法接收的是一组数据类型的参数,那么 Pig 会将它封装成一个 bag,这个数据包中每个 tuple 只会有一个这种类型的字段。因此如果有一个 Java 方法 com.yourcompany.Stats.stdev 参数是一组 double 类型的值,那么可能是通过如下方式使用的:

```
define stdev InvokeForDouble ('com.acme.State.stdev','double[]');
A = load'input'as(id:int,dp:double);
B = group A by id;
C = foreach B generate group,stdev(A.dp);
```

采用这种方式调用 Java 函数,需要付出一些性能上的代价,因为需要使用反射机制来找到和调用这些方法。

当输入的参数是 null 的时候,调用器函数会抛出一个 IllegalArgumentException 异常。因此调用之前需要放置一个过滤器来避免异常的出现。

第 10 章

SQL to Hadoop：Sqoop

10.1 Sqoop 概述

10.1.1 Sqoop 的产生背景

（1）多数使用 Hadoop 技术处理大数据业务的企业都有大量的数据存储在传统的关系型数据库（RDBMS）中。

（2）由于缺乏工具的支持，对 Hadoop 和传统数据库系统中的数据进行相互传输是一件十分困难的事情。

- 传统数据库的数据导入 Hadoop，便于廉价处理和分析。
- Hadoop 数据导入传统数据库，可利用强大的 SQL 进一步分析和展示。

（3）急需一个在 RDBMS 与 Hadoop 之间进行数据传输的项目。

10.1.2 Sqoop 是什么

Sqoop 即 SQL to Hadoop，是一款方便在传统型数据库与 Hadoop 之间进行数据迁移的工具，如图 10.1 所示，它充分利用 MapReduce 的并行特点以批处理的方式加快数据传输，发展至今，主要演化了两个大版本：Sqoop1 和 Sqoop2。

Sqoop 工具是 Hadoop 下连接关系型数据库和 Hadoop 的桥梁，支持关系型数据库和 Hive、HDFS、HBase 之间数据的相互导入，可以使用全表导入和增量导入。

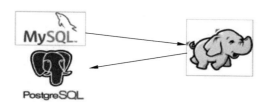

图 10.1　SQL to Hadoop

10.1.3　为什么选择 Sqoop

（1）高效可控的利用资源：任务并行度，超时时间；
（2）数据类型映射与转化：可自动进行，用户也可自定义；
（3）支持多种主流数据库：MySQL、Oracle、SQL Server、DB2 等。

10.1.4　Sqoop1 和 Sqoop2 的异同

（1）两个不同的版本，完全不兼容。
（2）版本号划分区别。
- Apache 版本：1.4.x(Sqoop1)、1.99.x(Sqoop2)。
- CDH 版本：Sqoop-1.4.3-cdh4(Sqoop1)、Sqoop2-1.99.2-cdh4.5.0(Sqoop2)。

（3）Sqoop2 比 Sqoop1 的改进之处。
- 引入 Sqoop Server，集中化管理 Connector 等。
- 多种访问方式：CLI、Web UI、REST API。
- 引入基于角色的安全机制。

10.1.5　Sqoop1 与 Sqoop2 的架构图

Sqoop1 的架构如图 10.2 所示。

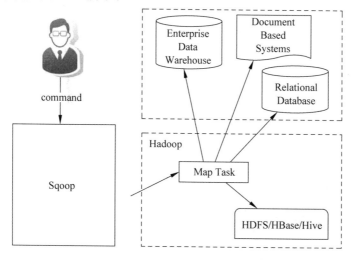

图 10.2　Sqoop1 架构图

Sqoop1 存在如下问题：

（1）基于命令行的操作方式易于出错，且不安全；

（2）数据传输和数据格式是紧耦合的，这使得 Connector 无法支持所有数据格式；

（3）安全密钥是暴露出来的，非常不安全；

（4）Sqoop 安装需要 root 权限；

（5）Connector 必须符合 JDBC 模型，并使用通用的 JDBC 词汇。

Sqoop2 的架构如图 10.3 所示。

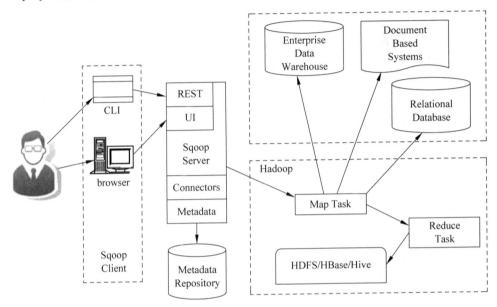

图 10.3　Sqoop2 架构图

Sqoop2 架构的优点如下：

（1）多种交互方式

- 命令行、Web UI、REST API。

（2）Connector 集中化安装。

所有 Connector 都安装在 Sqoop Server 上。

（3）权限管理机制

配置管理员、使用者等角色。

（4）Connector 规范化。

- 不再包含数据传输、格式转换以及与 Hive、HBase 交互等功能。
- 仅负责数据读写。

10.1.6　Sqoop1 与 Sqoop2 的优缺点

Sqoop1 与 Sqoop2 的优缺点比较如表 10.1 所示。

表 10.1 Sqoop1 与 Sqoop2 的优缺点对比

比较	Sqoop1	Sqoop2
架构	仅仅使用一个 Sqoop 客户端	引入了 Sqoop Server 集中化管理 Connector 以及 Rest Api、Web UI,并引入权限安全机制
部署	部署简单,安装需要 root 权限,connector 必须符合 JDBC 模型	架构稍复杂,配置部署更烦琐
使用	命令行方式容易出错,格式紧耦合,无法支持所有数据类型,安全机制不够完善,例如密码暴露	多种交互方式:命令行、Web UI、Rest API、Conncetor 集中化管理,所有的链接安装在 Sqoop Server 上,完善权限管理机制,Connector 规范化,仅仅负责数据的读写

10.2 Sqoop 安装部署

10.2.1 下载 Sqoop

(1)可以到 Sqoop 官网 http://sqoop.apache.org/,选择镜像下载地址 http://mirror.bit.edu.cn/apache/sqoop/下载一个稳定版本,如图 10.4 所示,下载支持 Hadoop1.X 的 1.4.6 版本包。

图 10.4 Sqoop 下载图

(2)在"/root/下载"文件夹中找到 Sqoop 安装包,如图 10.5 所示。

(3)打开终端,如图 10.6 所示选择"汉语-Pinyin"修改输入法。

(4)进入目录"/root/下载",找到 sqoop-1.4.6.bin_hadoop-2.0.4-alpha.tar.gz 安装包,解压该安装包并把解压后的文件移到/usr/sqoop 文件夹中。

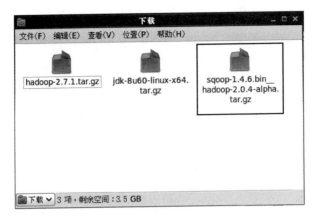

图 10.5 软件包

图 10.6 修改输入法

```
cd /root/下载
tar -zxvf sqoop-1.4.6.bin__hadoop-2.0.4-alpha.tar.gz
mkdir /usr/sqoop/
mv sqoop-1.4.6.bin__hadoop-2.0.4-alpha /usr/sqoop/sqoop-1.4.6
```

/usr/sqoop 文件夹中的内容如图 10.7 所示。

```
[root@master 下载]# ll /usr/sqoop
总用量 4
drwxr-xr-x. 9 root root 4096 4月  27 2015 sqoop-1.4.6
```

图 10.7 /usr/sqoop 文件夹中的内容

10.2.2 设置/etc/profile 参数

(1) 编辑~/.bashrc 文件,加入 Sqoop 的 Home 路径,在 PATH 中加入 bin 的路径(注意只需在原有的基础上添加红色字),如图 10.8 所示。

注:进入文本编辑界面后,单击键盘上的字母 i 键,进入插入状态,即可编写内容,此处介绍过,后面就不再具体指出。

```
vim ~/.bashrc
export SQOOP_HOME = /usr/sqoop/sqoop-1.4.6
export PATH = /usr/java/jdk1.8.0_60/bin:/usr/hadoop/hadoop-2.7.1/bin:/usr/sqoop/sqoop-1.4.6/bin:$PATH
```

图 10.8 编辑~/.bashrc 文件

(2) 编译配置文件~/.bashrc,并确认生效。

```
source ~/.bashrc
echo $PATH
```

10.2.3 设置 bin/configure-sqoop 配置文件

修改 bin/configure-sqoop 配置文件:

```
cd /usr/sqoop/sqoop-1.4.6/bin
vim configure-sqoop
```

注释掉 HBase 和 Zookeeper 等检查命令行(除非使用 HBase 和 Zookeeper 等 Hadoop 上的组件),如图 10.9 所示。

10.2.4 设置 conf/sqoop-env.sh 配置文件

(1) 如果不存在 sqoop-env.sh 文件,则复制 sqoop-env-template.sh 文件并将之重命名为 sqoop-env.sh。结果如图 10.10 所示。

```
## Moved to be a runtime check in sqoop.
#if [ ! -d "${HBASE_HOME}" ]; then
#   echo "Warning: $HBASE_HOME does not exist! HBase imports will fail."
#   echo 'Please set $HBASE_HOME to the root of your HBase installation.'
#fi

## Moved to be a runtime check in sqoop.
#if [ ! -d "${HCAT_HOME}" ]; then
#   echo "Warning: $HCAT_HOME does not exist! HCatalog jobs will fail."
#   echo 'Please set $HCAT_HOME to the root of your HCatalog installation.'
#fi

#if [ ! -d "${ACCUMULO_HOME}" ]; then
#   echo "Warning: $ACCUMULO_HOME does not exist! Accumulo imports will fail."
#   echo 'Please set $ACCUMULO_HOME to the root of your Accumulo installation.'
#fi
#if [ ! -d "${ZOOKEEPER_HOME}" ]; then
#   echo "Warning: $ZOOKEEPER_HOME does not exist! Accumulo imports will fail."
#   echo 'Please set $ZOOKEEPER_HOME to the root of your Zookeeper installation.'
#fi
```

图 10.9　注释掉 HBase 和 ZooKeeper 等

```
cd /usr/sqoop/sqoop-1.4.6/conf/
cp sqoop-env-template.sh sqoop-env.sh
```

```
[root@master conf]# ls
oraoop-site-template.xml   sqoop-env-template.cmd   sqoop-site-template.xml
sqoop-env.sh               sqoop-env-template.sh    sqoop-site.xml
```

图 10.10　创建 sqoop-env.sh 文件

（2）修改 sqoop-env.sh 配置文件：设置 Hadoop 运行程序所在路径和 hadoop-*-core.jar 路径（Hadoop1.X 需要配置）。结果如图 10.11 所示。

```
vim sqoop-env.sh
```

```
#Set path to where bin/hadoop is available
export HADOOP_COMMON_HOME=/usr/hadoop/hadoop-2.7.1

#Set path to where hadoop-*-core.jar is available
export HADOOP_MAPRED_HOME=/usr/hadoop/hadoop-2.7.1
```

图 10.11　修改 sqoop-env.sh 配置文件

（3）编译配置文件 sqoop-env.sh 使之生效。

10.2.5　验证安装完成

（1）先启动 Hadoop。

注：为了能够和 HDFS 之外的数据存储库进行交互，需要启动 Hadoop，如果只是验证 Sqoop 是否安装成功，可以不启动 Hadoop。

```
cd /usr/hadoop/hadoop-2.7.1/sbin
./start-all.sh
jps
```

执行 jps 命令的结果如图 10.12 所示。

出现图 10.12 所示结果，则 hadoop 启动成功，若图 10.12 中的进程未全部启动，则使用 stop-all.sh 关闭 Hadoop，之后再重新启动 Hadoop 直到出现图 10.12 中的所有内容。

```
[root@master sbin]# jps
15046 Jps
14823 ResourceManager
2153 VmServer.jar
14377 NameNode
14477 DataNode
14926 NodeManager
14670 SecondaryNameNode
```

图 10.12　启动 Hadoop

（2）下面的命令被用来验证 Sqoop 版本，如果正确安装 Sqoop，则会出现相应的版本信息，如图 10.13 所示。

```
sqoop - version
```

```
[root@master hadoop-2.7.1]# sqoop-version
16/08/16 11:16:49 INFO sqoop.Sqoop: Running Sqoop version: 1.4.6
Sqoop 1.4.6
git commit id c0c5a81723759fa575844a0a1eae8f510fa32c25
Compiled by root on Mon Apr 27 14:38:36 CST 2015
```

图 10.13　Sqoop 版本信息图

10.3　Sqoop 常用命令介绍

10.3.1　如何列出帮助

可以通过 Sqoop 的帮助命令查看相应的帮助信息，结果如 10.14 所示。

```
sqoop help
```

```
[root@master sqoop-1.4.6]# sqoop help
16/08/16 15:02:17 INFO sqoop.Sqoop: Running Sqoop version: 1.4.6
usage: sqoop COMMAND [ARGS]

Available commands:
  codegen            Generate code to interact with database records
  create-hive-table  Import a table definition into Hive
  eval               Evaluate a SQL statement and display the results
  export             Export an HDFS directory to a database table
  help               List available commands
  import             Import a table from a database to HDFS
  import-all-tables  Import tables from a database to HDFS
  import-mainframe   Import datasets from a mainframe server to HDFS
  job                Work with saved jobs
  list-databases     List available databases on a server
  list-tables        List available tables in a database
  merge              Merge results of incremental imports
  metastore          Run a standalone Sqoop metastore
  version            Display version information

See 'sqoop help COMMAND' for information on a specific command.
```

图 10.14　列出 Sqoop 帮助

10.3.2 Export

如图 10.15 所示，将数据从 Hadoop 导出到关系型数据库中。export 命令的基本选项如表 10.2 所示。

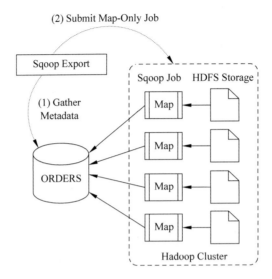

图 10.15　Export 图

表 10.2　export 基本选项

选　　项	含 义 说 明
--validate < class-name >	启用数据副本验证功能，仅支持单表复制，可以指定验证使用的实现类
--validation-threshold < class-name >	指定验证门限所使用的类
--direct	使用直接导出模式（优化速度）
--export-dir < dir >	导出过程中 HDFS 源路径
-m,--num-mappers < n >	使用 n 个 map 任务并行导出
--table < table-name >	导出的目的表名称
--call < stored-proc-name >	导出数据调用的指定存储过程名
--update-key < col-name >	更新参考的列名称，多个列名使用逗号分隔
--update-mode < mode >	指定更新策略，包括 updateonly（默认）、allowinsert
--input-null-string < null-string >	使用指定字符串，替换字符串类型值为 null 的列
--input-null-non-string < null-string >	使用指定字符串，替换非字符串类型值为 null 的列
--staging-table < staging-table-name >	在数据导出到数据库之前，数据临时存放的表名称
--clear-staging-table	清除工作区中临时存放的数据
--batch	使用批量模式导出

10.3.3 Import

将数据从关系型数据库导入 Hadoop 中，如图 10.16 所示。import 命令的基本选项如表 10.3 所示。import 和 export 命令的通用选项如表 10.4 所示。

第10章 SQL to Hadoop：Sqoop

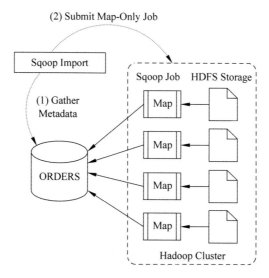

图 10.16　Import 图

表 10.3　Import 基本选项

选　　项	含　义　说　明
--append	将数据追加到 HDFS 上一个已存在的数据集上
--as-avrodatafile	将数据导入到 Avro 数据文件
--as-sequencefile	将数据导入到 SequenceFile
--as-textfile	将数据导入到普通文本文件（默认）
--boundary-query < statement >	边界查询，用于创建分片（InputSplit）
--columns < col,col,col… >	从表中导出指定的一组列的数据
--delete-target-dir	如果指定目录存在，则先删除掉
--direct	使用直接导入模式（优化导入速度）
--direct-aplit-size < n >	分割输入 stream 的字节大小（在直接导入模式下）
--fetch-size < n >	从数据库中批量读取记录数
--inline-lob-limit < n >	设置内联的 LOB 对象的大小
-m,--num-mappers < n >	使用 n 个 map 任务并行导入数据
-e,--query < statement >	导入的查询语句
--split-by < column-name >	指定按照哪个列去分割数据
--table < table-name >	导入的源表表名
--target-dir < dir >	导入 HDFS 的目标路径
--warehouse-dir < dir >	HDFS 存放表的跟路径
--where < where clause >	指定导出时所使用的查询条件
-z,compress	启用压缩
--compression-codec < c >	指定 Hadoop 的 codec 方式（默认 gzip）
--null-string < null-string >	如果指定列为字符串类型，使用指定字符串替换值为 null 的该类列的值
--null-non-string < null-string >	如果指定列为非字符串类型，使用指定字符串替换值为 null 的该类列的值

表 10.4　Import 和 Export 工具通用选项

选　　项	含 义 说 明
--connect < jdbc-uri >	指定 JDBC 连接字符串
--connection-manager < class-name >	指定要使用的连接管理器类
--driver < class-name >	指定要使用的 JDBC 驱动类
--hadoop-mapred-home < dir >	指定 $HADOOP_MAPRED_HOME 路径
--help	打印用法帮助信息
--password-file	设置用于存放密码信息文件的路径
-p	从控制台读取输入的密码
--password < password >	设置认证密码
--username < username >	设置认证用户名
--verbose	打印详细的运行信息
--connection-param-file < filename >	可选，指定存储数据库连接参数的属性文件

10.3.4　Job 作业

10.3.4.1　语法

以下是创建 Sqoop 作业的语法：

```
sqoop job (generic-args) (job-args)
    [-- [subtool-name] (subtool-args)]
sqoop-job (generic-args) (job-args)
    [-- [subtool-name] (subtool-args)]
```

10.3.4.2　创建作业(--create)

这里创建一个名为 myjob 的作业，可以将 employee 表的数据导入到 HDFS 作业。下面的命令用于创建一个从 db 数据库的 employee 表导入到 HDFS 文件的作业。

(1) 启动 Hadoop：

```
cd /usr/hadoop/hadoop-2.7.1/
sbin/start-all.sh
```

(2) 创建一个根目录：

```
hadoop fs -mkdir /user
hadoop fs -mkdir /user/root
```

(3) 输入下面命令创建作业 myjob：
注：-- import，中间有空格

```
sqoop job -- create myjob \
> -- import \
> -- connect jdbc:mysql://localhost/db \
> -- username root -- password root \
```

```
>-- table employee – m 1
```

10.3.4.3 验证作业（--list）

--list 参数用来验证保存的作业。下面的命令用来验证保存 Sqoop 作业的列表。

```
sqoop job -- list
```

它显示了保存的作业列表如图 10.17 所示。

```
[root@master work]# sqoop job --list
16/08/16 17:00:22 INFO sqoop.Sqoop: Running Sqoop version: 1.4.6
Available jobs:
  myjob
```

图 10.17　验证作业

10.3.4.4 检查作业（--show）

--show 参数用于检查或验证特定的工作及其详细信息。以下命令和样本输出用来验证一个名为 myjob 的作业。

```
sqoop job -- show myjob
Enter password:root
```

图 10.18 显示了工具以及作业使用的选择参数，这是在 myjob 中的作业情况。

```
[root@master work]# sqoop job --show myjob
16/08/16 17:01:58 INFO sqoop.Sqoop: Running Sqoop version: 1.4.6
16/08/16 17:01:59 INFO Configuration.deprecation: mapred.used.genericoptionspars
er is deprecated. Instead, use mapreduce.client.genericoptionsparser.used
Enter password:
Job: myjob
Tool: import
Options:
-------------------------
verbose = false
db.connect.string = jdbc:mysql://localhost/db
codegen.output.delimiters.escape = 0
codegen.output.delimiters.enclose.required = false
codegen.input.delimiters.field = 0
hbase.create.table = false
db.require.password = true
hdfs.append.dir = false
db.table = employee
codegen.input.delimiters.escape = 0
import.fetch.size = null
accumulo.create.table = false
codegen.input.delimiters.enclose.required = false
db.username = root
reset.onemapper = false
codegen.output.delimiters.record = 10
import.max.inline.lob.size = 16777216
hbase.bulk.load.enabled = false
hcatalog.create.table = false
db.clear.staging.table = false
codegen.input.delimiters.record = 0
enable.compression = false
hive.overwrite.table = false
```

图 10.18　myjob 的作业情况

10.3.4.5 执行作业（--exec）

--exec 选项用于执行保存的作业。下面的命令用于执行保存的名称为 myjob 的作业。

```
sqoop job -- exec myjob
Enter password:root
```

它会显示如下输出：

```
16/08/16 17:35:45 INFO tool.CodeGenTool: Beginning code generation
...
```

成功执行完成后，输入下面命令查看 HDFS 中应存在的表 employee，如图 10.19 所示。

```
hadoop fs -ls
```

```
[root@master sbin]# hadoop fs -ls
16/08/17 10:27:29 WARN util.NativeCodeLoader: Unable to load native-hadoop library f
or your platform... using builtin-java classes where applicable
Found 1 items
drwxr-xr-x   - root supergroup          0 2016-08-17 10:22 employee
```

图 10.19　查看 HDFS 中应存在表 employee

输入下面的命令查看表 employee 的内容：

```
hadoop fs -ls /user/root/employee
```

结果如图 10.20 和图 10.21 所示。

```
[root@master sbin]# hadoop fs -ls /user/root/employee
16/08/17 10:29:21 WARN util.NativeCodeLoader: Unable to load native-hadoop library f
or your platform... using builtin-java classes where applicable
Found 2 items
-rw-r--r--   1 root supergroup          0 2016-08-17 10:22 /user/root/employee/_SUCC
ESS
-rw-r--r--   1 root supergroup        229 2016-08-17 10:21 /user/root/employee/part-
m-00000
```

图 10.20　查看表 employee 的内容 1

```
hadoop fs -cat /user/root/employee/part-m-00000
```

```
[root@master sbin]# hadoop fs -cat /user/root/employee/part-m-00000
16/08/17 10:32:29 WARN util.NativeCodeLoader: Unable to load native-hadoop library f
or your platform... using builtin-java classes where applicable
1201, gopal, manager, 50000, TP
1202, manisha, preader, 50000, TP
1203, khalil, php dev, 30000, AC
1204, prasanth, php dev, 30000, AC
1205, kranthi, admin, 20000, TP
1206, satish p, grp des, 20000, GR
1207, Raju, UI dev, 15000, TP
1207, Raju, UI dev, 15000, TP
```

图 10.21　查看表 employee 的内容 2

因为之前做过"插入查询计算"操作,所以数据中会出现 ID 为 1207 的记录。

10.4 数据操作

10.4.1 MySQL 数据导入到 HDFS 中

10.4.1.1 下载 MySQL 驱动

(1)到 MySQL 官网进入下载页面 http://dev.mysql.com/downloads/connector/j/,选择所需要的版本进行下载,这里下载的是 zip 格式的文件,然后在本地解压,如图 10.22 所示。

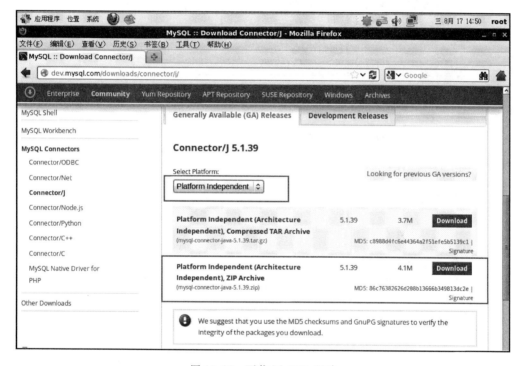

图 10.22 下载 MySQL 驱动

(2)找到 MySQL 驱动软件包 mysql-connector-java-5.1.39.zip 并解压,如图 10.23 所示。

```
cd /root/下载/
unzip mysql-connector-java-5.1.39.zip
cd mysql-connector-java-5.1.39
```

(3)将如图 10.24 所示 MySQL 驱动包,使用如下命令放到 Sqoop-1.4.6 的 lib 目录下:

```
cp mysql-connector-java-5.1.39-bin.jar /usr/sqoop/sqoop-1.4.6/lib
```

```
[root@master 下载]# ls
hadoop-2.7.1.tar.gz              mysql-connector-java-5.1.39.zip
jdk-8u60-linux-x64.tar.gz        sqoop-1.4.6.bin__hadoop-2.0.4-alpha.tar.gz
mysql-connector-java-5.1.39
[root@master 下载]# cd mysql-connector-java-5.1.39
[root@master mysql-connector-java-5.1.39]# ls
build.xml    COPYING    mysql-connector-java-5.1.39-bin.jar    README.txt
CHANGES      docs       README                                 src
```

图 10.23　解压 mysql-connector-java-5.1.39.zip

```
[root@master ~]# cd /usr/sqoop/sqoop-1.4.6/lib
[root@master lib]# ls
ant-contrib-1.0b3.jar              kite-data-mapreduce-1.0.0.jar
ant-eclipse-1.0-jvm1.2.jar         kite-hadoop-compatibility-1.0.0.jar
avro-1.7.5.jar                     mysql-connector-java-5.1.39-bin.jar
avro-mapred-1.7.5-hadoop2.jar      opencsv-2.3.jar
commons-codec-1.4.jar              paranamer-2.3.jar
commons-compress-1.4.1.jar         parquet-avro-1.4.1.jar
commons-io-1.4.jar                 parquet-column-1.4.1.jar
commons-jexl-2.1.1.jar             parquet-common-1.4.1.jar
commons-logging-1.1.1.jar          parquet-encoding-1.4.1.jar
hsqldb-1.8.0.10.jar                parquet-format-2.0.0.jar
jackson-annotations-2.3.0.jar      parquet-generator-1.4.1.jar
jackson-core-2.3.1.jar             parquet-hadoop-1.4.1.jar
jackson-core-asl-1.9.13.jar        parquet-jackson-1.4.1.jar
jackson-databind-2.3.1.jar         slf4j-api-1.6.1.jar
jackson-mapper-asl-1.9.13.jar      snappy-java-1.0.5.jar
kite-data-core-1.0.0.jar           xz-1.0.jar
kite-data-hive-1.0.0.jar
```

图 10.24　mysql-connector-java-5.1.39-bin.jar 包

10.4.1.2　启动 MySQL 服务

查看 MySQL 服务并查看状态，如图 10.25 所示表示已经启动。

```
[root@master lib]# service mysqld status
mysqld (pid  2659) 正在运行...
```

图 10.25　查看 MySQL 状态

如图 10.26 所示表示没有启动，此时需要启动服务。

```
[root@master ~]# service mysqld status
mysqld 已停
```

图 10.26　MySQL 没有启动图

```
service mysqld start
```

结果如图 10.27 所示。

```
[root@master ~]# service mysqld start
正在启动 mysqld：                                       [确定]
```

图 10.27　启动 MySQL

10.4.1.3　Sqoop 列出数据库

本节主要介绍如何使用 Sqoop 列出数据库。Sqoop 列表数据库工具解析并执行对数据库服务器的 SHOW DATABASES 查询。

以下语法用于 Sqoop 列表数据库命令：

```
sqoop list-databases (generic-args) (list-databases-args)
sqoop-list-databases (generic-args) (list-databases-args)
```

下面的命令用于列出 MySQL 数据库服务器的所有数据库：

```
sqoop list-databases \
> --connect jdbc:mysql://localhost/ \
> --username root --password root
```

如果命令成功执行，那么它会显示 MySQL 数据库服务器的数据库列表，如图 10.28 所示。

```
[root@master ~]# sqoop list-databases \
> --connect jdbc:mysql://localhost/ \
> --username root --password root
16/08/18 10:15:53 INFO sqoop.Sqoop: Running Sqoop version: 1.4.6
16/08/18 10:15:53 WARN tool.BaseSqoopTool: Setting your password on the command-
line is insecure. Consider using -P instead.
16/08/18 10:15:53 INFO manager.MySQLManager: Preparing to use a MySQL streaming
resultset.
information_schema
db
mysql
test
```

图 10.28　列出 MySQL 数据库服务器的所有数据库

10.4.1.4　Sqoop 列出所有表

本节介绍如何使用 Sqoop 列出 MySQL 数据库服务器的某个特定的数据库中的所有表。Sqoop 的 list-tables 工具解析并执行针对特定数据库的 SHOW TABLES 查询。

以下是使用 Sqoop 的 list-tables 命令的语法：

```
sqoop list-tables (generic-args) (list-tables-args)
sqoop-list-tables (generic-args) (list-tables-args)
```

10.4.1.5　示例查询

下面的命令用于列出 MySQL 数据库服务器的 db 数据库下的所有的表：

```
sqoop list-tables \
> --connect jdbc:mysql://localhost/db \
> --username root --password root
```

如果该指令执行成功，那么将显示 db 数据库中所有表，如图 10.29 所示。

10.4.1.6　查看 MySQL 中表的数据

进入 MySQL 数据库，选择有数据的一张表查看内容，比较导出结果是否正确，输入如下命令：

```
[root@master ~]# sqoop list-tables \
> --connect jdbc:mysql://localhost/db \
> --username root --password root
16/08/17 17:08:00 INFO sqoop.Sqoop: Running Sqoop version: 1.4.6
16/08/17 17:08:01 WARN tool.BaseSqoopTool: Setting your password on the command-
line is insecure. Consider using -P instead.
16/08/17 17:08:01 INFO manager.MySQLManager: Preparing to use a MySQL streaming
resultset.
employee
```

图 10.29　列出 MySQL 数据库服务器的 db 数据库下的所有的表

```
mysql - u root - p
Enter password:root
mysql> show databases;
```

结果如图 10.30 所示。

```
mysql> show databases;
+--------------------+
| Database           |
+--------------------+
| information_schema |
| db                 |
| mysql              |
| test               |
+--------------------+
4 rows in set (0.00 sec)
```

图 10.30　列出 MySQL 数据库服务器的所有数据库

执行如下命令，结果如图 10.31 所示。

```
mysql> use db;
```

```
mysql> use db;
Reading table information for completion of table and column names
You can turn off this feature to get a quicker startup with -A

Database changed
```

图 10.31　进入数据库 db

执行如下命令，结果如图 10.32 所示。

```
mysql> show tables;
```

```
mysql> show tables;
+--------------+
| Tables_in_db |
+--------------+
| employee     |
+--------------+
1 row in set (0.00 sec)
```

图 10.32　列出数据库 db 中的所有表

执行如下命令，结果如图 10.33 所示。

```
mysql > SELECT * FROM employee;
```

```
mysql> select *from employee;
+------+----------+-------------+--------+------+
| Id   | Name     | Designation | Salary | Dept |
+------+----------+-------------+--------+------+
| 1201 | gopal    | manager     | 30000  | TP   |
| 1202 | manisha  | preader     | 50000  | TP   |
| 1203 | khalil   | php dev     | 30000  | AC   |
| 1204 | prasanth | php dev     | 30000  | AC   |
| 1205 | kranthi  | admin       | 20000  | TP   |
| 1206 | satishp  | grp des     | 20000  | GR   |
| 1207 | Raju     | UI dev      | 15000  | TP   |
| 1207 | Raju     | UI dev      | 15000  | TP   |
+------+----------+-------------+--------+------+
8 rows in set (0.04 sec)
```

图 10.33 查看表 employee 的内容

10.4.1.7 把 MySQL 数据导入到 HDFS 中

使用如下命令列出 MySQL 中所有数据库：

```
sqoop list-databases --connect jdbc:mysql://localhost/db --username root --password root
```

结果如图 10.34 所示。

```
[root@master ~]# sqoop list-databases --connect jdbc:mysql://localhost/db --username root --password root
16/08/17 17:14:29 INFO sqoop.Sqoop: Running Sqoop version: 1.4.6
16/08/17 17:14:29 WARN tool.BaseSqoopTool: Setting your password on the command-line is insecure. Consider using -P instead.
16/08/17 17:14:29 INFO manager.MySQLManager: Preparing to use a MySQL streaming resultset.
information_schema
db
mysql
test
```

图 10.34 列出 MySQL 中所有数据库

要想把 MySQL 数据导入到 HDFS 中，首先需要在 MySQL 中创建一个数据表。
（1）在 shell 命令行下创建数据文件 empadd.txt，并输入如图 10.35 所示内容。

```
vim /root/empadd.txt
```

```
1201    288A    vgiri    jublee
1202    108L    aoc      sec-bad
1203    144Z    pgutta   hyd
1204    78B     old city          sec-bad
1205    720X    hitec    sec-bad
```

图 10.35 数据文件 empadd.txt 内容图

（2）进入 MySQL，创建表 empadd：

```
mysql -u root -p
Enter password:root
```

```
mysql> use db;
mysql> create table empadd(Id int,hno varchar(255),street varchar(255),city varchar(255));
mysql> load data local infile "/root/empadd.txt" into table empadd;
mysql> select * from empadd;
```

结果如图 10.36 所示。

图 10.36　创建表 empadd 并查看内容

执行以下命令：

```
mysql> exit
```

退出 MySQL 数据库。

（3）使用如下命令把 db 数据库的 empadd 表数据导入到 HDFS 中。

注：若此处没有启动 Hadoop，也没有创建根目录，那么先启动 Hadoop 并创建根目录，具体操作前面具体说过，此处不做详细介绍，若已经启动 Hadoop 并创建好根目录，那么就可以直接进行实验。

```
sqoop import -- connect jdbc:mysql://localhost/db -- username root -- password root -- table empadd -m 1
```

- --username 数据库用户名
- --password 连接数据库密码
- --table 表名
- -m 1 表示 map 数

结果如图 10.37 所示。

10.4.1.8　查看导入结果

使用如下命令查看导出数据到 HDFS 的结果，文件路径在当前用户的 Hadoop 目录下增加了 empadd 表目录：

```
hadoop fs -ls /user/root/empadd
```

结果如图 10.38 所示。
查看 part-m-00000 文件内容：

```
16/08/17 17:26:35 INFO mapreduce.Job:  map 0% reduce 0%
16/08/17 17:27:03 INFO mapreduce.Job:  map 100% reduce 0%
16/08/17 17:27:10 INFO mapreduce.Job: Job job_1471414656427_0001 completed succe
ssfully
16/08/17 17:27:14 INFO mapreduce.Job: Counters: 30
        File System Counters
                FILE: Number of bytes read=0
                FILE: Number of bytes written=132690
                FILE: Number of read operations=0
                FILE: Number of large read operations=0
                FILE: Number of write operations=0
                HDFS: Number of bytes read=87
                HDFS: Number of bytes written=116
                HDFS: Number of read operations=4
                HDFS: Number of large read operations=0
                HDFS: Number of write operations=2
        Job Counters
                Launched map tasks=1
                Other local map tasks=1
                Total time spent by all maps in occupied slots (ms)=28328
                Total time spent by all reduces in occupied slots (ms)=0
                Total time spent by all map tasks (ms)=28328
                Total vcore-seconds taken by all map tasks=28328
                Total megabyte-seconds taken by all map tasks=29007872
        Map-Reduce Framework
                Map input records=5
                Map output records=5
                Input split bytes=87
                Spilled Records=0
                Failed Shuffles=0
```

图 10.37　empadd 表导入到 HDFS 中

```
[root@master ~]# hadoop fs -ls /user/root/empadd
16/08/17 17:30:23 WARN util.NativeCodeLoader: Unable to load native-hadoop librar
y for your platform... using builtin-java classes where applicable
Found 2 items
-rw-r--r--   1 root supergroup          0 2016-08-17 17:27 /user/root/empadd/_SUC
CESS
-rw-r--r--   1 root supergroup        116 2016-08-17 17:26 /user/root/empadd/part
-m-00000
```

图 10.38　查看 employee 表目录

hadoop fs – cat /user/root/empadd/part – m – 00000

结果如图 10.39 所示。

```
[root@master ~]# hadoop fs -cat /user/root/empadd/part-m-00000
16/08/17 17:31:46 WARN util.NativeCodeLoader: Unable to load native-hadoop librar
y for your platform... using builtin-java classes where applicable
1201,288A,vgiri,jublee
1202,108L,aoc,sec-bad
1203,144Z,pgutta,hyd
1204,78B,old city,sec-bad
1205,720X,hitec,sec-bad
```

图 10.39　查看 part-m-00000 文件内容

10.4.1.9　Sqoop 导入所有表

本节介绍如何将数据库服务器的所有表导入到 HDFS。每个表的数据存储在一个单独的目录中，目录名与表名相同。

以下语法用于导入所有表：

sqoop import – all – tables (generic – args) (import – args)
sqoop - import – all – tables (generic – args) (import – args)

(1) 进入 MySQL，创建一个数据库和用于测试的表。

```
mysql -u root -p
Enter password:root
mysql> create database mydb;
mysql> use mydb;
mysql> create table mytable(TBL_ID int,CREATE_TIME long,DB_ID int,OWNER varchar(255),TBL_NAME varchar(255),TBL_TYPE varchar(255));
mysql> create table youtable(TBL_ID int,CREATE_TIME long,DB_ID int,OWNER varchar(255),TBL_NAME varchar(255),TBL_TYPE varchar(255));
mysql> insert into mytable value(6,1433862566,1,'shiyanlou','test','MANAGED_TABLE');
mysql> insert into youtable value(7,1433868888,2,'jiaoxuelou','test','MANAGED_TABLE');
mysql> SELECT * FROM mytable;
mysql> SELECT * FROM youtable;
```

结果如图 10.40 和图 10.41 所示。

```
mysql> SELECT * FROM mytable;
+--------+------------+-------+-----------+----------+---------------+
| TBL_ID | CREATE_TIME | DB_ID | ONER      | TBL_NAME | TBL_TYPE      |
+--------+------------+-------+-----------+----------+---------------+
|      6 | 1433862566 |     1 | shiyanlou | test     | MANAGED_TABLE |
+--------+------------+-------+-----------+----------+---------------+
1 row in set (0.00 sec)
```

图 10.40　创建表 mytable 并查看内容

```
mysql> select * from youtable;
+--------+------------+-------+------------+----------+---------------+
| TBL_ID | CREATE_TIME | DB_ID | OWNER      | TBL_NAME | TBL_TYPE      |
+--------+------------+-------+------------+----------+---------------+
|      7 | 1433868888 |     2 | jiaoxuelou | test     | MANAGED_TABLE |
+--------+------------+-------+------------+----------+---------------+
1 row in set (0.00 sec)
```

图 10.41　创建表 youtable 并查看内容

(2) 执行如下命令，可以看到数据库 mydb 包含表的列表如图 10.42 所示。

```
mysql> show tables;
```

```
mysql> show tables;
+----------------+
| Tables_in_mydb |
+----------------+
| mytable        |
| youtable       |
+----------------+
2 rows in set (0.00 sec)
```

图 10.42　数据库 mydb 包含表的列表图

(3) 退出 MySQL 数据库：

```
exit
```

(4) 下面的命令用于从 mydb 数据库中导入所有的表：

```
sqoop import-all-tables \
>--connect jdbc:mysql://localhost/mydb \
>--username root --password root -m 1
```

(5) 查看导入的所有表：

```
hadoop fs -ls
```

结果如图 10.43 所示。

图 10.43　查看 HDFS 导入的所有表

10.4.2　HDFS 数据导入到 MySql 中

本节介绍如何将数据从 HDFS 导出到 db 数据库。目标表必须存在于目标数据库中。以下是 export 命令语法：

```
sqoop export (generic-args) (export-args)
sqoop-export (generic-args) (export-args)
```

现在需要创建一个员工联系方式的数据文件 emp_contact，并将该数据文件导入 HDFS 中，最后再使用命令将 HDFS 目录/user/root/下的数据文件 emp_contact 导出到 MySQL 中。

(1) 执行如下命令，创建 emp_contact 数据文件，结果如图 10.44 所示（数据之间使用逗号分隔）。

```
vim emp_contact
```

```
1201,2356742,gopal@tp.com
1202,1661663,manisha@tp.com
1203,8887776,khalil@ac.com
1204,9988774,prasanth@ac.com
1205,1231231,kranthi@tp.com
```

图 10.44　创建 empcontact 数据文件

(2) 执行如下命令，将文件 emp_contact 导入 HDFS 中：

```
hadoop fs -put emp_contact emp_contact
hadoop fs -ls
```

结果如图 10.45 所示。

```
[root@master ~]# hadoop fs -ls
16/08/17 17:48:44 WARN util.NativeCodeLoader: Unable to load native-hadoop librar
y for your platform... using builtin-java classes where applicable
Found 4 items
-rw-r--r--   1 root supergroup        138 2016-08-17 17:48 emp_contact
drwxr-xr-x   - root supergroup          0 2016-08-17 17:27 empadd
drwxr-xr-x   - root supergroup          0 2016-08-17 17:39 mytable
drwxr-xr-x   - root supergroup          0 2016-08-17 17:40 youtable
```

图 10.45　文件 emp_contact 导入 HDFS

（3）它是强制性的，该表手动导出创建，并且存在于要导出的数据库中。下面的查询被用来创建 MySQL 命令行表 empcontact。

```
mysql – u root – p
Enter password:root
mysql > USE mydb;
mysql > CREATE TABLE empcontact (
    - > id INT,
    - > phno long,
    - > email VARCHAR(255));
Query OK,0 rows affected (0.01 sec)

mysql > exit
```

（4）下面的命令用来导出表数据（这是在 HDFS 中的 emp_contact 文件）到 MySQL 数据库服务器 mydb 数据库的 empcontact 表中。

```
sqoop export \
> -- connect jdbc:mysql://localhost/mydb \
> -- username root -- password root \
> -- table empcontact \
> -- export-dir /user/root/emp_contact
```

（5）下面的命令是用来验证表 mysql 的命令行。

```
mysql – u root – p
Enter password:root
mysql > USE mydb;
mysql > select * from empcontact;
```

（6）如果给定的数据存储成功，那么可以在如图 10.46 所示的 empcontact 表中找到数据。

```
mysql> select * from empcontact;
+------+---------+-----------------+
| id   | phno    | email           |
+------+---------+-----------------+
| 1201 | 2356742 | gopal@tp.com    |
| 1202 | 1661663 | manisha@tp.com  |
| 1205 | 1231231 | kranthi@tp.com  |
| 1204 | 9988774 | prasanth@ac.com |
| 1203 | 8887776 | khalil@ac.com   |
+------+---------+-----------------+
5 rows in set (0.01 sec)
```

图 10.46　查看 MySQL 中 empcontact 表的内容

实 战 篇

第 11 章 项目实战

11.1 项目背景与数据情况

11.1.1 项目概述

11.1.1.1 数据来源

项目的数据日志来源于国内某技术学习论坛,构建项目的目的在于通过对该技术论坛的日志进行分析,计算该论坛的一些关键指标,以供进行决策时参考。

项目的数据日志如图 11.1 所示。

```
27.19.74.143 - - [30/May/2013:17:38:20 +0800] "GET /static/image/common/faq.gif HTTP/1.1" 200 1127
110.52.250.126 - - [30/May/2013:17:38:20 +0800] "GET /data/cache/style_1_widthauto.css?y7a HTTP/1.1" 200 1292
27.19.74.143 - - [30/May/2013:17:38:20 +0800] "GET /static/image/common/hot_1.gif HTTP/1.1" 200 680
27.19.74.143 - - [30/May/2013:17:38:20 +0800] "GET /static/image/common/hot_2.gif HTTP/1.1" 200 682
27.19.74.143 - - [30/May/2013:17:38:20 +0800] "GET /static/image/filetype/common.gif HTTP/1.1" 200 90
110.52.250.126 - - [30/May/2013:17:38:20 +0800] "GET /source/plugin/wsh_wx/img/wsh_zk.css HTTP/1.1" 200 1482
110.52.250.126 - - [30/May/2013:17:38:20 +0800] "GET /data/cache/style_1_forum_index.css?y7a HTTP/1.1" 200 2331
110.52.250.126 - - [30/May/2013:17:38:20 +0800] "GET /source/plugin/wsh_wx/img/wx_jqr.gif HTTP/1.1" 200 1770
27.19.74.143 - - [30/May/2013:17:38:20 +0800] "GET /static/image/common/recommend_1.gif HTTP/1.1" 200 1030
110.52.250.126 - - [30/May/2013:17:38:20 +0800] "GET /static/image/common/logo.png HTTP/1.1" 200 4542
27.19.74.143 - - [30/May/2013:17:38:20 +0800] "GET /data/attachment/common/c8/common_2_verify_icon.png HTTP/1.1" 200 582
110.52.250.126 - - [30/May/2013:17:38:20 +0800] "GET /static/js/logging.js?y7a HTTP/1.1" 200 603
8.35.201.144 - - [30/May/2013:17:38:20 +0800] "GET /uc_server/avatar.php?uid=29331&size=middle HTTP/1.1" 301 -
110.52.250.126 - - [30/May/2013:17:38:20 +0800] "GET /data/cache/common_smilies_var.js?y7a HTTP/1.1" 200 3184
27.19.74.143 - - [30/May/2013:17:38:20 +0800] "GET /static/image/common/pn.png HTTP/1.1" 200 592
```

图 11.1 项目数据日志

11.1.1.2 需要用到的技术

(1) Linux Shell 编程。
(2) HDFS、MapReduce。

(3) Hive、Sqoop 框架。

11.1.2 项目分析指标

本项目旨在对以下四个项目指标进行分析。

11.1.2.1 浏览量 PV

定义：页面浏览量即为 PV(Page View)，是指所有用户浏览页面的总和，一个独立用户每打开一个页面就被记录一次。

分析：网站总浏览量，可以考核用户对于网站的兴趣，就像收视率对于电视剧一样。但是对于网站运营者来说，更重要的是每个栏目下的浏览量。

计算公式：记录计数，从日志中获取访问次数，又可以细分为各个栏目下的访问次数。

11.1.2.2 注册用户数

分析：该论坛的用户注册页面为 member.php，而当用户点击注册请求时，将会访问 member.php？mod＝register 的 url，如下所示：

```
GET /member.php?mod = register HTTP/1.1
```

计算公式：对访问 member.php？mod＝register 的 url 进行计数。

11.1.2.3 网站访问 IP 数

定义：一天之内访问网站的不同独立 IP 个数进行求和。其中同一 IP 无论访问了几个页面，独立 IP 数均为 1。

分析：这是我们最熟悉的一个概念，无论同一个 IP 上有多少计算机，或者其他用户，从某种程度上来说，独立 IP 的多少，是衡量网站推广活动好坏最直接的数据。

计算公式：对不同的访问者 IP 进行计数。

11.1.2.4 跳出率

定义：只浏览了一个页面便离开了网站的访问次数占总的访问次数的百分比，即只浏览了一个页面的访问次数/全部的访问次数汇总。

分析：跳出率是非常重要的访客黏性指标，它显示了访客对网站的兴趣程度。跳出率越低说明流量质量越好，访客对网站的内容越感兴趣，这些访客越可能是网站的有效用户、忠实用户。

该指标也可以衡量网络营销的效果，指出有多少访客被网络营销吸引到宣传产品页或网站上之后，又流失掉了。比如，网站在某媒体上打广告推广，分析从这个推广来源进入的访客指标，其跳出率可以反映出选择这个媒体是否合适，广告语的撰写是否优秀，以及网站入口页的设计是否用户体验良好。

计算公式：统计一天内只出现一条记录的 IP，称为跳出数。

跳出率＝跳出数/PV。

11.1.3 项目开发步骤

11.1.3.1 上传日志文件至 HDFS

把日志数据上传到 HDFS 中进行处理,可以分为以下几种情况:

(1) 如果是日志服务器数据较小、压力较小,可以直接使用 shell 命令把数据上传到 HDFS 中;

(2) 如果是日志服务器数据较大、压力较大,则使用 NFS 在另一台服务器上上传数据;

(3) 如果日志服务器非常多、数据量大,则使用 flume 进行数据处理。

本项目中使用第一种方式进行操作,即直接使用 shell 命令把数据上传到 HDFS 中。

11.1.3.2 数据清洗

由于原始数据中会有一些残缺的无效数据,因此在进行数据统计之前需要对数据进行筛选,剔除原始数据集中的无效数据。

使用 MapReduce 对 HDFS 中的原始数据进行清洗,以便后续进行统计分析。

11.1.3.3 数据统计分析

使用 Hive 对清洗后的数据进行统计分析。

11.1.3.4 将统计分析结果导入 MySQL

使用 Sqoop 把 Hive 产生的统计结果导出到 MySQL 中。

11.1.4 表结构设计

我们使用 MySQL 存储关键指标的统计分析结果,MySQL 表结构设计如表 11.1 所示。

表 11.1 MySQL 数据表结构

汇 总 表	
日期	acc_data
浏览量	pv
新用户注册数	newer
独立 IP	iip
跳出数	jumper

11.2 环境搭建

11.2.1 MySQL 的安装

1. 添加 YUM 源

检测系统:

```
yum list installed | grep mysql
yum -y remove mysql-libs.x86_64
```

去官网下载对应操作系统版本的 rpm 文件：

```
wget http://dev.mysql.com/get/mysql-community-release-el6-5.noarch.rpm
```

安装 mysql-community-release-el6-5.noarch.rpm：

```
yum localinstall mysql-community-release-el6-5.noarch.rpm
yum repolist all | grep mysql
yum-config-manager --disable mysql55-community
yum-config-manager --disable mysql56-community
yum-config-manager --enable mysql57-community-dmr
yum repolist enabled | grep mysql
```

查看 yum 源列表：

```
yum repolist
```

或者进入 /etc/yum.repos.d/ 目录下面查看：

```
cd /etc/yum.repos.d/
```

发现多了文件 mysql-community.repo 和 mysql-community-source.repo。

查看 mysql-community.repo 文件的内容，如图 11.2 所示。

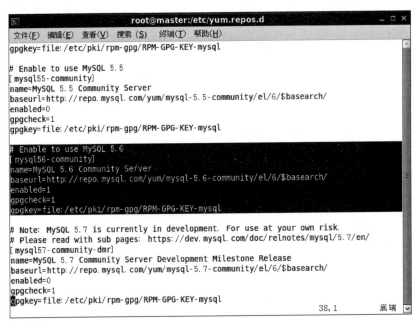

图 11.2　mysql-community.repo 文件内容

2. 安装 MySQL

安装 MySQL 执行命令如下：

```
yum install mysql-community-server
```

3. 修改初始密码

停止 MySQL 服务：

```
/etc/init.d/mysqld stop
```

配置 mysql 服务的配置文件：

```
vim /etc/my.cnf
```

在文档中加入下面的语句，再登录 MySQL 时就不需要密码了：

```
skip-grant-tables
```

修改之前如图 11.3 所示。

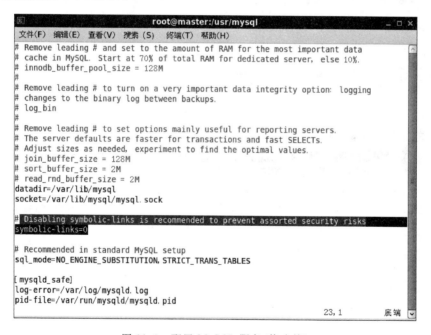

图 11.3　配置 MySQL 服务（修改前）

修改之后如图 11.4 所示。

启动 MySQL 服务：

```
/etc/init.d/mysqld start
```

登录进入 MySQL：

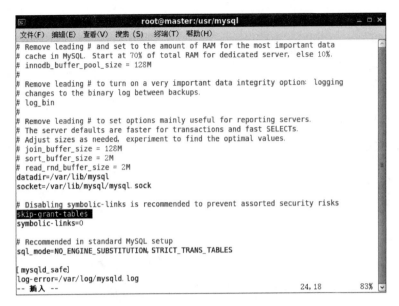

图 11.4 配置 MySQL 服务（修改后）

```
mysql
```

接下来修改用户密码为 root，执行命令如下：

```
use mysql;
update user set password = password('root') where user = 'root';
```

将/etc/my.cnf 中 skip-grant-tables 注释掉，恢复配置文件的原样。
修改之后如图 11.5 所示。

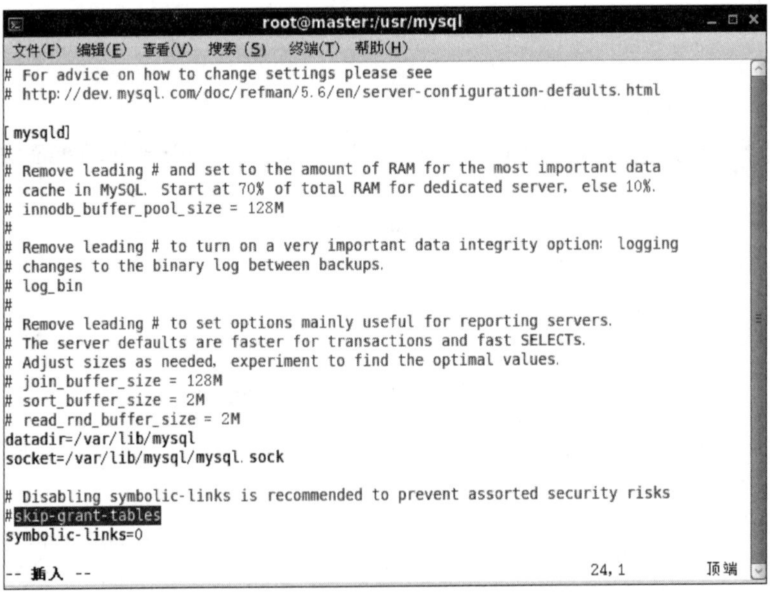

图 11.5 恢复配置文件

4. 检验 MySQL 是否能启动

重新启动 MySQL：

```
/etc/init.d/mysqld restart
```

登录 MySQL：

```
mysql -u root -p
```

注：输入密码不可见。

若出现如图 11.6 所示界面，则 MySQL 服务安装成功。

```
[root@master mysql]# mysql -u root -p
Enter password:
Welcome to the MySQL monitor.  Commands end with ; or \g.
Your MySQL connection id is 2
Server version: 5.6.26 MySQL Community Server (GPL)

Copyright (c) 2000, 2015, Oracle and/or its affiliates. All rights reserved.

Oracle is a registered trademark of Oracle Corporation and/or its
affiliates. Other names may be trademarks of their respective
owners.

Type 'help;' or '\h' for help. Type '\c' to clear the current input statement.

mysql>
```

图 11.6　验证 MySQL 服务安装成功

11.2.2　Eclipse 的安装

应下载安装包，将安装包解压缩，假设解压到了 /usr/eclipse 目录下。
使符号链接目录，执行命令如下：

```
ln -s /usr/eclipse/eclipse /usr/bin/eclipse
```

为了方便使用，可以创建桌面的快捷方式，步骤如下：
（1）在桌面右击，选择"创建启动器"命令，如图 11.7 所示。
（2）进入项目页面，如图 11.8 所示。

图 11.7　创建启动器

图 11.8　进入项目页面

单击图 11.8 中左上角的图标,然后找到 Eclipse 的安装路径,如图 11.9 所示。

图 11.9　找到 Eclipse 图标

在"名称"文本框填写快捷名称,单击"浏览"按钮,如图 11.10 所示。

图 11.10　单击"浏览"按钮

找到 Eclipse 启动图标,并单击,如图 11.11 所示。

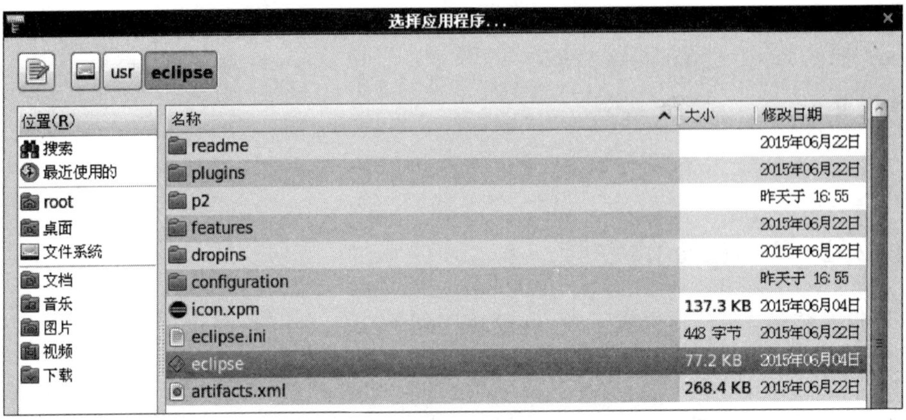

图 11.11　找到 Eclipse 启动图标

如图 11.12 所示,已经完成所有配置。

图 11.12　完成配置

如图 11.13 所示,桌面已经生成 Eclipse 的快捷图标。
在桌面上直接单击图标,即可启动 Eclipse。

图 11.13　显示快捷图标

11.3　数据清洗

11.3.1　数据分析

11.3.1.1　数据情况回顾

图 11.14 展示了该日志数据的记录格式，其中每行记录有 5 个部分：访问者 IP、访问时间、访问资源、访问状态（HTTP 状态码）、本次访问流量。

图 11.14　项目数据日志

11.3.1.2　要清理的数据

（1）根据 11.1.2 节的关键指标的分析，我们所要统计分析的数据均不涉及访问状态（HTTP 状态码）以及本次访问的流量，于是首先可以将这两项记录清理掉；

（2）根据日志记录的数据格式，需要将日期格式转换为平常所见的普通格式，如

20160325，于是可以写一个类将日志记录的日期进行转换；

(3) 由于静态资源的访问请求对我们的数据分析没有意义，于是可以将“GET /staticsource/”开头的访问记录过滤掉，又因为 GET 和 POST 字符串也没有意义，因此也可以将其省略掉。

11.3.2 数据清洗流程

在进行数据操作之前，首先需要启动 Hadoop 集群，执行如下命令，结果如图 11.15 所示。

```
strat-all.sh
```

```
[root@master ~]# start-all.sh
This script is Deprecated. Instead use start-dfs.sh and start-yarn.sh
16/12/13 15:00:52 WARN util.NativeCodeLoader: Unable to load native-hadoop libra
ry for your platform... using builtin-java classes where applicable
Starting namenodes on [master]
master: starting namenode, logging to /usr/hadoop/hadoop-2.7.1/logs/hadoop-root-
namenode-master.out
master: starting datanode, logging to /usr/hadoop/hadoop-2.7.1/logs/hadoop-root-
datanode-master.out
Starting secondary namenodes [master]
master: starting secondarynamenode, logging to /usr/hadoop/hadoop-2.7.1/logs/had
oop-root-secondarynamenode-master.out
16/12/13 15:01:15 WARN util.NativeCodeLoader: Unable to load native-hadoop libra
ry for your platform... using builtin-java classes where applicable
starting yarn daemons
starting resourcemanager, logging to /usr/hadoop/hadoop-2.7.1/logs/yarn-root-res
ourcemanager-master.out
master: starting nodemanager, logging to /usr/hadoop/hadoop-2.7.1/logs/yarn-root
-nodemanager-master.out
[root@master ~]#
```

图 11.15 启动集群

需要确保如图 11.16 所示进程全部启动。

```
[root@master ~]# jps
3040 NodeManager
2771 SecondaryNameNode
2582 DataNode
2935 ResourceManager
3354 Jps
2043 VmServer.jar
2479 NameNode
[root@master ~]#
```

图 11.16 查看进程

11.3.2.1 上传日志到 HDFS 文件系统

可以直接使用 shell 命令把数据上传到 HDFS 中，在“论坛访问日志分析_模板 1”中，已经将相应的日志数据文件 access.log 放到了本地磁盘的/usr 目录下。

在 HDFS 文件系统上创建/usr/project 目录：

```
hadoop fs -mkdir /user/project
```

在 HDFS 文件系统上创建 /usr/project/input 目录：

```
hadoop fs -mkdir /user/project/input
```

通过以下命令将 access.log 日志数据上传到 HDFS 文件系统的 /usr/project/input 目录下：

```
hadoop fs -put /usr/access.log /user/project/input
```

通过以下命令查看 HDFS 中的数据是否已经正确上传：

```
hadoop fs -ls /user/project/input
```

结果如图 11.17 所示。

图 11.17　查看文件

11.3.2.2　创建项目

双击桌面上的 Eclipse 图标打开 Eclipse。

（1）单击左上角的 File→New 命令，新建一个 Java Project，命名为 MyProject，如图 11.18～图 11.22 所示。

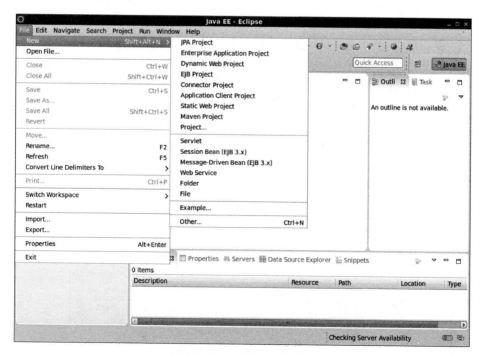

图 11.18　创建项目（一）

Hadoop 核心技术与实战

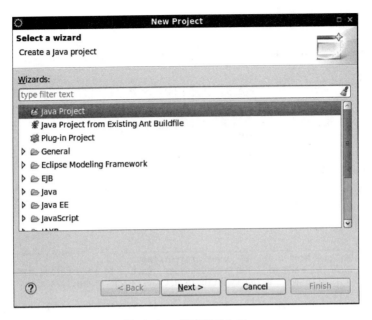

图 11.19　创建项目（二）

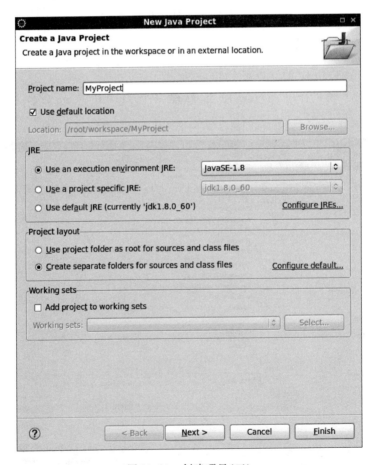

图 11.20　创建项目（三）

第11章 项目实战

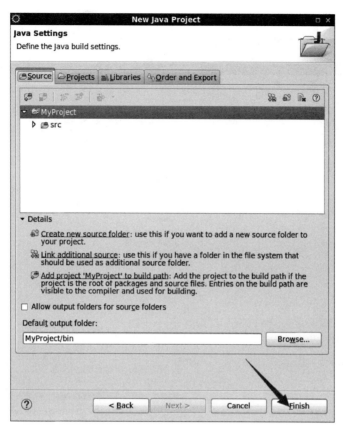

图 11.21 创建项目(四)

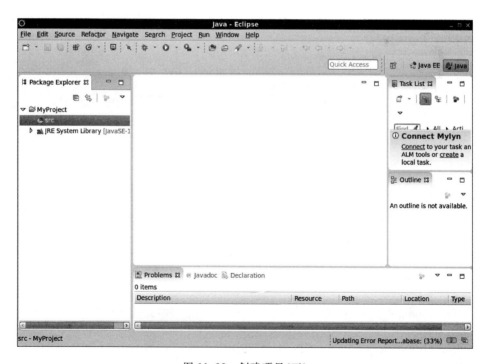

图 11.22 创建项目(五)

（2）在项目的 src 上右击，选择 New→Package 命令，新建一个 package，命名为 myPackage，如图 11.23 和图 11.24 所示。

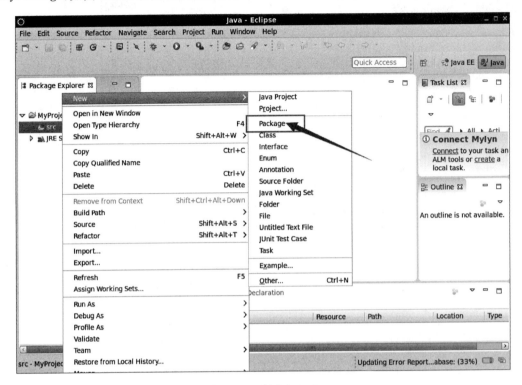

图 11.23　创建包（一）

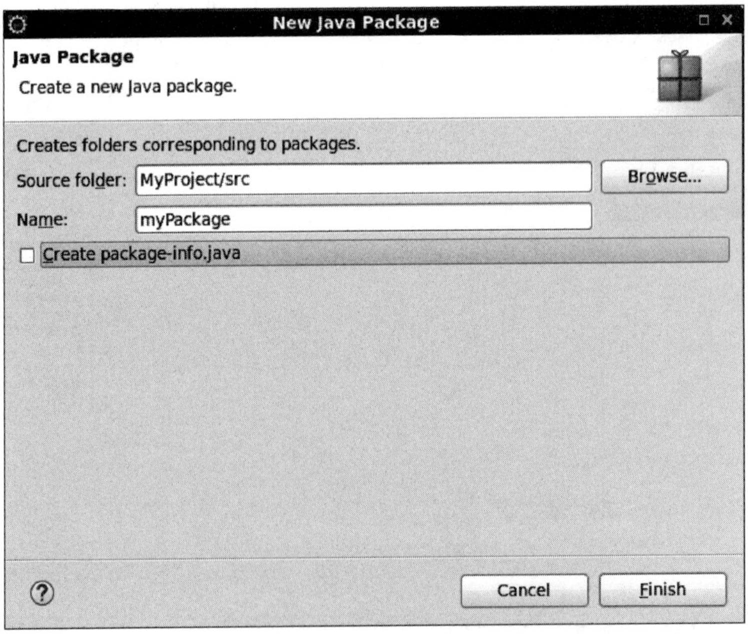

图 11.24　创建包（二）

(3) 在新建的包 myPackage 下添加类文件，命名为 DataClean，如图 11.25 和图 11.26 所示。

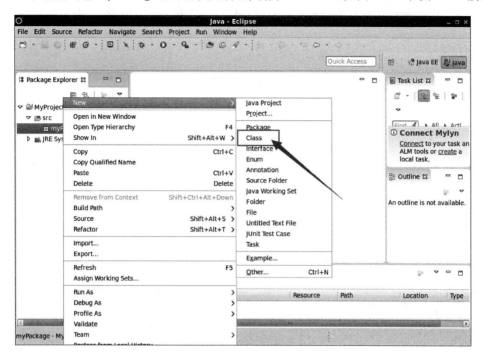

图 11.25　创建类（一）

图 11.26　创建类（二）

（4）导入项目所需要的 Hadoop jar 包。

在模板中 Hadoop 的安装目录为/usr/hadoop/hadoop-2.7.1/，因此需要导入的 jar 包包括：

/usr/hadoop/hadoop-2.7.1/share/hadoop/common/ * .jar

/usr/hadoop/hadoop-2.7.1/share/hadoop/common/lib/ * .jar

/usr/hadoop/hadoop-2.7.1/share/hadoop/hdfs/ * .jar

/usr/hadoop/hadoop-2.7.1/share/hadoop/hdfs/lib/ * .jar

/usr/hadoop/hadoop-2.7.1/share/hadoop/mapreduce/ * .jar

/usr/hadoop/hadoop-2.7.1/share/hadoop/mapreduce/lib/ * .jar

/usr/hadoop/hadoop-2.7.1/share/hadoop/tools/ * .jar

/usr/hadoop/hadoop-2.7.1/share/hadoop/yarn/ * .jar

/usr/hadoop/hadoop-2.7.1/share/hadoop/yarn/lib/ * .jar

以上路径下的 jar 包必须被导入，如果没有正确导入，程序将无法正常运行。

jar 包导入方式如图 11.27～图 11.29 所示。

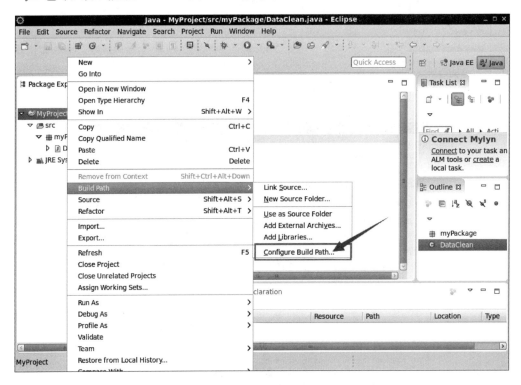

图 11.27　导入依赖文件（一）

至此，项目搭建完毕，接下来进行程序代码的编写。

11.3.2.3　编写 MapReduce 程序清理日志

（1）编写日志解析类对每行记录的五个组成部分进行单独的解析。

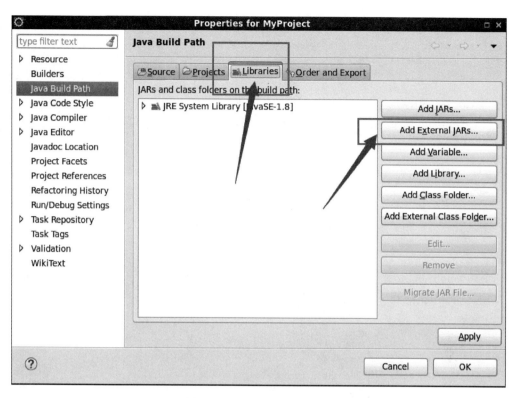

图 11.28　导入依赖文件（二）

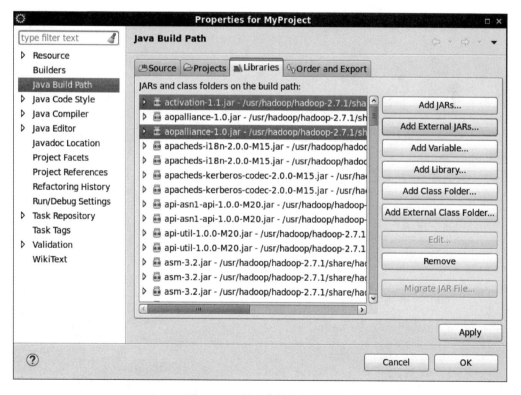

图 11.29　导入依赖文件（三）

```java
static class LogParser {
    public static final SimpleDateFormat FORMAT = new SimpleDateFormat(
            "d/MMM/yyyy:HH:mm:ss", Locale.ENGLISH);
    public static final SimpleDateFormat dateformat1 = new SimpleDateFormat(
            "yyyyMMddHHmmss");
    public static void main(String[] args) throws ParseException {
        final String S1 = "27.19.74.143 - - [30/May/2013:17:38:20 +0800] \"GET /static/image/common/faq.gif HTTP/1.1\" 200 1127";
        LogParser parser = new LogParser();
        final String[] array = parser.parse(S1);
        System.out.println("样例数据: " + S1);
        System.out.format(
                "解析结果: ip=%s, time=%s, url=%s, status=%s, traffic=%s",
                array[0], array[1], array[2], array[3], array[4]);
    }

    /**
     * 解析英文时间字符串
     *
     * @param string
     * @return
     * @throws ParseException
     */
    private Date parseDateFormat(String string) {
        Date parse = null;
        try {
            parse = FORMAT.parse(string);
        }
        catch (ParseException e) {
            e.printStackTrace();
        }
        return parse;
    }

    /**
     * 解析日志的行记录
     *
     * @param line
     * @return 数组含有五个元素,分别是ip、时间、url、状态、流量
     */
    public String[] parse(String line) {
        String ip = parseIP(line);
        String time = parseTime(line);
        String url = parseURL(line);
        String status = parseStatus(line);
        String traffic = parseTraffic(line);
        return new String[] { ip, time, url, status, traffic };
    }
    private String parseTraffic(String line) {
        final String trim = line.substring(line.lastIndexOf("\"") + 1)
                .trim();
        String traffic = trim.split(" ")[1];
        return traffic;
    }
```

```java
    private String parseStatus(String line) {
        final String trim = line.substring(line.lastIndexOf("\"") + 1)
                .trim();
        String status = trim.split(" ")[0];
        return status;
    }
    private String parseURL(String line) {
        final int first = line.indexOf("\"");
        final int last = line.lastIndexOf("\"");
        String url = line.substring(first + 1, last);
        return url;
    }
    private String parseTime(String line) {
        final int first = line.indexOf("[");
        final int last = line.indexOf(" +0800]");
        String time = line.substring(first + 1, last).trim();
        Date date = parseDateFormat(time);
        return dateformat1.format(date);
    }
    private String parseIP(String line) {
        String ip = line.split(" - - ")[0].trim();
        return ip;
    }
}
```

(2) 编写 MapReduce 程序对指定日志文件的所有记录进行过滤。

Mapper 类：

```java
static class MyMapper extends Mapper<LongWritable, Text, LongWritable, Text> {
    LogParser logParser = new LogParser();
    Text outputValue = new Text();

        protected void map(LongWritable key, Text value, Context context) throws java.io.IOException,
InterruptedException {
            final String[] parsed = logParser.parse(value.toString());
            // step1.过滤掉静态资源访问请求
             if (parsed[2].startsWith("GET /static/")|| parsed[2].startsWith("GET /uc_
server")) {
                return;
            }
            // step2.过滤掉开头的指定字符串
            if (parsed[2].startsWith("GET /")) {
                parsed[2] = parsed[2].substring("GET /".length());
            } else if (parsed[2].startsWith("POST /")) {
                parsed[2] = parsed[2].substring("POST /".length());
             }
            // step3.过滤掉结尾的特定字符串
            if (parsed[2].endsWith(" HTTP/1.1")) {
                parsed[2] = parsed[2].substring(0, parsed[2].length() - " HTTP/1.1".
length());
            }
            // step4.只写入前三个记录类型项
```

```
            outputValue.set(parsed[0] + "\t" + parsed[1] + "\t" + parsed[2]);
            context.write(key, outputValue);
        }
    }
```

Reducer 类：

```
static class MyReducer extends Reducer<LongWritable, Text, Text, NullWritable>
{
    protected void reduce(LongWritable k2, Iterable<Text> v2s, Context context) throws java.io.IOException, InterruptedException {
        for (Text v2 : v2s) {
            context.write(v2, NullWritable.get());
        }
    };
}
```

（3）DataClean.java 的完整示例代码如下：

```
package myPackage;
import java.text.ParseException;
import java.text.SimpleDateFormat;
import java.util.Date;
import java.util.Locale;

import org.apache.hadoop.conf.Configuration;
import org.apache.hadoop.conf.Configured;
import org.apache.hadoop.fs.FileSystem;
import org.apache.hadoop.fs.Path;
import org.apache.hadoop.io.LongWritable;
import org.apache.hadoop.io.NullWritable;
import org.apache.hadoop.io.Text;
import org.apache.hadoop.mapreduce.Job;
import org.apache.hadoop.mapreduce.Mapper;
import org.apache.hadoop.mapreduce.Reducer;
import org.apache.hadoop.mapreduce.lib.input.FileInputFormat;
import org.apache.hadoop.mapreduce.lib.output.FileOutputFormat;
import org.apache.hadoop.util.Tool;
import org.apache.hadoop.util.ToolRunner;

public class DataClean extends Configured implements Tool {

    public static void main(String[] args) {
        Configuration conf = new Configuration();
        try {
            int res = ToolRunner.run(conf, new DataClean(), args);
            System.exit(res);
        } catch (Exception e) {
            e.printStackTrace();
        }
```

```java
    }

    @Override
    public int run(String[] args) throws Exception {
        Configuration conf = new Configuration();
        conf.set("fs.defaultFS", "hdfs://master:9000");

        Job job = new Job(conf);
        // 设置为可以打包运行
        job.setJarByClass(DataClean.class);
        job.setMapperClass(MyMapper.class);
        job.setMapOutputKeyClass(LongWritable.class);
        job.setMapOutputValueClass(Text.class);
        job.setReducerClass(MyReducer.class);
        job.setOutputKeyClass(Text.class);
        job.setOutputValueClass(NullWritable.class);
        FileInputFormat.setInputPaths(job, args[0]);

        // 清理已存在的输出文件
        FileSystem fs = FileSystem.get(conf);
        Path outPath = new Path(args[1]);
        if (fs.exists(outPath)) {
            fs.delete(outPath, true);
        }
        FileOutputFormat.setOutputPath(job, new Path(args[1]));
        boolean success = job.waitForCompletion(true);
        if(success){
            System.out.println("Clean process success!");
        }
        else{
            System.out.println("Clean process failed!");
        }
        return 0;
    }

    static class MyMapper extends Mapper<LongWritable, Text, LongWritable, Text> {
        LogParser logParser = new LogParser();
        Text outputValue = new Text();

        protected void map(LongWritable key, Text value, Context context) throws java.io.IOException, InterruptedException {
            final String[] parsed = logParser.parse(value.toString());
            // step1.过滤掉静态资源访问请求
            if      (parsed[2].startsWith("GET /static/")|| parsed[2].startsWith("GET /uc_server")) {
                return;
            }
            // step2.过滤掉开头的指定字符串
            if (parsed[2].startsWith("GET /")) {
                parsed[2] = parsed[2].substring("GET /".length());
            } else if (parsed[2].startsWith("POST /")) {
                parsed[2] = parsed[2].substring("POST /".length());
            }
```

```java
            // step3.过滤掉结尾的特定字符串
            if (parsed[2].endsWith(" HTTP/1.1")) {
                parsed[2] = parsed[2].substring(0, parsed[2].length() - " HTTP/1.1".length());
            }
            // step4.只写入前三个记录类型项
            outputValue.set(parsed[0] + "\t" + parsed[1] + "\t" + parsed[2]);
            context.write(key, outputValue);
        }
    }

    static class MyReducer extends Reducer<LongWritable, Text, Text, NullWritable> {
        protected void reduce(LongWritable k2, Iterable<Text> v2s, Context context) throws java.io.IOException, InterruptedException {
            for (Text v2 : v2s) {
                context.write(v2, NullWritable.get());
            }
        };
    }

    /*
     * 日志解析类
     */
    static class LogParser {
        public static final SimpleDateFormat FORMAT = new SimpleDateFormat(
                "d/MMM/yyyy:HH:mm:ss", Locale.ENGLISH);
        public static final SimpleDateFormat dateformat1 = new SimpleDateFormat(
                "yyyyMMddHHmmss");

        public static void main(String[] args) throws ParseException {
            final String S1 = "27.19.74.143 - - [30/May/2013:17:38:20 +0800] \"GET /static/image/common/faq.gif HTTP/1.1\" 200 1127";
            LogParser parser = new LogParser();
            final String[] array = parser.parse(S1);
            System.out.println("样例数据: " + S1);
            System.out.format(
                    "解析结果: ip=%s, time=%s, url=%s, status=%s, traffic=%s",
                    array[0], array[1], array[2], array[3], array[4]);
        }

        /**
         * 解析英文时间字符串
         *
         * @param string
         * @return
         * @throws ParseException
         */
        private Date parseDateFormat(String string) {
            Date parse = null;
            try {
                parse = FORMAT.parse(string);
            }
            catch (ParseException e) {
```

```java
            e.printStackTrace();
        }
        return parse;
    }

    /**
     * 解析日志的行记录
     *
     * @param line
     * @return 数组含有五个元素,分别是 ip、时间、url、状态、流量
     */
    public String[] parse(String line) {
        String ip = parseIP(line);
        String time = parseTime(line);
        String url = parseURL(line);
        String status = parseStatus(line);
        String traffic = parseTraffic(line);
        return new String[] { ip, time, url, status, traffic };
    }

    private String parseTraffic(String line) {
        final String trim = line.substring(line.lastIndexOf("\"") + 1)
                .trim();
        String traffic = trim.split(" ")[1];
        return traffic;
    }

    private String parseStatus(String line) {
        final String trim = line.substring(line.lastIndexOf("\"") + 1)
                .trim();
        String status = trim.split(" ")[0];
        return status;
    }

    private String parseURL(String line) {
        final int first = line.indexOf("\"");
        final int last = line.lastIndexOf("\"");
        String url = line.substring(first + 1, last);
        return url;
    }

    private String parseTime(String line) {
        final int first = line.indexOf("[");
        final int last = line.indexOf(" +0800]");
        String time = line.substring(first + 1, last).trim();
        Date date = parseDateFormat(time);
        return dateformat1.format(date);
    }

    private String parseIP(String line) {
        String ip = line.split("- -")[0].trim();
        return ip;
    }
}
```

(4) 将项目导出为 jar 包,导出到/root 目录下,命名为 dataClean.jar,具体过程如图 11.30~图 11.38 所示。

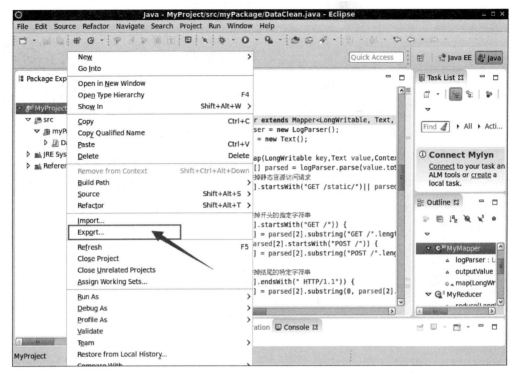

图 11.30 导出 jar 包(一)

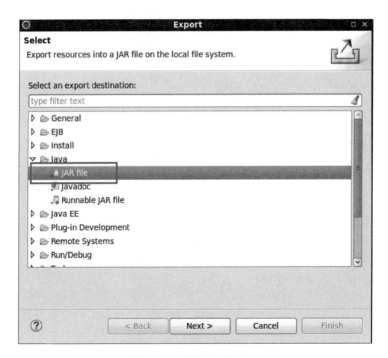

图 11.31 导出 jar 包(二)

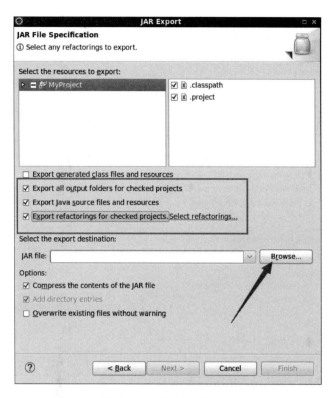

图 11.32　导出 jar 包（三）

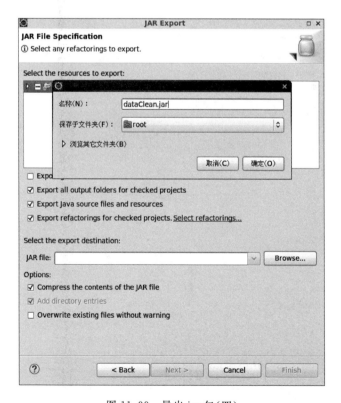

图 11.33　导出 jar 包（四）

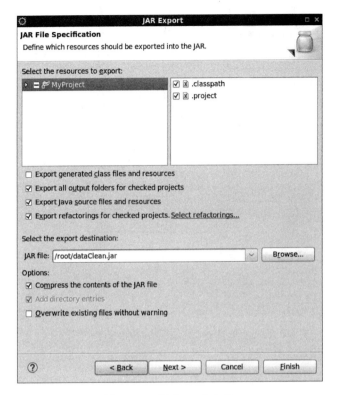

图 11.34　导出 jar 包（五）

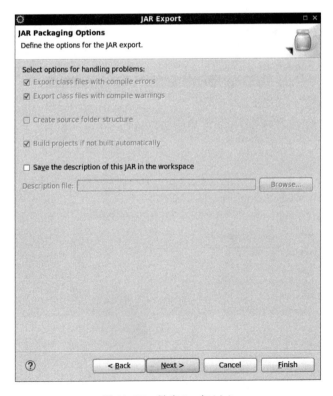

图 11.35　导出 jar 包（六）

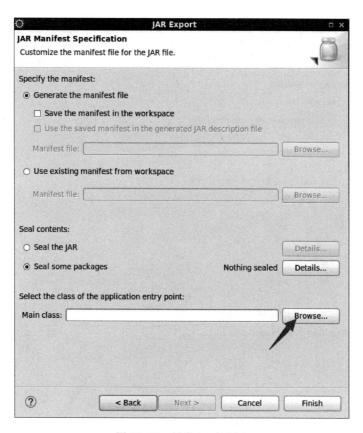

图 11.36　导出 jar 包(七)

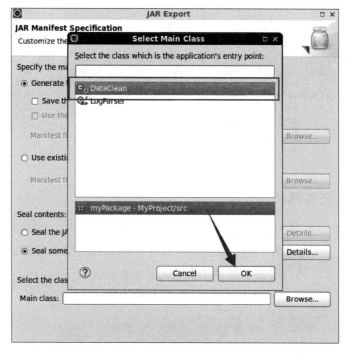

图 11.37　导出 jar 包(八)

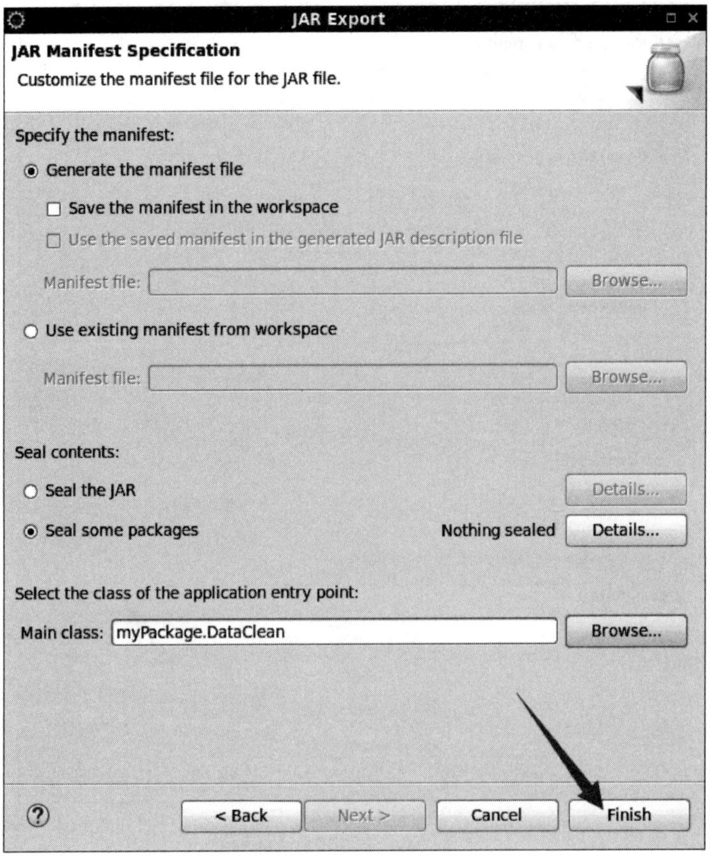

图 11.38　导出 jar 包（九）

导出完成后在 root 目录下将会看到导出的 dataClean.jar 文件，如图 11.39 所示。

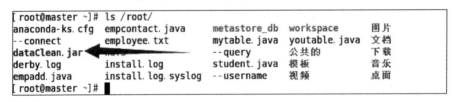

图 11.39　查看文件

11.3.2.4　任务执行

通过以下命令在集群上执行我们编写的 MapReduce 任务进行数据清洗（程序的执行需要指定两个参数，分别是集群上输入数据文件的存放地址和清洗后的数据在集群上的输出地址，输入地址设置为我们之前上传的数据地址 /user/project/input/access.log，数据输出地址设置为 /user/project/output），命令如下，执行结果如图 11.40 所示。

```
hadoop jar /root/dataClean.jar /user/project/input /user/project/output
```

通过以下命令可以验证数据清洗后的结果：

```
                Reduce input records=502404
                Reduce output records=502404
                Spilled Records=1004808
                Shuffled Maps =2
                Failed Shuffles=0
                Merged Map outputs=2
                GC time elapsed (ms)=3210
                CPU time spent (ms)=52150
                Physical memory (bytes) snapshot=577241088
                Virtual memory (bytes) snapshot=6170796032
                Total committed heap usage (bytes)=378470400
        Shuffle Errors
                BAD_ID=0
                CONNECTION=0
                IO_ERROR=0
                WRONG_LENGTH=0
                WRONG_MAP=0
                WRONG_REDUCE=0
        File Input Format Counters
                Bytes Read=157200725
        File Output Format Counters
                Bytes Written=38015669
Clean process success!
```

图 11.40　任务执行

```
hadoop fs -ls /user/project/output
```

结果如图 11.41 所示。

```
[root@master ~]# hadoop fs -ls /user/project/output
16/12/13 16:12:28 WARN util.NativeCodeLoader: Unable to load native-hadoop libra
ry for your platform... using builtin-java classes where applicable
Found 2 items
-rw-r--r--   1 root supergroup          0 2016-12-13 16:11 /user/project/output/
_SUCCESS
-rw-r--r--   1 root supergroup   38015669 2016-12-13 16:11 /user/project/output/
part-r-00000
[root@master ~]#
```

图 11.41　查看结果

默认会在输出路径下生成以 part-r-xxxxxx 命名的文件，即为清洗后的数据文件，可以通过以下命令将文件下载到本地系统：

```
hadoop fs -get /user/project/output/part-r-00000 /root/
```

结果如图 11.42 所示。

```
[root@master ~]# ls /root/
anaconda-ks.cfg    empcontact.java     metastore_db      --username   视频   桌面
--connect          employee.txt        mytable.java      workspace    图片
dataClean.jar      hdfs                part-r-00000      youtable.java 文档
derby.log          install.log         --query           公共的        下载
empadd.java        install.log.syslog  student.java      模板          音乐
[root@master ~]#
```

图 11.42　导出文件

通过以下命令查看文件内容，因为文件内容过多，因此在此只查看前 10 行数据：

```
head – n 10 /root/part – r – 00000
```

结果如图 11.43 所示。

图 11.43 查看文件内容

可见执行清洗后，文件中只剩下了我们所需要的三类数据：访问 IP 地址、访问时间以及访问网址。

11.4 数据统计分析

11.4.1 建立分区表

为了能够借助 Hive 进行统计分析，首先需要将清洗后的数据存入 Hive 中，那么需要先建立一张表。这里选择分区表，以日期作为分区的指标，建表语句如下（这里关键之处就在于确定映射的 HDFS 位置，这里是/user/project/output/part-r-00000，即清洗后的数据存放的位置）：

1. 启动 hive

在终端输入 hive 命令，即可启动：

```
hive
```

结果如图 11.44 所示。

```
[root@master ~]# hive
Logging initialized using configuration in jar: file:/usr/hive/lib/hive-common-1.2.1.jar!/hive-log4j.properties
hive>
```

图 11.44 启动 Hive

2. 创建分区表

命令如下：

```
hive> CREATE EXTERNAL TABLE mydata(ip string, atime string, url string) PARTITIONED BY (logdate string) ROW FORMAT DELIMITED FIELDS TERMINATED BY '\t';
```

结果如图 11.45 所示。

```
hive> CREATE EXTERNAL TABLE mydata(ip string, atime string, url string) PARTITIO
NED BY (logdate string) ROW FORMAT DELIMITED FIELDS TERMINATED BY '\t';
OK
Time taken: 2.566 seconds
hive>
```

图 11.45　创建分区表

3. 增加分区，并向分区中装载数据（即将清洗后的数据 /user/project/output/ part-r-00000 导入分区）

执行如下命令：

```
hive> ALTER TABLE mydata ADD PARTITION(logdate = '2016_11_01') LOCATION '/user/project/output';
```

结果如图 11.46 所示。

```
hive> ALTER TABLE mydata ADD PARTITION(logdate='2016_11_01') LOCATION '/user/pro
ject/output';
OK
Time taken: 0.432 seconds
hive>
```

图 11.46　增加分区

11.4.2　使用 HQL 统计关键指标

1. 关键指标一：PV 量

页面浏览量即为 PV（Page View），是指所有用户浏览页面的总和，一个独立用户每打开一个页面就被记录 1 次。这里只需要统计日志中的记录个数即可，HQL 代码如下：

```
select count(*) from mydata where logdate = '2016_11_01';
```

结果如图 11.47 所示。

```
MapReduce Total cumulative CPU time: 5 seconds 650 msec
Ended Job = job_1481612482869_0004
MapReduce Jobs Launched:
Stage-Stage-1: Map: 1  Reduce: 1   Cumulative CPU: 5.65 sec   HDFS Read: 3802273
0 HDFS Write: 7 SUCCESS
Total MapReduce CPU Time Spent: 5 seconds 650 msec
OK
502404
Time taken: 32.898 seconds, Fetched: 1 row(s)
hive>
```

图 11.47　统计指标

创建 mydata_pv 表存储 PV 量：

```
hive> CREATE TABLE mydata_pv AS SELECT COUNT(*) AS PV FROM mydata WHERE logdate = '2016_11_01';
```

验证 mydata_pv 表：

```
select * from mydata_pv;
```

结果如图 11.48 所示。

```
hive> select * from mydata_pv;
OK
502404
Time taken: 0.14 seconds, Fetched: 1 row(s)
hive>
```

图 11.48　查看结果

2. 关键指标二：注册用户数

该论坛的用户注册页面为 member.php，而当用户点击注册时请求的是 member.php?mod=register 的 URL。这里只需要统计出日志中访问的 URL 是 member.php?mod=register 的数目即可，因此创建 mydata_regis 表存储用户注册数，HQL 代码如下：

```
hive> CREATE TABLE mydata_regis AS SELECT COUNT(*) AS REGUSER FROM mydata WHERE logdate = '2016_11_01' AND INSTR(url,'member.php?mod=register')>0;
```

验证 mydata_regis 表：

```
select * from mydata_regis;
```

结果如图 11.49 所示。

```
hive> select * from mydata_regis;
OK
523
Time taken: 0.1 seconds, Fetched: 1 row(s)
hive>
```

图 11.49　查看结果

3. 关键指标三：独立 IP 数

一天之内，访问网站的不同独立 IP 个数加和。其中同一 IP 无论访问了几个页面，独立 IP 数均为 1。这里只需要统计日志中处理的独立 IP 数即可，在 SQL 中可以使用 DISTINCT 关键字，在 HQL 中也可以使用这个关键字，创建 mydata_ip 表存储独立 IP 数：

```
hive> CREATE TABLE mydata_ip AS SELECT COUNT(DISTINCT ip) AS IP FROM mydata WHERE logdate = '2016_11_01';
```

验证 mydata_ip 表：

```
select * from mydata_ip;
```

结果如图 11.50 所示。

```
hive> select * from mydata_ip;
OK
24635
Time taken: 0.102 seconds, Fetched: 1 row(s)
hive>
```

图 11.50　查看结果

4. 关键指标四：跳出用户数

只浏览了一个页面便离开了网站的访问次数，即只浏览了一个页面便不再访问的访问次数。这里可以通过用户的 IP 进行分组，如果分组后的记录数只有一条，那么即为跳出用户。将这些用户的数量相加，就得出了跳出用户数，创建 mydata_jump 表存储跳出用户数，HQL 代码如下：

> hive > CREATE TABLE mydata_jump AS SELECT COUNT(1) AS jumper FROM (SELECT COUNT(ip) AS times FROM mydata WHERE logdate = '2016_11_01' GROUP BY ip HAVING times = 1)?e;

验证 mydata_jump 表：

> select * from mydata_jump;

结果如图 11.51 所示。

```
hive> select * from mydata_jump;
OK
8454
Time taken: 0.089 seconds, Fetched: 1 row(s)
hive>
```

图 11.51　查看结果

跳出率是指只浏览了一个页面便离开了网站的访问次数占总的访问次数的百分比，即只浏览了一个页面的访问次数/全部的访问次数汇总。也可以将这里得出的跳出用户数除以 PV 数即可得到跳出率。

5. 将所有关键指标放入一张汇总表中以便于通过 Sqoop 导出到 MySQL

为了方便通过 Sqoop 统一导出到 MySQL，这里借助一张汇总表 mydata_all 将刚刚统计到的结果整合起来，通过表连接结合，HQL 代码如下：

> hive > CREATE TABLE mydata_all AS SELECT '2016_11_01', a.pv, b.reguser, c.ip, d.jumper FROM mydata_pv a JOIN mydata_regis b ON 1 = 1 JOIN?mydata_ip c ON 1 = 1 JOIN mydata_jump d ON 1 = 1;

验证 mydata_all 表：

> select * from mydata_all;

结果如图 11.52 所示。

```
hive> select * from mydata_all;
OK
2016_11_01        502404    523     24635    8454
Time taken: 0.085 seconds, Fetched: 1 row(s)
hive>
```

图 11.52　查看结果

11.4.3　使用 Sqoop 将数据导入到 MySQL 数据表

（1）在 Linux shell 终端运行如下命令进入 MySQL：

```
mysql -u root -proot
```

结果如图 11.53 所示。

```
[root@master ~]# mysql -u root -proot
Welcome to the MySQL monitor.  Commands end with ; or \g.
Your MySQL connection id is 2
Server version: 5.1.73 Source distribution

Copyright (c) 2000, 2013, Oracle and/or its affiliates. All rights reserved.

Oracle is a registered trademark of Oracle Corporation and/or its
affiliates. Other names may be trademarks of their respective
owners.

Type 'help;' or '\h' for help. Type '\c' to clear the current input statement.

mysql>
```

图 11.53　进入 MySQL

（2）在 MySQL 中创建结果汇总表。

执行如下命令创建一个新数据库 project：

```
mysql> create database project;
```

结果如图 11.54 所示。

```
mysql> create database project;
Query OK, 1 row affected (0.00 sec)

mysql>
```

图 11.54　创建数据库

切换到 project 数据库：

```
mysql> use project;
```

创建一张新数据表 logs_stat。

```
mysql> create table logs_stat(logdate varchar(10) primary key, pv int, reguser int, ip int, jumper int);
```

（3）通过 Sqoop 的 export 命令将统计后的数据导入到 MySQL 的 logs_stat 数据表中：

```
sqoop export -- connect jdbc:mysql://master:3306/project -- username root -- password root
-- table logs_stat -- fields - terminated - by '\001' -- export - dir '/user/hive/warehouse/
mydata_all'
```

结果如图 11.55 所示。

图 11.55　任务执行

其中--export-dir 是指定的 hive 目录下的汇总表所在位置，这里是/user/hive/warehouse/mydata_all。

执行如下命令查看导出结果：

```
mysql > select * from logs_stat;
```

结果如图 11.56 所示。

图 11.56　查看结果

11.5　定时任务处理

上面只是对一个固定的日志文件进行处理分析，而在实际情况中，日志文件是每天都会产生的，因此需要对每天的所有日志数据文件进行统计分析。

如果对每天的日志数据文件进行手动处理，那任务量是比较大的，不切实际。可以通过设置一个定时任务,在某个指定时间自动对前一天产生的日志文件进行统计分析。

11.5.1 日志数据定时上传

因为日志文件是每天产生的,因此可以设置一个定时任务,在每天的 1 点钟自动将前一天产生的日志文件上传到 HDFS 的指定目录中。

可以通过 shell 脚本结合 crontab 创建一个定时任务 project_core.sh,内容如下:

```
vi /root/project_core.sh
```

假设每天产生的日志文件存储在本地的/usr/local/files/apache_logs/目录下,并以 access_日期.log 形式进行命名,我们需要将其上传到 HDFS 文件系统的/project/data 目录下进行处理。

```
#!/bin/sh
# get yesterday format string
yesterday=$(date --date='1 days ago' +%Y_%m_%d)
# upload logs to hdfs
hadoop fs -put /usr/local/files/apache_logs/access_${yesterday}.log /project/data
```

输入以下命令,编辑 crontab 文件:

```
crontab -e
```

在文件中输入以下内容(其中 1 代表每天 1:00,/root/project_core.sh 为要执行的脚本文件在本地文件系统中的绝对路径):

```
* 1 * * * /root/project_core.sh
```

通过以下命令可以查看已经设置的定时任务:

```
crontab -l
```

结果如图 11.57 所示。

```
[root@master ~]# crontab -l
* 1 * * * /root/project_core.sh
[root@master ~]#
```

图 11.57 查看定时任务文件

11.5.2 日志数据定期清理

在之前编写的定时任务脚本中添加以下内容(其中加粗部分为需要添加的内容),将自动执行清理的 MapReduce 程序加入脚本中,内容如下(其中/root/dataClean.jar 为前面编写的 MapReduce 程序导出的 jar 包的绝对路径):

```sh
#!/bin/sh
# get yesterday format string
yesterday=$(date --date='1 days ago' +%Y_%m_%d)
# upload logs to hdfs
hadoop fs -put /usr/local/files/apache_logs/access_${yesterday}.log /project/data

# clean log data
hadoop jar /root/dataClean.jar /project/data/access_${yesterday}.log /project/cleaned/${yesterday}
```

这段脚本的作用是将日志文件上传到 HDFS 后，执行数据清理程序对已存入 HDFS 的日志文件进行过滤，并将过滤后的数据存入 cleaned 目录下。

11.5.3 数据定时统计分析

数据上传与清理之后，应实现自动化的统计分析并导出数据，所以需要改写定时任务脚本文件。

在之前编写的定时任务脚本中添加以下内容（其中加粗部分为需要添加的内容）：

```sh
#!/bin/sh
# get yesterday format string
yesterday=$(date --date='1 days ago' +%Y_%m_%d)
# upload logs to hdfs
hadoop fs -put /usr/local/files/apache_logs/access_${yesterday}.log /project/data

# clean log data
hadoop jar /root/dataClean.jar /project/data/access_${yesterday}.log /project/cleaned/${yesterday}

# alter hive table and then add partition
hive -e "ALTER TABLE mydata ADD PARTITION(logdate='${yesterday}') LOCATION '/project/cleaned/${yesterday}';"
# create hwive table everyday
hive -e "CREATE TABLE mydata_pv_${yesterday} AS SELECT COUNT(1) AS PV FROM hmbbs WHERE logdate='${yesterday}';"
hive -e "CREATE TABLE mydata_regis_${yesterday} AS SELECT COUNT(1) AS REGUSER FROM hmbbs WHERE logdate='${yesterday}' AND INSTR(url,'member.php?mod=register')>0;"
hive -e "CREATE TABLE mydata_ip_${yesterday} AS SELECT COUNT(DISTINCT ip) AS IP FROM hmbbs WHERE logdate='${yesterday}';"
hive -e "CREATE TABLE mydata_jump_${yesterday} AS SELECT COUNT(1) AS jumper FROM (SELECT COUNT(ip) AS times FROM mydata WHERE logdate='${yesterday}' GROUP BY ip HAVING times=1) e;"
hive -e "CREATE TABLE mydata_all_${yesterday} AS SELECT '${yesterday}', a.pv, b.reguser, c.ip, d.jumper FROM mydata_pv_${yesterday} a JOIN mydata_regis_${yesterday} b ON 1=1 JOIN mydata_ip_${yesterday} c ON 1=1 JOIN mydata_jump_${yesterday} d ON 1=1;"
# delete hive tables
hive -e "drop table mydata_pv_${yesterday};"
hive -e "drop table mydata_regis_${yesterday};"
hive -e "drop table mydata_ip_${yesterday};"
hive -e "drop table mydata_jump_${yesterday};"
# export to mysql
```

```
sqoop export -- connect jdbc:mysql://master:3306/project -- username root -- password root
-- table logs_stat -- fields-terminated-by '\001' --export-dir '/user/hive/warehouse/
mydata_all_${yesterday}'
# delete hive table
hive -e "drop table mydata_all_${yesterday};"
```

这段脚本的作用是对过滤后的数据文件进行统计分析,同时将分析后的结果存入 MySQL 数据表。

思考与延伸:

如果一个网站已经生成了很多天的日志以后日志分析系统才上线,那么此时如何有效对之前已经产生的数据进行分析呢?

此时需要写一个初始化脚本任务,来对之前每天的日志进行统计分析并导出结果。这里新增一个 project_init.sh 脚本文件,内容如下:

```
#!/bin/sh

# create external table in hive
hive -e "CREATE EXTERNAL TABLE mydata(ip string, atime string, url string) PARTITIONED BY
(logdate string) ROW FORMAT DELIMITED FIELDS TERMINATED BY '\t';"

# compute the days between start date and end date
s1=`date --date="$1" +%s`
s2=`date +%s`
s3=$((($s2-$s1)/3600/24))

# excute project_core.sh $3 times
for ((i=$s3; i>0; i--))
do
   logdate=`date --date="$i days ago" +%Y_%m_%d`
   /root/project_core.sh $logdate
done
```

通过以下命令执行任务即可(需要指定一个"日志生成开始日期的时间戳"参数):

```
project_init.sh 日志生成开始日期的时间戳
```

时间戳是按照从格林威治时间 1970 年 1 月 1 日起计算的一个秒数。

通过以上的介绍,网站的日志分析工作基本完成,当然还有很多没有完成的内容,但是大体上的思路已经明了,后续的工作只需要在此基础上稍加分析即可完成。可以通过 JSP 或 ASP.NET 读取 MySQL 的分析结果表来开发关键指标查询系统,供网站运营决策者查看和分析。

参 考 文 献

［1］ 朱杰.大数据架构详解：从数据获取到深度学习［M］.北京：电子工业出版社，2016.
［2］ 林子雨.大数据技术原理与应用［M］.2版.北京：人民邮电出版社，2017.
［3］ Tom White.Hadoop权威指南［M］.3版.华东师范大学数据科学与工程学院，译.北京：清华大学出版社，2015.
［4］ 张良均，樊哲，李成华，等.Hadoop大数据分析与挖掘实战［M］.北京：机械工业出版社，2015.
［5］ 范东来.Hadoop海量数据处理技术详解与项目实战［M］.2版.北京：人民邮电出版社，2016.
［6］ Garry Turkington.Hadoop基础教程［M］.张治起，译.北京：人民邮电出版社，2014.
［7］ Benoy Antony，Cazen Lee，Kai Sasaki.Hadoop大数据解决方案.北京：清华大学出版社，2017.
［8］ Sandeep Karanth.精通Hadoop［M］.北京：人民邮电出版社，2016.
［9］ 翟周伟.Hadoop核心技术［M］.北京：机械工业出版社，2015.
［10］ Jonathan R. Owens.Hadoop实战手册［M］.北京：人民邮电出版社，2014.
［11］ 刘刚.Hadoop应用开发技术详解［M］.北京：机械工业出版社，2014.
［12］ Edward Capriolo，等.Hive编程指南［M］.北京：人民邮电出版社，2013.
［13］ Lars George.HBase权威指南［M］.北京：人民邮电出版社，2013.
［14］ Nick Dimiduk，Amandeep Khurana.HBase实战［M］.谢磊，译.北京：人民邮电出版社，2013.
［15］ 马延辉，孟鑫，李立松.HBase.企业应用开发实战［M］.北京：机械工业出版社，2014.

图书资源支持

感谢您一直以来对清华版图书的支持和爱护。为了配合本书的使用,本书提供配套的资源,有需求的读者请扫描下方的"书圈"微信公众号二维码,在图书专区下载,也可以拨打电话或发送电子邮件咨询。

如果您在使用本书的过程中遇到了什么问题,或者有相关图书出版计划,也请您发邮件告诉我们,以便我们更好地为您服务。

我们的联系方式:

地　　址:北京市海淀区双清路学研大厦 A 座 714

邮　　编:100084

电　　话:010-83470236　010-83470237

客服邮箱:2301891038@qq.com

QQ:2301891038(请写明您的单位和姓名)

资源下载:关注公众号"书圈"下载配套资源。

资源下载、样书申请

书　圈

获取最新书目

观看课程直播